主审 吕志涛

Introduction to 土木工程导论 Civil Engineering

主 编 刘荣桂 胡白香
副主编 李琮琦 韩 豫 滕 斌 朱 炯

江苏大学出版社
JIANGSU UNIVERSITY PRESS

镇 江

图书在版编目(CIP)数据

土木工程导论 / 刘荣桂,胡白香主编. —镇江：
江苏大学出版社，2013.8(2013.11 重印)
ISBN 978-7-81130-479-4

Ⅰ.①土… Ⅱ.①刘… ②胡… Ⅲ.①土木工程—高
等学校—教材 Ⅳ.①TU

中国版本图书馆 CIP 数据核字(2013)第 198250 号

土木工程导论
Tumu Gongcheng Daolun

主　　编/刘荣桂　胡白香
责任编辑/李菊萍　张璐
出版发行/江苏大学出版社
地　　址/江苏省镇江市梦溪园巷 30 号(邮编：212003)
电　　话/0511-84446464(传真)
网　　址/http：//press. ujs. edu. cn
排　　版/镇江新民洲印刷有限公司
印　　刷/丹阳市兴华印刷厂
经　　销/江苏省新华书店
开　　本/787 mm×1 092 mm　1/16
印　　张/23.25
字　　数/595 千字
版　　次/2013 年 8 月第 1 版　2013 年 11 月第 2 次印刷
书　　号/ISBN 978-7-81130-479-4
定　　价/52.00 元

如有印装质量问题请与本社营销部联系(电话：0511-84440882)

序

几千年以来,土木工程历经了几个划时代的发展过程。从古埃及金字塔、古希腊雅典卫城、印度泰姬陵、中国长城,再到法国埃菲尔铁塔、澳大利亚悉尼歌剧院和中国的三峡工程、青藏铁路等,建成了无数伟大的工程。随着社会生产力的不断发展,人类对建筑功能的要求也日益复杂多样,大量不同类型的结构体系,与之相伴的设计、计算理论、实验与测试方法、施工技术与工程管理方法等都在变革、创新的驱动力作用下不断更新、发展与进步。如今土木工程已成为国民经济发展的主要支柱产业之一。为了让本专业及对本专业感兴趣的中外学生了解土木工程的概况,作者决定编写双语版《土木工程导论》一书。

《土木工程导论》是以全国土木工程专业指导委员会最新专业指导意见(2012年)为指南,为土木工程及相关专业的中外大学生编写的了解土木工程学科及其分支领域历史沿革与发展状况的概论性教材,采用中英文双语编写。

本教材以"引领专业学习,激发专业兴趣"为主旨,立足于反映近20年土木工程领域的主要理论与应用成果以及未来发展趋势,通过系统地介绍土木工程(含部分工程管理知识)学科的基本框架、土木工程专业的基本概念、主要特点、基础理论与技术方法等,帮助学生在进入专业学习之前了解土木工程行业及学科的基本情况与发展态势、执业资格认证体系及就业导向等,最终激发学生对本专业学习、研究的兴趣,为今后专业学习奠定基础;同时,通过"知识拓展"、"相关链接"和"小贴士"等形式,使用网络等技术,扩展知识量,提供学生自学通道,为学生的学业生涯设计、就业核心竞争力的提升提供专业支撑。

作者及编写组的多数成员从事土木工程专业教学都在20年以上,并一直活跃在土木工程科研和社会服务一线,理论与实践经验比较丰富。本书尽作者所能,力图使编写内容从整体上反映土木工程学科的综合性、理论性、技术性和实用性,重点展现我国土木工程大建设、大发展、大提升的最新成果与未来发展趋势,并与国际接轨。

本教材充分借鉴国内外土木工程概论类教材和课程教学的先进经验,如美国爱荷华州立大学工程学院开设的"土木工程的伟大成就(The great achievements of civil engineering)"课程等,贯穿式地介绍了土木工程在推动人类社会进步过程中的巨大作用、土木工程建设与可持续发展的关系、工程全寿命期管理等。本书内容丰富,观点独特,写作严谨认真,力求文字精准,图文并茂。本书的出版将弥补目前有关土木工程导论(或概论)类教程编写的不足,也可为专科生、研究生等不同层次读者了解土木工程学科提供参考。

中国工程院院士 吕志涛

2013 年 4 月

前 言

　　《土木工程导论》是以全国土木工程专业指导委员会最新(2012 年)专业指导意见为指南,为土木工程及相关专业的中外大学生编写的,旨在介绍土木工程学科及其分支领域历史沿革与发展状况的概论性教材,采用中英文双语编写。

　　本教材以"引领专业学习,激发专业兴趣"为主旨,立足于反映近 20 年土木工程领域的基本概况、国内外主要工程应用成果以及未来发展趋势;力图通过系统地介绍土木工程(含部分工程管理内容)学科的基本框架和关键知识,让学生形成对本学科和相关专业的宏观了解,激发学生对本专业学习、研究的兴趣,为今后专业学习奠定基础。教材通过"链接"牵引等形式,打开学生自学的通道,为学生的学业生涯设计、就业核心竞争力的提升提供专业支撑。此外,本教材对研究生(尤其是海外留学研究生)也有一定的参考作用。

　　本教材中,土木工程内容约占 75% ,工程管理内容约占 25% 。编者力求从整体上反映土木工程学科的综合性、理论性、技术性和实用性,重点展现我国土木工程大建设、大发展、大提升的最新成果与未来发展趋势。本书的部分内容融合了编者及其科研团队最新的研究成果,写作风格力求文字精准,图文并茂。

　　本教材由绪论统领全书,以土木工程及其各子学科和分支体系以及土木工程实施中的关键管理任务为基本架构展开介绍,最终通过系统和全面的学科展望,力求提升学生的专业素养,建立起宏观的土木工程及相关知识体系框架,完成引领后续专业课程学习的任务。

　　参加编写本书人员的具体分工如下(以下未注工作单位的人员均为江苏大学教师):刘荣桂、李琼琦(扬州大学),胡凤庆(扬州大学)编写绪论;陆春华、谢桂华、滕斌(金土木建设集团)编写第1章,殷杰、滕斌、朱炯(徐州工程学院)编写第2章,胡白香、谢甫哲、沈圆顺、杨帆编写第3章与第4章,延永东、杨帆、滕斌、朱炯编写第5章,徐荣进、滕斌、朱炯编写第6章,李琮琦、胡凤庆、操礼林、苏波、郭兴龙编写第7章,李琮琦、胡凤庆、蔡东升、滕斌编写第8章,韩豫、朱炯编写第9章,温修春、韩豫、朱炯编写第10章,孙莹、韩豫编写第11章。刘荣桂制定了编写大纲并对全书进行了最后统稿,韩豫为本书出版做了大量的联系、协调工作,东南大学吕志涛院士对本书进行了主审。

感谢东南大学、浙江大学、江苏省建筑科学研究院、扬州大学、徐州工程学院等兄弟单位的技术帮助与支持;感谢编者课题组的老师与研究生为本书编写所做的工作;感谢江苏大学出版社李锦飞教授及汪再非、李菊萍、常钰、张璐及其同仁为本书的顺利出版所做出的不懈努力;最后要特别感谢东南大学的吕志涛院士对本书出版的指导与支持。

土木工程导论作为一本带有科普、引导性教程,涉及的问题很多且较为复杂,尚有许多问题亟待完善。希望使用本书进行教学的各位同仁多提宝贵意见,以便再版时修改与完善。同时由于编者水平有限,书中难免存在不足之处,恳请读者批评指正。

刘荣桂

2013 年 6 月

Preface

Introduction to Civil Engineering is a textbook written in both Chinese and English. It is compiled for Chinese and foreign students majoring in civil engineering or the relative specialties, according to the latest professional guidance given by National Steering Committee of Civil Engineering Specialty in 2012. Students can learn about the historical evolution and development of the civil engineering and its branches from this book.

The aim of this textbook is to lead students to professional fields and arouse their professional interest. Therefore, this book is designed to report the general situation of civil engineering in the past 20 years, the main achievements in engineering application both at home and abroad, and the development trend in the future. The framework and the key knowledge of civil engineering (including some project management knowledge) are introduced systematically to help students get the basic informatiom of civil engineering and the relative fields, arouse students' interest in professional learning and research, and lay the foundation for professional learning in the future. The book provides students with a way of studying professional knowledge by themselves, and making plans for their academic career and improving the core competitiveness in getting jobs. Besides, the book is also a useful reference for graduates (especially for the overseas graduates).

In this textbook, civil engineering accounts for 75 percent, and the project management covers 25 percent. As a whole, the content reflects the comprehensive, theoretical, technical and practical properties of civil engineering discipline, while the latest achievements of great constructions in China, the improvement, and the development trend in the future in civil engineering field are emphasized. Part of the book is the latest research results of the author and his team. The author aims for accuracy in writing, and abundant pictures are included in the book.

This textbook is guided by the introduction and extended by the basic framework of civil engineering, its various sub-disciplines and branch systems combined with the key management tasks in civil engineering implementation. Through systematic and comprehensive prospects for civil engineering, this textbook will enhance students' professional knowledge, help them set up a macroscopic knowledge system and framework for civil engineering and the related knowledge, and lead them to the following specialized courses.

The following people participated in the writing(unspecified participants are all the teachers in Jiangsu University): the introduction is completed by Liu Ronggui, Li Congqi(Yangzhou University), Hu Fengqing(Yangzhou University). Chapter 1 is written by Lu Chunhua, Xie Guihua, Teng Bin(Jiangsu Gold nstruction Group Co. Ltd.). Chapter 2 is completed by Yin Jie, Teng Bin, Zhu Jiong(Xuzhou Institute of Technology). Chapter 3 and Chapter 4 are written by Hu Baixiang, Xie Fuzhe, Shen Yuanshun, Yang Fan. Chapter 5 is written by Yan Yongdong, Yang Fan, Teng Bin, Zhu Jiong. Chapter 6 is written by Xu Rongjin, Teng Bin, Zhu Jiong. Chapter 7 is written by Li Congqi, Hu Fengqing, Cao Lilin, Su Bo, Guo Xinglong. Chapter 8 is written by Li Congqi, Hu Fengqing, Cai Dongsheng, Teng Bin. Chapter 9 is written by Han Yu, Zhu Jiong. Chapter 10 is written by Wen Xiuchun, Han Yu, Zhu Jiong. Chapter 11 is written by Sun Ying, Han Yu. Liu Ronggui made the outline of the book and compiled the integrated manuscript. Han Yu did much work on contacting and coordinating for the publishment. Academician Lv Zhitao from Southeast University reviewed the book chiefly.

Thanks for the technical assistance and support from Southeast University, Zhejiang University, Jiangsu Institute of Building Research, Yangzhou University, Xuzhou Institute of Technology and other fraternal units. Thanks to Professor Li Jinfei and his colleagues from Jiangsu University Press for their tireless efforts for the publication of this book. The graduates of the author made a great contribution for the book, to whom I owe many thanks. Finally, special thanks to Academician Lv Zhitao who guided and supported the publishment of this book.

As an educational and instructive textbook, *Introduction to Civil Engineering* involves a lot of complex problems to be resolved. Criticism and suggestions from the readers will be highly welcomed and appreciated. Due to the limtited knowledge of the authors, defects in the textbook are unavoidable. We deeply appreciate your criticism and correction.

Liu Ronggui

June 2013

目 录

Contents

绪 论

Introduction

本章主要介绍土木工程的基本概念及土木工程专业涉及的主要技术领域。通过本章的学习,可以加深对土木工程的含义、类型、发展历史的认识,了解土木工程未来的发展趋势,建立对土木工程的学习兴趣,明确土木工作者的责任。

This article mainly introduces the basic concepts of civil engineering and its related technological fields. From this article, people will have a deeper understanding of the concepts, types, development history and the developing tendency of civil engineering. Besides, people can build their interests in civil engineering and realize their responsibilities.

0.1 土木工程的内涵

0.1 Connotations of civil engineering

0.1.1 土木工程

土木工程的英文是 civil engineering, 直译为民用工程,是各种建造工程的统称。它既指建设的对象,即建造在地上、地下、水中的工程设施,也指应用的材料设备和进行的勘测、设计施工、保养维修等专业技术。

1828 年,英国土木工程师协会在其皇家宪章中给出这样的定义:"土木工程是利用伟大的自然资源为人类造福的艺术。"其作为国内生产和交通的实现手段,

0.1.1 Civil engineering

The literal meaning of civil engineering in English is non-military engineering, which is the general term of the construction of various projects. It not only refers to its objects such as the engineering facilities constructed on the ground, underground and underwater, but also includes the materials and the professional technology of investigation, construction and maintenance.

In 1828, in its Royal Charter, the Institution of Civil Engineers defined civil engineering as the art of benefiting human beings by using great natural resources. Civil engineering works as the means of production

进行国内外贸易。例如,应用于道路、桥梁、渡槽、运河、内河航运和码头的建设,而这些建设又服务于内部交流和交换;应用于港口、码头、防波堤和灯塔的建设;应用于以商业为目的,以人为的力量进行的航海行为;应用于机械的建设和使用;应用于城市和乡镇的排水系统。

"土木"来源于拉丁文"公民"一词。1782 年,英国人 John Smeaton 为了把他的非军事工程工作区别于当时占优势地位的军事工程师的工作而采用该名词。自那时起,"土木工程"一词常被从事公共设施建设的工程师所应用,尽管其包含的领域更为广阔。"土木"在中国是一个古老的术语,意指建造房屋,而古代建房主要依靠"土"(包括岩石、沙、泥土、石灰以及由土烧制成的砖、瓦等)和"木"(包括木材、茅草、竹子、藤条等),故将"civil engineering"译为土木工程。

随着时代的发展和技术的进步,土木工程被不断注入新鲜血液,显示出勃勃生机,其中工程材料的变革和力学理论的发展起着最为重要的推动作用。现代土木工程早已不是传统意义上的砖、瓦、灰、砂石,而是由新理论、新材料、新技术、新方法武装起来的为众多领域和行业所不可缺少的大型综合性学科群,是一个古老而年轻的学科。

and transportation for external and internal trade. For example, it can be applied to the construction of roads, bridges, aqueducts, canals, river navigation and docks for internal intercourse and exchange; it can also be applied to the construction of ports, harbors, moles, breakwaters, lighthouses and the navigation by artificial power for the purposes of commerce, the construction and application of machinery, as well as the drainage system both in cities and rural areas.

The word "civil" derives from the Latin word "citizen". In 1782, Englishman John Smeaton used this term to differentiate his non-military engineering work from that of the military engineers who predominated at that time. Since then, the term civil engineering has often been used among engineers who build public facilities, although the field of civil engineering is supposed to be much broader. "Civil" is an ancient term in Chinese, which means construction. In ancient times, "soil" (including rock, sand, clay, lime , the brick and tile made of soil) and "wood" (including wood, thatch, bamboo and rattan) have been mainly used in the construction. Therefore, the meaning of "civil engineering" in Chinese includes the meaning of wood and soil.

With the development of technology and the times, civil engineering is developing quickly, in which the revolution of engineering material and the development of mechanical theory play the most important roles. The modern civil engineering is not just the brick and stone in the traditional sense, but an indispensable and comprehensive subject clot armed by the new theory, new materials, new technologies and new methods. It is a young subject with a long history of development.

0.1.2 土木工程的范畴

土木工程是工程学科分支之一,旨在为人们提供舒适而安全的生活。人们生活离不开衣、食、住、行。其中"住"是人们最基本的生活需求之一,它与土木工程直接相关,而供水及灌溉工程的合理规划与设计可有效提高粮食产量。从古老的金字塔到今天的薄壳结构,所有的工程奇迹都是土木工程不断发展的结果,而公路、铁路、桥梁等运输线路也是土木工程师的工作成果。

随着近现代工程建设和科学技术的迅猛发展,土木工程逐渐划分成一些专门学科,如结构工程、岩土工程、交通工程、环境工程、水利工程、建设工程、材料科学、测量学、城市工程等,其包含的内容和涉及的范围非常广泛。土木工程不仅为人类生存与发展建造了单体的建筑、桥梁、隧道、大坝等,也创造了城市、乡村、厂矿等综合的生态与环境。

若业主拥有充足的资金,那么建设项目的所有步骤可按图 0.1 所示的流程完成。

0.1.2 The scope of civil engineering

Civil engineering is the branch of engineering, which aims to provide the comfortable and safe life for people. People cannot live without clothing, food, shelter and transportation. Shelter, one of the primary needs of mankind, directly relates to civil engineering. The efficient planning of water supply and irrigation system can increase the food production. The engineering marvels in the world, starting from the ancient pyramids to thin shell structures, are the results of the development in civil engineering. Besides, the transport routes like roads, railways, bridges are also the products of civil engineers.

With the rapid development of the modern engineering and the technology, civil engineering has been gradually subdivided into some specific disciplines, including structural engineering, geotechnical engineering, transportation engineering, environmental engineering, hydraulic engineering, construction engineering, etc., materials science, surveying science and urban engineering, etc. The content and the scope of civil engineering are much broader. Civil engineering not only includes the monomer building for human survival and development, such as bridges, tunnels and dams, but also contains the ecology and environment in cities, villages, factories and so on.

Assuming that the house owner has enough money, all steps of the construction project for housing or industry can be implemented as the following(Fig. 0.1).

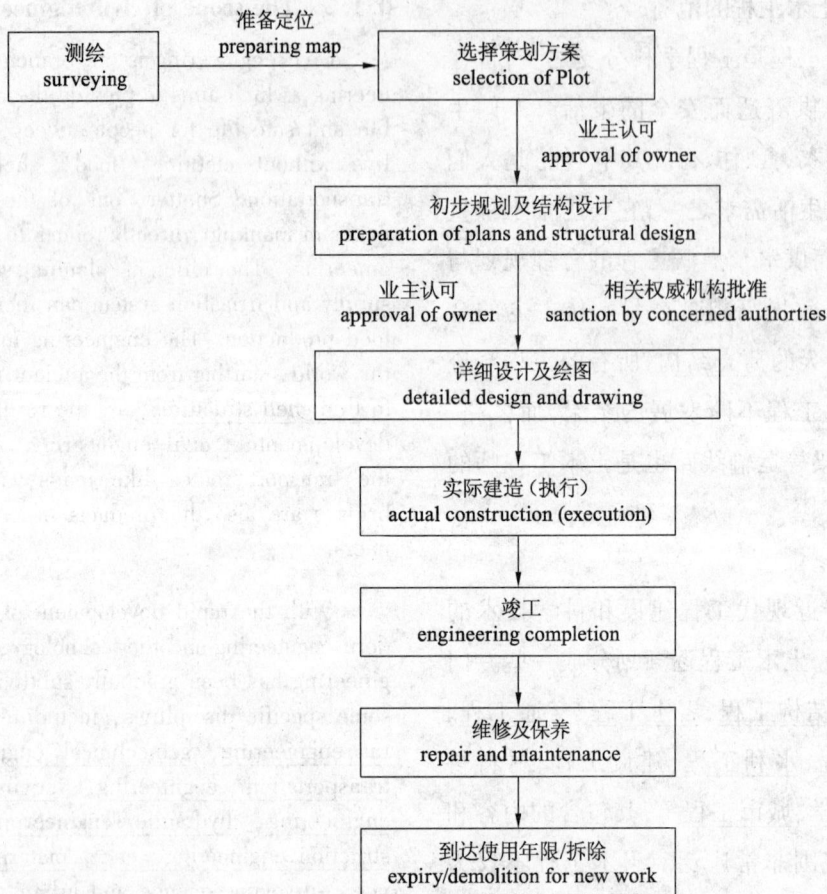

```
测绘                准备定位                选择策划方案
surveying  ──────  preparing map  ──→   selection of Plot

                                              │ 业主认可
                                              │ approval of owner
                                              ↓

                              初步规划及结构设计
                      preparation of plans and structural design

          业主认可                              相关权威机构批准
          approval of owner                    sanction by concerned authorties
                              ↓

                              详细设计及绘图
                          detailed design and drawing
                              ↓

                              实际建造（执行）
                          actual construction (execution)
                              ↓

                              竣工
                          engineering completion
                              ↓

                              维修及保养
                          repair and maintenance
                              ↓

                              到达使用年限/拆除
                          expiry/demolition for new work
```

图 0.1　工程建设概要

Fig. 0.1　Outline of the construction activity

一个项目开工之初,土木工程师要对场地进行测绘,定位有用的布置,如地下水水位、下水道和电力线等。岩土工程专家则进行土力学试验以确定土壤能否承受工程荷载。环境工程专家研究工程对当地的影响,包括可能对空气和地下水产生的污染,对当地动植物生活的影响,以及如何让工程设计满足政府对环境保护的要求等。交通工程专家确定必需的不同种类设施,以减轻整个工程对当地公路和其他交通网络的负担。同时,结构工程

When a project begins, the site should be surveyed and mapped by civil engineers. The engineers should locate the utility placement, such as the stage of underwater, sewer and power lines. Geotechnical specialists will perform soil experiments to determine whether the earth can bear the weight of the project. Environmental specialists will study the impact of the project on the local area: the potential pollution for air and groundwater, the impact on local animals and plants, and how the project can be designed to meet government's requirements aimed at protecting the environment. Transportation specialists will determine what kinds of facilities are needed to ease the burden on local roads and

专家利用初步数据对工程作详细规划、设计和说明。

从项目开始到结束,对这些土木工程专家的工作进行监督和调配的则是施工管理专家。根据其他专家所提供的信息,施工管理专家计算材料和人工的数量和开支,确定所有工作的进度表,订购工作所需要的材料和设备,雇佣承包商和分包商,还要做些额外的监督工作以确保工程能按时按质完成。

0.2 土木工程发展简史

土木工程的发展可以分为 3 个阶段:古代土木工程、近代土木工程和现代土木工程。

0.2.1 古代土木工程

古代土木工程的历史跨度很长,大致从旧石器时代(约公元前 5 000 年)到 17 世纪中叶。在这一时期内,人们修建各种设施时主要依靠经验,没有设计理论指导,所运用的材料也大多取自于自然,如石块、草筋、土坯等,大约在公元前 1 000 年才采用烧制的砖。这一时期,所用的工具也很简单,只有斧、锤、刀、铲和石夯等手工工具。尽管如此,古人还是以他们卓越的智慧建造了许多具有历史价值的建筑。

other transportation networks. Meanwhile, structural specialists will use preliminary data to make detailed designs, plans and specifications for the project.

From the beginning to the end of the project, the construction management specialists will be in charge of the supervision and coordination of the civil engineering specialists. Based on the information supplied by the other specialists, construction management specialists will estimate quantities and costs of materials and labor. They will also schedule all tasks, order materials and equipment, hire contractors and subcontractors, and perform other supervisory work to ensure the project will be completed on time and as specified.

0.2 Brief history of civil engineering

The development history of civil engineering can be divided into three stages: ancient civil engineering, modern civil engineering and contemporary civil engineering.

0.2.1 Ancient civil engineering

The ancient civil engineering had a long time span, roughly from the Paleolithic Age (5 000 B. C.) to the mid-17th century. During this period, people built all kinds of facilities mainly depending on experience without any guidance. The materials are mostly taken from nature, such as rocks, grass and adobe. Fired brick was not adopted until 1 000 B.C. During this period, the used tools were very simple, only include stone axes, hammers, knives; sickles, stone ram and other hand tools. Even without advanced tools, the outstanding ancients still built lots of buildings with historical value.

人类历史上最早的土木工程大约出现在公元前 4 000 年到公元前 2 000 年的古埃及和美索不达米亚，那时人类开始放弃游牧生活方式，因而需要有作为庇护所的建筑。

古巴比伦人和亚述人曾致力于解决包括水坝、大堤和运河在内的水利工程问题。他们通过考虑直角三角形的三边关系来解决问题，同时也能解一些简单的代数方程。他们能计算土地面积、砖石砌体体积以及开挖运河所必需的土方量等。亚述帝国完成了第一次有组织的道路建设，而第一座具有技术含量的桥梁则于公元前 6 世纪建造在幼发拉底河上。

古埃及人用最原始的力学原理和工具建造了众多的庙宇和金字塔，至今仍有许多不朽建筑屹立不倒，包括吉萨的大金字塔和卡纳克的阿蒙-拉神庙（如图 0.2 所示）。146.6 m 高的大金字塔由 225 万块平均重量超过 1.4 t 的巨石建成。这种标志性建筑动用了大量的人力来修建。此外，埃及人也使用青铜作为切削工具，切割巨石建造方尖碑，其中一些巨石的重量超过 900 t。

公元前 6 世纪到公元前 3 世纪，古希腊人在将理论引入工程方面取得了巨大的进展，并发展了线、角、面和体的抽象概念。古希腊建筑工程中的几何学基础包括像正方形、矩形和三角形这样的图形。古希腊的工长通常既是设计者也是建造

The earliest practice of civil engineering may appear between 4 000 B.C. and 2 000 B.C. in ancient Egypt and Mesopotamia, when human beings started to abandon their nomadic way of life. Instead, they needed one kind of construction as the shelter.

Babylonians and Assyrians struggled with problems of hydraulic engineering involving dams, levees and canals. They solved problems by concerning the sides of right triangle, and they also solved simple algebraic equations. They calculated the areas of land, volumes of masonry, the necessary cubic contents of excavation for canals. The first organized road building was finished in the Assyrian Empire, and the first high-tech bridge was constructed over the Euphrates River in the 6th century B.C.

In ancient Egypt, the simplest mechanical principles and devices were used to construct many temples and pyramids that are still standing today, including the Great Pyramid at Giza and the temple of Amon-ra in Karnak (Fig. 0.2). The Large Pyramid, 146.6 meters high, was made of 2.25 million stone blocks with an average weight of more than 1.4 tons. Large number of labor was used during the construction of such monuments. The Egyptians also built obelisks by huge stone blocks with tools made of hard bronze. Some huge stones even weight more than 900 tons.

The Greeks made great strides in introducing theory into engineering during the 6th century B.C. to the 3rd century B.C. They developed an abstract knowledge of lines, angles, surfaces and solids rather than referring to specific objects. The geometric bases of Greek building construction includes figures such as the square, rectangle and triangle. The Greek architect was usually the designer, as

者,他们带领工匠、泥瓦匠和雕刻匠建造了很多杰出的工程。在古希腊时代,所有重要的房屋都由石灰石或大理石建成,例如帕台农神庙就是由大理石建成的。这是一座公元前 5 世纪建造在雅典卫城之上的供奉雅典娜的神殿(如图 0.3 所示)。

well as the builder of the architectural and engineering masterpieces. Craftsmen, masons and sculptors worked under his supervision. In the classical period of Greece, all important buildings were built of limestone or marble. The Parthenon, for example, was built of marble. It was a temple of Athena, built in the 5th century B.C. on the Acropolis of Athens(Fig. 0.3).

(a)大金字塔
The Great Pyramid

(b)卡纳克阿蒙-拉神庙
The temple of Amon-ra in Karnak

图 0.2 古埃及建筑
Fig. 0.2 Egyptian architecture

图 0.3 帕台农神庙
Fig. 0.3 The Parthenon

公元 2 世纪,罗马帝国进入全盛期,统治了从苏格兰至波斯的大片疆域。当罗马人征服其他民族时,那些战败国的工程理论和实践也为其所拥有和继承。其中,古埃及建筑对古罗马建筑影响最大。

In its heyday of the 2nd century A.D., Rome ruled the world from Scotland to Persia. As the Romans conquered other nations, they borrowed their captives' ideas and practices about engineering. The influence of Egyptian building is especially noticeable. However, the Roman arch construction crea-

但是,古罗马的工程师发明了带有拱顶石的半圆拱结构,这表明他们当时已熟知砖石砌体受压的事实,尽管他们没有书面或正式的关于力平衡的知识。古罗马的建筑师是技术专家,他们设计并建造了桥梁、沟渠、公路以及公共建筑。古罗马时期的筑路技术达到了古代土木工程的最高水平。除了著名的阿皮亚古道(如图0.4所示),古罗马人还修建了隧道、渡槽、石拱桥、海港、码头以及灯塔等。

ted a central keystone at the top, which indicated that Roman engineers were familiar with masonry under compression, although they had no written or formal knowledge about equilibrium of forces. The Roman technical experts designed and constructed bridges, aqueducts, highway and buildings for public uses. The art of road building in ancient Rome reached its highest level in civil engineering filed. Besides the Via Appia(Fig. 0.4b), Romans also built tunnels for roadways, aqueducts(Fig. 0.4a), stone arch bridges, harbors, dock and lighthouse.

(a) 渡槽
The aqueducts

(b) 阿皮亚古道
The Via Appia

图 0.4 古罗马工程
Fig. 0.4 The ancient Roman engineering

在我国黄河流域的仰韶文化遗址和西安半坡遗址发现了供居住的浅穴和直径为 5~6 m 的圆形房屋,这说明在新石器时代出现了原始的房屋建筑。进入封建社会后,我国开始了秦砖汉瓦的土木工程技术时代。秦朝开始修建的万里长城、大型宫殿、陵墓,战国时代建造的都江堰等,都是这个时期不朽工程的代表。中国古代的建筑多采用木结构,并逐渐形成与此相适应的风格(如图 0.5 所示)。公元14 世纪建造的北京故宫是世界上最大、最

In China, the shallow holes for living and the round houses with the diameter of 5 to 6 meters have been found at Yangshao culture sites of the Yellow River basin and Xi'an Banpo Ruins. It shows that the original construction appeared in the Neolithic Age. The Great Wall, large palaces, tombs built in Qin Dynasty and the Dujiang Dam built in the Warring States are immortal engineering representatives in the period of the feudal society. Most Chinese ancient architectures (Fig. 0.5) were built with wooden structure, and gradually formed unique styles to match it. The Imperial Palace in Beijing, constructed in the 14th century, is the largest and most complete ancient wooden

完整的古代木结构宫殿建筑群;建于辽代的应县木塔是世界上最高的木建筑,并成为世界木结构建筑的典范。

structure buildings in the world. The Yingxian Wooden Tower built in Liao Dynasty is the tallest wooden building in the world, and might be considered as the excellent example of wooden constructions.

(a)长城
The Great Wall

(b) 都江堰
The Dujiang Dam

(c) 故宫
The Imperial Palace

(d) 应县木塔
The Yingxian wooden tower

图 0.5 中国古代工程
Fig. 0.5 Chinese ancient engineering

0.2.2 近代土木工程

近代土木工程的历史通常认为是从 17 世纪中期到二战后总共超过 300 年的时间。在这段时期,土木工程成为一门独立学科,并进入了定量分析阶段。一些理论的发展,新材料的出现,新工具的发明,都使土木工程学科日渐完善和成熟。

在力学理论方面,1638 年意大利学者伽利略发表了《关于两门新科学的对话》

0.2.2 Modern civil engineering

The history of modern civil engineering is generally regarded as the period from the middle of 17th century to the post-second world war. During this period, civil engineering was becoming an independent discipline, and stepped into a quantitative analysis stage. The development of some theories, the emergence of new materials and the invention of new tools made the civil engineering perfect and mature gradually.

In mechanical theory, Italian scholar Galileo published *Dialogues Concerning Two*

一文,首次用公式表述了梁的设计理论。随后,在 1687 年牛顿总结出力学三大定律,为土木工程奠定了力学分析的基础。1825 年法国的维纳在材料力学、弹性力学和材料强度理论的基础上,建立了土木工程中结构设计的容许应力法。

在材料方面,1824 年波特兰水泥和 1867 年钢筋混凝土的发明成为土木工程历史上的重大事件。水泥和钢铁的大批量生产和供应,使得土木工程师可以利用这些材料实现大规模复杂工程设施的建设。在近代及现代建筑中,高耸、大跨、巨型、复杂的工程结构,绝大多数应用了钢结构或钢筋混凝土结构。

1825 年,英国修建了世界上第一条长 21 km 的铁路;1863 年,伦敦修建了世界上第一条地铁;1889 年,法国建成了高达 300 m 的埃菲尔铁塔(如图 0.6 所示),该塔总重 8 500 t,现已成为巴黎乃至法国的标志性建筑;1890 年,英国在爱丁堡附近修建了福斯桥(如图 0.7 所示),它是一座由众多钢管弦杆构件组成的双伸臂梁铁路桥,被后人公认为桥梁史上的里程碑之一;1931 年美国纽约建成 102 层的帝国大厦(如图 0.8 所示),378 m 高的钢骨架总重超过 50 000 t,这一建筑高度保持世界纪录长达 40 年;1936 年美国旧金山建成了金门大桥(如图 0.9 所示),该桥主跨 1 280 m,是世界上第一座主跨超过 1 000 m

New Sciences in 1638, in which the design formula of the beam theory was put forward for the first time. In 1687, Newton summed up the three laws of mechanics, which laid the basis for the analysis of civil engineering. Then, the allowable stress method was established by the Frenchman Navier in 1825 for civil engineering structural design, based on material mechanics, theory of elasticity and strength of materials.

As for materials, the invention of Portland cement in 1824 and reinforced concrete in 1867 became the historical events in civil engineering. As the mass production of concrete and steel, civil engineers can make use of these materials to construct large and complex engineering facilities. In modern times, steel structures and reinforced concrete structures are applied to the vast majority of the tall, large span, huge and complicated engineering structure of modern buildings.

In 1825, the first railway in the world, 21 km long, was built in Britain. In 1863, the first underground railway was built in London. The Eiffel Tower (Fig. 0. 6), as high as 300 m, was built in 1889. The total weight of this building is 8 500 t, and it has become the landmark building for Paris even for France. In 1890, the British built the Forth Bridge (Fig. 0. 7) near Edinburgh, which is a railway bridge of double cantilever beam consisted of many steel chord members. It is recognized as one of the milestones in the history of bridge. The Empire State Building (Fig. 0. 8) which was built in 1931 in New York is 378 meters high, and the total weight of its steel skeleton is more than 50 000 t. The building height had kept the world record for 40 years. In 1936, the Golden Gate Bridge was built in San Francisco (Fig. 0. 9), with main span of 1 280 m, which is the first bridge of main span more than 1 000 m in the world.

的桥梁。

图 0.6　埃菲尔铁塔
Fig. 0.6　The Eiffel Tower

图 0.7　福斯桥
Fig. 0.7　The Forth Bridge

图 0.8　帝国大厦
Fig. 0.8　The Empire State Building

图 0.9　金门大桥
Fig. 0.9　The Golden Gate Bridge

在这一时期，我国由于近代历史原因，土木工程的发展长期处于落后状态，直到洋务运动后，才开始学习西方现代技术，并建造了一批有影响力的土木工程。例如，1909 年詹天佑主持修建的京张铁路，全长 200 km，它的建成在我国近代土木工程史上具有重要的历史意义；1934 年建成的上海国际饭店一直是这座城市的中心（如图 0.10 所示），作为上海最高楼的纪录保持了半个世纪之久；1937 年，茅以升主持建造了钱塘江大桥（如图 0.11

During this period, due to the historical reasons in China, the development of civil engineering was backward for a long time. After the westernization movement, Chinese people began to learn western modern technologies, and a group of influential civil engineering was built. For example, Beijing-Zhangjiakou Railway, with the total length of 200 km, was built by Zhan Tianyou in 1909, and it has important historical significance in the history of modern civil engineering. Since its opening in 1934, the Park Hotel in Shanghai (Fig. 0. 10) has been regarded as the focus and the tallest building in this city for more than half a century. The Qiantang

所示），它是公路、铁路两用的双层钢结构桥，是我国近代土木工程的优秀成果。

River Bridge （Fig. 0. 11） built by Mao Yi-sheng in 1937 is a double layer steel structure bridge both for road and rail, and it is the outstanding achievement in the modern civil engineering in China.

图 0.10　上海国际饭店
Fig. 0.10　The Park Hotel in Shanghai

图 0.11　钱塘江大桥
Fig. 0.11　The Qiantang River Bridge

0.2.3　现代土木工程

　　二战后，战后恢复的需要和现代科学的高速发展为土木工程提供了更为雄厚的物质基础和理论指导，一个崭新的"现代土木工程"的辉煌时代到来了。这一时期的土木工程有以下几个特点。

（1）功能要求多样化、复杂化

　　现代土木工程的特征之一就是工程设施同它的使用功能或生产工艺紧密地结合在一起。公共建筑和住宅建筑要求建筑结构与水、暖、电、气的供应及室内温、湿度的自动控制等现代化设备相结合，并且与环境相协调。悉尼歌剧院因其独树一帜的帆状屋顶成为澳大利亚的象征，也被公认为是世上最与众不同的现代建筑（如图 0.12 所示）。工业建筑要求恒

0.2.3　Contemporary civil engineering

　　After the Second World War, the postwar recovery needs and the development of modern technology provided the material basis and theoretical guidance for the development of civil engineering. A new age for civil engineering was coming. During that period, the features of civil engineering were as following.

（1）The diversification and complication of the function

　　One of the features of modern civil engineering is that the construction facilities are closely combined with their functions or manufacturing techniques. Public buildings and residential building require the combination of structures and the supply of water, heating, electricity and gas, also the automatic control of indoor temperature and humidity. Besides, They should be in harmony with the environment. The Sydney Opera House（Fig. 0.12）is as representative of Australia as the pyramids of Egypt, because of the distinguishing sails of the roof, and it is considered to

温、恒湿、防微振、防腐蚀、防辐射、防磁、除尘，并向大跨度、超重型、分隔灵活的方向发展。同时，具有特殊功能要求的各类特种工程结构，则对土木工程提出高标准的要求，如水利枢纽工程（如图0.13所示）、核电站中的安全壳要求具有极高的安全度，而海洋平台（如图0.14所示）则因其功能多样、使用环境恶劣、荷载复杂、施工困难而对工程结构提出了更高的要求。

be one of the most recognizable images of the modern world. Industrial constructions need to be constructed in constant temperature and humidity. They also ask for anti-corrosion, anti-slight shock, radiation protection, anti-magnetism, dedusting, to the large span, and flexible segmentation. Besides, all constructions of special structures ask for higher standards of civil engineering. For example, the water control project (Fig. 0.13) and the containment of nuclear power station require high safety degree, while the ocean platform (Fig. 0.14) focus on the construction structure based on its function diversity, bad operating environment, complicated load, and its difficulty in construction.

图 0.12 悉尼歌剧院
Fig. 0.12 Sydney Opera House

图 0.13 三峡水利枢纽工程
Fig. 0.13 Three Gorges Project

（2）城市建设立体化

随着经济的发展和人口的膨胀，城市人口密度快速增长，造成城市用地紧张，交通拥挤，地价昂贵，迫使城市建设向立体化发展，表现为高层建筑的大量兴起，地下工程的高速发展和城市高架公路、立交桥的大量涌现。现代化城市建设在地面、空中、地下同时展开，形成了立体化发展的局面。位于阿拉伯联合酋长国迪拜的哈利法塔（如图0.15所示），是目前最高的人工建筑，共169层，总高度828 m，历时5年建成，被称为"一座垂直而立的

（2）Three-dimensional development in cities

With the development of economy and population, the population density in cities has been grown quickly, which causes the tension of land use, traffic jam, high price of land and finally the three-dimensional development of city constructions. Many tall buildings, underground projects, urban elevated road and flyovers appear fast. Constructions are operated on the ground, underground and in the sky. The Burj Khalifa Tower (Fig. 0.15) in Dubai, United Arab Emirates is now the highest artificial building in the world. It has 169 layers and 828 m in height, and it was finished in 5 years, and named as "a vertical city". The building

城市"。日本东京八重洲地下商业街建筑面积近 70 000 m², 分 3 层。

area of Japan's Tokyo Yaesu underground commercial street is nearly 70 000 m², and it has three layers.

图 0.14　海洋平台
Fig. 0.14　Ocean Platform

图 0.15　哈利法塔
Fig. 0.15　Burj Khalifa Tower

（3）交通工程快速化

　　经济的繁荣和发展对运输系统提出了快速、高效的要求,同时现代技术的进步也为满足这种要求提供了条件。交通运输的高速化让世界变得越来越小,主要表现为高速公路的大规模修建、铁路电气化的形成和大量长距离海底隧道的出现。现在,高速公路的里程数已成为衡量一个国家现代化程度的标志之一。我国高速公路由 2002 年底的 2.51 万公里增加到 2011 年底的 8.49 万公里,跃居世界第二。2006 年建成通车的青藏铁路东起青海西宁,西至拉萨,全长 1 956 公里,是世界海拔最高、线路最长的高原铁路（如图 0.16 所示）。1993 年,贯通英吉利海峡的英法海底隧道（如图 0.17 所示）实现通车,人们仅用 35 分钟就可以从欧洲大陆穿越英吉利海峡到达英国本土。

(3) Fast development of traffic engineering

　　The development of economy requires fast and efficient work of transportation, and the advanced technology also provides condition for such development. The world has became smaller with the fast development of transportation, such as the construction of highways, the electrification of railways and the appearance of long distance channel tunnel. Now the mileage of highway has become the symbol of a country's modernization degree. At the end of 2002, the mileage of highway in China was 25 100 km, while it was increased to 84 900 km at the end of 2011, which ranks secord in the world. The Qinghai-Tibet railway (Fig. 0. 16), completed in 2006 with the total length of 1 956 km, is from Xining in Qinghai to Lhasa. It is the highest and longest plateau railway in the world . In 1993, the Channel Tunnel (Fig. 0. 17) went into trial operation. People can travel from the continental Europe to England in only 35 minutes.

图 0.16　青藏铁路
Fig. 0.16　Qinghai-Tibet railway

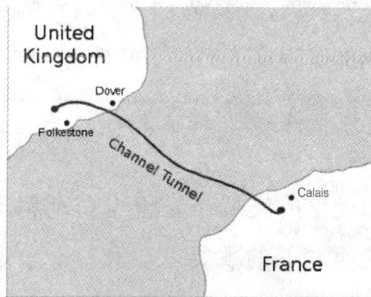

图 0.17　海底隧道
Fig. 0.17　Channel tunnel

（4）工程材料轻质高强化

工程材料中,混凝土材料发展迅速,由普通混凝土向轻骨料混凝土、加气混凝土和高性能混凝土的方向发展,既降低了混凝土的重度,又提高了其强度,同时其他性能也得到了很大改善。此外,诸如低合金、高强度钢材的发展,铝合金、建筑塑料等轻质高强材料的应用为建筑材料的轻质高强化创造了条件。纤维增强材料的发展与应用,也是 21 世纪新材料在土木工程中应用的重要标志。如碳纤维（CFRP）、芳纶纤维（AFRP）、玄武岩纤维（BF）、金属纤维以及混杂纤维等,它们既可以作为增强材料或智能材料直接掺入混凝土中,也可以制成片材或棒材作为结构构件的补强或加筋材料。

（5）施工过程工业化、装配化

在工厂成批生产房屋、桥梁的各种构配件、组合体,以在现场进行拼装的生产方式极大地加快了施工速度,减少了户外工作时间。此外,各种先进的施工手段也

（4）Lighter and more qualified engineering materials

In engineering materials, the concrete material developed quickly. Normal concrete now tends to be lightweight aggregate concrete, aerated concrete and high performance concrete, which not only decreases the weight of concrete, but also advances its intensity and improved other properties greatly. The development of low content alloy, high strength steel and the application of aluminum alloys and building plastic make the engineering materials lighter and more qualified. The development of fiber reinforced material also reflects the development of new materials in the 21st century. Materials such as carbon fiber reinforced plastic (CFRP), aramid fibers reinforced plastic (AFRP), basalt fiber (BF), metal fiber and hybrid fiber can not only be mixed into concrete as strength materials, but also be made as slice and stick materials.

（5）The industrialization and assemblage of construction

It makes the construction faster by producing the components of houses and bridges in factories while assembling in site. Besides, various advanced construction ways have been developed quickly, which makes the existence of complex tall buildings pos-

得到了很大发展,为复杂、大型、高耸的建(构)筑物的工程施工提供了条件。

(6) 设计理论精确化、科学化

结构理论的发展与完善也是现代土木工程快速发展的重要基础和标志。现代力学和分析方法以及计算机技术的发展,使得土木工程学科的理论基础得到了迅速发展,结构设计方法实现了从经验方法、安全系数法到可靠度设计方法的过渡。进入 21 世纪,基于性能设计理论、抗连续倒塌设计理论、结构耐久性理论、结构的振动控制理论、结构实验技术等又有了重大发展,为建(构)筑物的安全、可靠、合理、经济提供了设计保证。

0.3 土木工程专业及其知识构成

0.3.1 土木工程专业简介

土木工程专业是为培养土木工程专门技术人才而设的。早期的土木工程师很少受过正规教育。最早针对土木工程师的正规训练计划由法国国立路桥学院提供,该学院是 1716 年为了路桥建设的科学发展而组建的,旨在为国家路桥兵团准备人才。

我国土木工程教育则始于 1895 年的天津大学。新中国成立前,我国工科学科的设置基本上是仿效英美的,土木工程没有明确的专业和统一的教学计划,更没有

sible.

(6) The precision and scientization of design theories

The development of structure theories also helps modern civil engineering. Modern mechanics, analytical methods and the development of computer technologies accelerate the development of theoretical basis of civil engineering. Instead of experience method and safe coefficient method, reliability design method has been more accepted. From the 21st century, performance-based design theory, continuous collapse resistance design theory, durability of structure theory, the vibration control theory of structure and the structural experimental technology have been used widely, which guarantee the safety, reliability, rationality and economy of the buildings.

0.3 Major civil engineering and its knowledge composition

0.3.1 Brief introduction to major civil engineering

The major civil engineering is set up for training civil engineering specialists. Early civil engineers, as a general rule, had very little formal education. The earliest formal training program for civil engineers was offered by the Ecole Nationale Des Ponts et Chaussees, which was formed in 1716 for the scientific advancement of bridge building and road building.

The education of civil engineering in China started from Tianjing University in 1895. Before the founding of New China, the settings-up of engineering subjects in China was learnt from Europe and America.

教学大纲,各校土木系开课很不一致,开设的课程很广泛。新中国成立后,特别是改革开放以来,我国土木工程专业的教育有了很大的发展。针对专业划分过细、专业范围过窄、门类之间专业重复设置等问题,我国分别在 1982 年、1993 年、1997 年进行了三次专业目录的调整,坚持拓宽专业口径、增强适应性原则,专业主要按学科划分,使培养的人才具有较宽广的适应性。自 1997 年起建筑工程、交通土建、地下工程等近十个专业合并成为目前的"土木工程专业"。

按人才培养的层次分,土木工程专业培养的人才有专科、本科(工学学士)、硕士(工学硕士)、博士(工学博士)等几个层次。在本科教育阶段,土木工程专业属于一级学科专业,下设建筑工程、道路工程、桥梁工程、铁路工程、隧道工程等多个专业方向。进入硕士或博士教育阶段则具体分二级学科专业,如岩土工程、结构工程、防灾减灾与防护工程、桥隧工程、市政工程等。

0.3.2 土木工程专业知识构成

土木工程专业的工程对象及业务范围非常广,每个工程对象中的工作内容又分勘察、设计、施工、监理、管理等多个方面和环节。因此,专业培养中应贯彻"大土木"的人才培养理念,在这一理念下,应坚持"宽口径、厚基础"的培养原则,这样才能培养出既符合现实工程需要又符合

There were no clear subjects, no formal teaching plan and no syllabus in civil engineering department. After 1949, civil engineering developed quickly especially after the opening up to the out-side world. China changed the catalogue of specialty in 1982, 1993 and 1997, aiming to solve the problems such as the too narrow professional range or repeated majors. The new plan has aiming to cultivate students' adaptability. Since 1997, the major civil engineering has combined nearly ten majors, such as constructional engineering, traffic engineering and underground construction.

Civil engineering major aims to cultivate students with three-year college, bachelor diploma, master degree and doctor degree. Civil engineering is regarded as first discipline major in the stage of undergraduate education. It comprises constructional engineering, road engineering, bridge engineering, railway engineering, tunnel engineering, etc. In the stage of post graduate and doctoral education, civil engineering is seen as second discipline subject. It comprises geotechnical engineering, structural engineering, disaster prevention, bridge and tunnel engineering, municipal engineering, etc.

0.3.2 The professional knowledge in civil engineering

Civil engineering has a wide range of project objects and businesses. Every task includes reconnaissance, design, construction, supervision and management. Therefore, people should regard this major as "big civil engineering", learn more and set up a good foundation. By doing so, we will train students as promising specialist with strong adaptability.

土木工程未来发展要求的适应能力强的专门人才。

土木工程专业的知识体系由 4 部分组成,分别为工具性知识、人文社会科学知识、自然科学知识和专业知识。每个知识体系所涵盖的知识领域见表 0.1。

Knowledge in civil engineering can be divided into four parts: instrumental knowledge, humanities and social science knowledge, natural science knowledge and professional knowledge. The following chart is the knowledge hierarchy in each part (Table 0.1).

表 0.1　土木工程专业知识体系和知识领域

Table 0.1　Professional knowledge hierarchy and knowledge areas of civil engineering

序号 No.	知识体系 Knowledge hierarchy	知识领域 Knowledge area
1	工具性知识 instrumental knowledge	外国语、信息科学基础、计算机技术与应用 foreign languages, information science, computer technology and application
2	人文社会科学知识 humanities and social science knowledge	政治、历史、伦理学与法律、心理学、管理学、体育运动 politics, history, ethics and law, psychology, management, sports
3	自然科学知识 natural science knowledge	工程数学、普通物理学、普通化学、环境科学基础 engineering mathematics, general physics, general chemistry, environmental science basis
4	专业知识 professional knowledge	力学原理和方法、材料科学基础、专业技术相关基础、工程项目经济与管理、结构基本原理和方法、施工原理和方法、计算机应用技术 mechanics principle and method, material science basis, professional technology, economic and project management, basic principle and method of structure, construction principle and method, computer application technology

0.4　土木工程师的能力素质及职业发展

0.4　The competence and career development of civil engineers

0.4.1　土木工程师的能力素质

21 世纪是高科技时代,土木工程将引进更多的高新技术,不断提高、创新和发展,以满足人们日渐提高的社会生活和生产需求。为了适应新时期土木工程的发展,作为一名土木工程师必须具备扎实的

0.4.1　The competence of civil engineers

21st century is an high-tech age. Civil engineering will bring in more high technologies to keep development and innovation in this century, and fulfill people's increasing production needs. To adapt the development of civil engineering, a civil engineer should have solid foundation of professional know-

土木工程导论

专业基础、较强的实践能力和较高的综合素质。

（1）土木工程师的专业技能

土木工程是一个应用性的学科，具有很强的个性和综合性，大量问题需要依靠工程师的经验和工程实例来解决。土木工程师想要把在学校里学到的专业基础知识、专业知识和实践技能应用到工程项目中去，就必须依靠他们自身的各种能力，包括工程能力、科技开发能力、组织管理能力、表达能力和公关能力等专业技能。

工程能力就是土木工程技术人员在从事土木工程工作时应用工程技术知识和技能的能力。科技开发能力就是在现有的设计方法和施工技术的基础上，对设计方法和施工技术提出改进设想并予以实施的能力。土木工程是一项群体性的工作，对于土木工程师，要求其具有必要的组织管理能力，这里的管理包括人力资源管理、投资管理、进度管理、质量管理、安全管理、工程项目管理、各工种工作的协调管理等；土木工程师需要具有文字、图纸和口头的表达能力，以及社会活动、人际交往和公关的能力。

（2）土木工程师的综合素质

土木工程师的综合素质一般包括4个方面的内容：个人修养、心理和体魄、自然科学知识、土木工程专业知识。土木工

ledge, strong practical abilities and competence.

（1）The professional ability of civil engineers

Civil engineering is an applied science, which has its own characteristics and comprehensiveness. A lot of problems need to be solved by engineers' experience. Civil engineers should combine knowledge and practice in actual constructional tasks which need their abilities, such as constructional ability, science development capability, management ability, expression ability and public relation ability.

The constructional ability is the ability to apply engineering technology when operating objects; science development capability is the ability to improve the designed methods and construction skills. Civil engineering is a group work. Civil engineers should have the necessary management include the human resources, investment, quality control, schedule quality, safety control and coordination. Civil engineers should also have an excellent expression abilities both literally and pictorially, as well as the communication abilities in social public.

（2）The competence of civil engineers

The competence of civil engineers normally includes personal cultivation, physical healthy and mental health, scientific knowledge and professional knowledge. Civil engineers should have good personalities, social

程师在个人修养方面,应具有良好的思想品德、社会公德,具有高尚的科学人文素养和精神;应具有健康的心理和良好体魄,能保持乐观、积极向上;应能了解当代科学技术发展的主要方面,学会科学思维的方法;在土木工程专业方面,土木工程师要有良好的职业道德、强烈的社会责任感、正确的设计思想和创新意识,以及深入实践的愿望和本领。

(3) 土木工程师的创新意识

土木工程师的创新能力和品质,是土木工程这个有"创造力的专业"永葆青春的基本保证。但创新意识不可能孤立地培养,土木工程师除了应有扎实的知识结构、良好的实践技能和完善的能力结构外,还必须结合各自工作自觉地培养创新意识。在大学学习期间,学生首先应结合自身特点,逐步明确今后的发展方向,做好学业规划。其次,在学习的不同阶段,循序渐进地培养自学能力。同时,接受艺术和美学方面的熏陶,对激发土木工程师的创新意识大有好处。

(4) 土木工程师的法律意识

工程建设活动通常具有建设周期长、涉及面广、人员流动性大、技术要求高等特点,因此,工程建设活动应确保工程建设的质量与安全,应当符合国家的工程建设安全标准,应当遵守法律、法规的强制性规定,不得损害社会公共利益和他人的

morality, healthy body and mind. They should be aware of the new technologies in civil engineering field. Civil engineers should be able to understand the development of contemporary science and technology, and the scientific thought. They also should have good professional ethics, social responsibility, the right design thought, innovation consciousness, the desire and ability of practice.

(3) The innovation consciousness of civil engineers

Innovaiton consciousness and creativity are the essence of civil engineering. Innovation consciousness cannot be cultivated singly. Civil engineers should not only have the solid foundation of academic knowledge, excellent practice skills and perfect knowledge structure, but also be aware of the coming tendency of civil engineering. What is more, civil engineers should study all the time and accept the nurture of art to inspire their innovation abilies.

(4) The legal consciousness of civil engineers

Some construction tasks require a long duration, as well as widely related, staff mobilized and high-tech requirement, so construction progress should be safe and qualified. Civil engineers should follow the nation's law, regulations and disciplines. The construction law is aiming to rule and guide the constructional behaviors, to protect the legal constructional tasks and punish illegal

合法权益,而依法进行的建设活动也不得受到任何单位和个人的妨碍和阻挠。建设法规的作用主要表现为:规范与指导建设行为,保护合法建设行为,处罚违法建设行为。

engineering behaviors.

(5) 土木工程师的风险意识

建设工程项目在设计、施工和竣工验收等各个阶段中都大量存在未确定因素,这些未确定因素会不断变化,由此而造成的风险直接威胁工程项目的顺利实施和完工。为了降低风险,土木工程师在项目建设以前应进行风险评估,即在风险识别和风险估测的基础上把握风险发生的概率、损失严重程度,然后根据评估结果制订出完整的、切实可行的风险控制计划。

(5) The risk consciousness of civil engineers

Because of many possible unsure factors, constructional projects are facing changes during the period of design, building and final check of construction. Any tiny problem will make the task unsuccessful. In order to reduce the risk, civil engineers need to estimate the risk before applying the project. Engineers need to be aware of the probability of the risk and work out a complete and feasible risk management plan.

(6) 土木工程师的可持续发展意识

当前可持续发展已成为国际社会的共识。任何土木工程都要占据一定的自然空间并直接或间接地消耗大量的物质资源。为解决可持续发展理论中最基本的"资源有限"问题,土木工程师在价值观念、理论基础、方法原理和技术手段等方面应进行一系列的变革,以最大限度地提高自然资源的利用率,保护、恢复自然生态环境。

(6) The sustainable development awareness of civil engineers

The sustainable development thought has been accepted by people all over the world. Any civil engineering projects will occupy some natural space and consume lots of resources directly or indirectly. In order to save resources as many as possible, civil engineers should take actions to efficiently make use of the resources, which will be helpful in the environment protection.

0.4.2 土木工程师的职业发展

由于自身的专业特点,土木工程专业的就业范围比较广泛,同时随着城市建设和公路建设的不断升温,土木工程专业的

0.4.2 The career development of civil engineering

In light of its specialties, the major civil engineering has a wide range of career choices, and as the urban and road construction is heating up constantly, its career ex-

就业形势持续走好。随着我国执业资格认证制度的不断完善,土木工程师不但需要精通专业知识和技术,还需要取得必要的执业资格证书。总体来说,土木工程专业的主要就业方向可以概括为以下 3 个方面(见表 0.2)。

pectation has been on a good trend in recent years. With the improvement of the professional qualification certification system in China, civil engineers not only need to be excellent in professional knowledge and technique, but also should acquire some necessary professional qualifications certificates. The career objectives of civil engineering, by and large, can be summarized to three aspects as follows (Table 0.2).

表 0.2 土木工程专业的主要就业方向
Table 0.2 Career objectives of civil engineering

就业方向 Career objectives	代表职位 Representative titles	代表行业 Representative industries	执业资格认证 Qualification certificates
工程技术 technical engineering	施工员、建筑工程师、结构工程师、岩土工程师、建造师、技术经理、项目经理等 construction crew, architectural engineer, construction engineer, geotechnical engineer, constructor, technical manager and project manager	建筑施工企业、房地产开发企业、路桥施工企业等 construction enterprises, real estate enterprises, road and bridge construction enterprises	全国一、二级注册建筑师,全国一、二级注册结构工程师,全国一、二级注册建造师,注册土木工程师(岩土)等 First or Second Class National Certified Architect, First or Second Class National Certified Construction Engineer, First or Second Class National Certified Constructor and Certified Civil Engineer (Geotechnical field)
设计、规划及预算 design, plan and budget	项目设计师、城市规划师、预算工程师等 project designer, city planner and budgetary engineer	工程勘察设计单位、房地产开发企业、交通或市政工程类、工程造价咨询机构等。 project design, investigation and survey; real estate development;transportation and public work; project budget consulting	全国一、二级注册建筑师,全国一、二级注册结构工程师,注册土木工程师(岩土),注册造价工程师 First or Second Class National Certified Architect, First or Second National Class Certified Construction Engineer, Certified Civil Engineer (Geotechnical field) and Certified Cost Engineer
质量监督及工程监理 quality control and engineering supervision	监理工程师 supervising engineer	建筑、路桥监理公司、工程质量检测监督部门等 building, road and bridge quality supervision company, construction quality detection and supervision sector	注册监理工程师 Certified Supervising Engineer

0.5 土木工程的发展趋势

0.5.1 工程材料向轻质、高强、多功能化发展

土木工程材料是新型结构出现与发展的基础，而新型结构的出现又是新材料出现的驱动力。在结构材料方面，高强、高性能混凝土已在工程中广泛应用。目前世界上研究的混凝土抗压强度可达 300 MPa，而且混凝土的各种性能如施工性、耐久性等显著改善。但是，混凝土的生产消耗了大量自然资源。因此，从人类社会可持续发展的前景出发，混凝土也要坚持可持续发展的方向。近年来提出的发展"绿色混凝土"或"生态混凝土"正是上述出发点的集中体现，这不仅贯穿从生产到使用的全过程，还包括材料的再循环使用问题等。近年来，钢材的性能与加工工艺得到了显著改善和提高，而耐腐蚀、耐高温、易焊接、易加工的新型超级钢铁材料的研究已成为多国科学家关注的热点。

为了满足工程结构保温、隔热、隔声、耐高温、耐高压、耐磨、耐火等方面的需求，化学合成高分子材料将会广泛应用于土木工程结构中，其应用范围将扩展至大面积围护结构以及抗力结构。

0.5 The development tendency of civil engineering

0.5.1 The engineering materials tend to be lighter, stronger and multi-functionalization

Civil engineering materials are also the basis of the development of any new structures, which are the motivation of the occurrence of new materials. Now more and more high strength concretes are used. At present, the compressive strength of concretes is 300 MPa, and the application property and durability are getting better. However, the production of concretes consumes a lot of resources. Therefore, concretes industry should develop with the sustainable consciousness. The "Green Concrete" and "ecological concrete" clearly define the core idea of the above mentioned opinion, which should be applied not only into every detail in the construction, but also into the material recycle. Nowadays, the performance and processing technology of the steel have been quickly developed. New types of steel that are corrosion/heat resisting and easy to be produced are mainly concerned by scientists in many countries.

In order to meet the requirements of heat preservation, heat/noise insulation, resistance to high temperature and high pressure, abrasion resistance, fire resistance, etc., chemical synthetic polymer materials will be widely used in the structure of civil engineering. Its application scope will be extended to large areas, such as palisade structure and resistance structure.

0.5.2 工程设计向工程全生命周期综合决策发展

所谓工程的"全生命周期"是指包括工程建造、使用和老化的全过程。在不同阶段,工程的风险来源不完全相同。建造阶段的风险来自于对未完成结构及其支撑系统的分析不完全,以及对人为错误的失控;老化阶段的风险来自于结构或材料长期在自然环境和使用环境中功能的逐渐退化。相对而言,工程使用阶段的平均风险率是最低的,其主要危险来自于自然灾害和可能出现的人为灾害。以往的工程设计有的仅考虑使用阶段的工程安全,而现在除了要考虑使用阶段的安全,还要考虑安全以外的内容,如结构的功能能否得到保证以及耐久性问题等。对一些重大的高坝、水库、桥梁、高层建筑、海洋工程和港口工程,目前比较科学的做法是综合考虑建造、使用、老化三个阶段后,再做最后决策。

0.5.3 土木工程向海洋、太空、荒漠开拓

地球表面只有约30%的面积为陆地,而这其中又有大约1/3为沙漠或荒漠地区。随着地球上人口的不断增长,资源逐渐枯竭,随之带来的人类生存问题已迫在眉睫。因此,人类大力开发海洋、荒漠甚至太空资源已成为一种趋势。

现在世界各国已有许多这类成功案例,如关西国际机场是建在人工岛上的少数东亚机场的典范,可抵挡地震和台风的

0.5.2 The engineering design tends to develop in the whole life cycle project

The project "life cycle" means the whole process that includes the construction, application and maturing. The project faces different risks in different stages. In the construction process, people may make a wrong judgement of the support system of an unfinished project, and materials will degenerate in the long operating and natural environment in the maturing stage. In comparison, there will be fewer risks in the application stage, natural and man-made disasters are the main causes of the risks. Now people should not only concern about the safety in the application process, but also other problems such as the guarantee of the performances of the structures and the durability. As for those important projects such as high dams, reservoir, bridges, tall buildings, ocean engineering and harbor engineering, it is better to consider those three stages carefully and then make possible decisions.

0.5.3 Civil engineering has been exploited towards ocean, outer space and deserts

Only about 30% area of land can be used on the earth, and about 1/3 of the land are deserts. With the growing population, people are facing serious living problems due to lack of resources. It is a new tendency to open up more space from ocean, desert and outer space.

Now we have a lot of successful examples in the world. Such as Kansai International Airport, which was built on man-made

侵袭，中国香港大屿山国际机场劈山填海，而荷兰1/5 的土地面积是 800 年来通过填海得到的。在中国西北部，利用兴修水利、种植固沙植物、改良土壤等方法，使一些沙漠变成了绿洲。这些都是成功造福人类的宏大工程。人类已有的探测资料显示，外太空星球上拥有丰富的资源，有些星球甚至有可能适宜人类居住，届时移民外太空将成为可能，这也将很大程度上扩大人类的生存空间。

0.5.4 工程信息化与智能技术的发展

信息、计算机、智能化技术在土木工程领域得到了愈来愈广泛的应用，并且将是今后相当长时间内的重要发展方向。

智能化技术的应用主要体现在智能建筑与智能交通系统的发展与推广上。智能建筑的特征是所有设备都是用计算机的先进管理系统进行监测与控制的，并能通过优化控制来满足使用者对舒适、安全、能源利用率和可靠性的需求。智能交通系统是将先进的信息技术、通信技术、传感技术、控制技术以及计算机技术等有效地集成运用于整个交通运输管理体系，从而建立起实时、准确、高效、综合的运输和管理系统。

建筑信息建模（BIM）作为一种创新的工具与生产方式，是信息化技术在建筑业的直接应用，自 2002 年被提出后，已在欧

island could resist the attack of earthquake and typhoon. The Hong Kong International Airport was built on the sea, and 1/5 lands in Netherlands were obtained by filling the sea within 800 years. In northwestern China, people make some deserts into oasis by water conservancy, planting sand binder and improving soil. Some outer spaces may be suitable for human habitation, which will be helpful for human to expanse their living spaces.

0.5.4 The development of engineering information and intelligent technology

Information, computer and intelligent technology have been widely used in civil engineering, and they will become an important development direction for quite a long time.

The application of intelligent technology is mainly reflected in the development of intelligent building and intelligent transportation system. By using highly controlled and supervised machines, the features of intelligent technology meet the needs of the users, which aim for comfort, safty, energy efficiency and reliability. Intelligent transportation system is to combine information technology with communication technology, sensor technology, control technology and computer technology, aiming to build up an accurate and efficient management system.

Building information modeling(BIM) is an innovative tool and production process, which is a typical application of information technology in construction industry. The ra-

美等发达国家引发了建筑业的巨大变革。BIM 技术通过建立数字化的 BIM 参数模型,涵盖与项目相关的大量信息,服务于建设项目的设计、建造安装、运营等整个生命周期,在提高生产效率、保证生产质量、节约成本、缩短工期等方面发挥巨大的作用。虽然我国的 BIM 应用还处于雏形阶段,但是认识并发展 BIM,实现行业的信息化转型已是势不可挡的趋势。

随着计算机技术的迅猛发展,计算机在土木工程中的应用已从早期的数值计算发展到了工程设计的各个阶段和许多环节。在土木工程领域中,计算机辅助设计(computer-aided design,CAD)已经被应用到规划、设计、施工、维修和加固等各个方面,CAD 技术朝着标准化、集成化、智能化等方面发展。通过土木工程的计算机仿真分析,可在计算机上模拟原形大小的工程结构在灾害荷载作用下从变形到倒塌的全过程,从而揭示结构不安全的部位和因素。地下工程开挖全过程的计算机模拟仿真可以预示和防止出现土体失稳或管涌、潜蚀、流沙、土洞等现象。

0.5.5 土木工程的可持续发展

随着人口增长,生态失衡、资源枯竭,人类面临越来越严峻的生态环境恶化问

pid development of BIM has brought a gigantic revolution in construction industries of Europe, the United States and other developed countries since 2002. With BIM technology, a BIM parameter mode is constructed digitally, which contains the relevant data needed to support the design, construction, fabrication and procurement activities through the whole life cycle of the project. When properly implemented, the BIM users can gain a range of benefits that include improving productivity, enhancing quality, saving cost, shorting construction period and so on. Although the BIM application is still in its infancy in China, recognizing and developing BIM to realize the reforming of Chinese construction industry becomes an irresistible trend.

With the increasing development of computer technology, computer has been used in many aspects in engineering design. The computer-aided design (CAD) has been used in design, construction, maintenance and many other aspects of civil engineering. The development of CAD technology is tend to be standardized, integrated and intellectualized. The computer simulation analysis in civil engineering can simulate the whole process of the collapse of projects in disaster. This simulation will help people find the parts and factors of any dangerous structures. The computer simulation of underground projects will forecast and prevent disasters, such as soil instability or piping, shallow corrosion, sand drifting and soil cave.

0.5.5 Sustainable development of civil engineering

People are faced with more and more severe environmental degradation, population growth, ecological imbalance and resource

题。世界环境与发展委员会（World Commission on Environment and Development, WCED）在1987年提出了"可持续发展"的概念。经过几十年的努力,可持续发展的观念已成为世界各国的共识,我国更是把"可持续发展"原则定为国策之一,并在这方面做了很大的努力。

土木工程要消耗大量资源(一般占总资源消耗的25%)。为了得到土地,有时会毁林建设,围湖造地,导致环境失衡与恶化,可见土木工程师在贯彻可持续发展原则方面负有重大的责任。在工程项目建设中,土木工程师应遵循"安全、经济、适用和可持续发展"的原则,综合考虑以下方面:尽量少占耕地,提高土地利用率;城市应加快地下空间的开发利用;尽量利用可再生资源,提高废物回收再利用的效率;建设过程中应采取措施减少对环境的影响;推广节能建筑;提倡循环用水等。

根据可持续发展原则,人类居住环境应保持与自然环境的协调和谐,为此专家提出了"绿色建筑"的理念。对"绿色建筑"的理解各国学者已普遍形成如下共识:在建筑物的全生命周期中,最低程度地占有和消耗地球资源,用量最小且效率最高地使用能源,最少量地产生废弃物及排放有害环境物质,使之成为与自然和谐

depletion. World Commission on Environment and Development (WCED), put forward the concept of "sustainable development" in 1987. After decades of effort, the concept of "sustainable development" has been accepted all over the world. China has even designed "sustainable development" as one of the national policy, and made great efforts in this aspect.

Civil engineering consumes a large amount of resources (typically account for 25% of the consumption resources). To occupy the land, sometimes people need to deforest and turn lake into land, which lead to environmental imbalance and deterioration, so civil engineer should be more responsible in implementing the principles of sustainable development. In construction projects, civil engineers should follow the principle of "safe, economy, applicable and sustainable development", which means they should take the following aspects into consideration: using the arable lands as little as possible, improving land utilization; speeding up the development and utilization of underground; maximizing the use of renewable resources, improving the efficiency of waste recycling; taking measures to reduce the environmental impact during the construction process; promoting the energy-efficient buildings, water recycling and so on.

According to the principle of sustainable development, people's living environment should maintain coordination and harmony with the natural environment. The experts put forward the concept of "green building". As to the "green building", scholars have generally reached consensus as follows: minimizing the possession and consumption of the resources during the full life cycle of the buildings; using the energy in minimum amount and most efficiency; producing the least waste and discharging the least harmful

共生、有利于生态系统与人居系统共同安全,健康且满足人类功能需求、心理需求、生理需求及舒适度需求的宜居可持续建筑物。可见,绿色建筑的目标是通过人类的建设行为,达到人与自然安全、健康、和谐共生,满足人类追求适宜生存居所的需求和愿望。

绿色建筑综合了大量的实践、技术和技能,以减少并最终消除建筑对环境和人类的影响。绿色建筑要根据地理条件,充分利用环境提供的天然可再生能源,如设置太阳能装置,利用植物建造绿色屋顶或雨水花园以减少雨水径流。为了补充地下水,可以使用砾石或透水混凝土来代替普通混凝土或沥青等传统路面材料。达姆施塔特工业大学设计的亚热带全生态减碳屋(如图 0.18 所示),将太阳能作为唯一的能源。

substance to the environment; becoming a sustainable building which is in harmony with nature, conductive to the common safety and healthy for ecosystem and human settlements system; satisfying the human functioning requirement, psychological needs, physical needs and comfort needs. Therefore, the purpose of green buildings is to meet the needs of safety, healthy and harmony between human and nature through the construction, and to satisfy people's demand of appropriate living condition.

Green buildings bring together a vast array of practice, techniques and skills to reduce and ultimately eliminate the impacts of buildings on the environment and human being. Take advantage of renewable resources is emphasized, e. g., using sunlight through passive solar, active solar, and photovoltaic equipment, using plants and trees through green roofs, rain gardens to reduce rainwater run-off. Many other techniques are used, such as low-impact building materials, packed gravel or permeable concrete instead of conventional concrete or asphalt to enhance replenishment of ground water. Solar energy is the only energy in the subtropical ecosystem carbon reduction house designed by Darmstadt University of Technology(Fig. 0.18).

图 0.18 达姆施塔特工业大学设计的亚热带全生态减碳屋
Fig. 0.18 Subtropical ecosystem carbon reduction house designed by Darmstadt University of Technology

绿色建筑的标准在不同国家和地区都存在差异，但绿色建筑的设计都必须包含以下方面：选址和结构的能效、能耗、水资源的能效、材料能效、室内环境改善、运行和维护优化、废弃物减排等。绿色建筑的本质是以上一个或多个原则的优化。通过适当的协同设计，各个绿色建筑技术可以共同工作以发挥更大的累积效应。

While the standard for green buildings is constantly evolving and may differ from region to region. The fundamental principles are as follows：siting and structure design efficiency, energy efficiency, water efficiency, materials efficiency, indoor environmental quality enhancement, operations and maintenance optimization, waste and toxics reduction. The essence of green building is an optimization of one or more of these principles. With the proper synergistic design, individual green building technologies may work together to produce a greater cumulative effect.

中国地域广大、人口众多、环境条件恶化、资源匮乏且存在分布不合理的现象，经济与社会条件差异性突出，因此，发展绿色建筑具有特殊的科学地位、政治作用、社会意义和经济价值。

China has a vast area with a large population, deteriorating environmental condition, insufficient resource and irrational distribution. The differences of economy and social condition are highlighted. Therefore, the development of green building has the special scientific status, political role, social significance and economic value.

注：本章插图 0.2 ~ 0.18 来源于网络。
Note：the illustrations in this chapter Fig. 0.2 ~ 0.18 are from the network.

知识拓展
Learning More

相关链接　Related Links

（1）中国土木工程网 http://www.civil.edu.cn/

（2）美国土木工程师协会官网 http://www.asce.org/

（3）美国混凝土学会 ACI 官网 http://www.concrete.org/

（4）中国土木网 http://www.zgtm.com/

（5）中国土木科技网 http://www.tumukeji.com/

（6）中国土木工程学会官网 http://www.cces.net.cn/

（7）美国发现工程网 http://www.discover engineering.org/

思考题　Review Questions

（1）什么是土木工程？思考土木工程在人类社会发展中的作用。

What is civil engineering? Think about the role of civil engineering in the development of human society.

（2）了解和认识土木工程的历史、现状和未来。

Learn and understand the history, the present situation and the future of civil engineering.

（3）土木工程专业的工程对象与业务范畴有哪些？

What are the project objects and scopes of civil engineering?

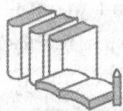

参考文献
References

［1］全国土木工程专业指导委员会：《本科教育培养目标和培养方案及课程教学大纲》，中国建筑工业出版社，2002年。

［2］高等学校土木工程学科专业指导委员会：《高等学校土木工程本科指导性专业规范》，中国建筑工业出版社，2011年。

［3］叶列平：《土木工程科学前沿》，清华大学出版社，2006年。

［4］赵鸿佐，胡鹤钧：《中国土木建筑百科辞典》，中国建筑工业出版社，1999年。

［5］叶志明：《土木工程概论》（第3版），高等教育出版社，2009年。

［6］周新刚：《土木工程概论》（第1版），中国建筑工业出版社，2011年。

［7］BHAVIKATTI. S. S. Basic Civil Engineering. New Age International Ltd. 2010.

［8］丁大钧，蒋永生：《土木工程概论》（第2版），中国建筑工业出版社，2010年。

［9］［印］M S帕拉理查米：《土木工程概论》（第3版），机械工业出版社，2005年。

［10］Nitsure,S. P. Pawar A. D. Basic Civil Engineering. 1st ed. Technical Publications Pune，2006.

［11］《绿色建筑》教材编写组：《绿色建筑》，中国计划出版社，2008年。

第1章　土木工程材料

Chapter 1　Civil Engineering Material

　　土木工程材料是土木工程建(构)筑物所使用的各种材料及制品的总称。它是一切土木工程的物质基础,决定了建筑结构形式及其施工方法。土木工程材料的发展与当前科技水平、工业化水平以及环境保护要求密切相关。为了适应建筑工业化、提高工程质量、保护生态环境、实现可持续发展的要求,土木工程界不断涌现出各种新型材料;新材料的出现,又促使建筑形式发生变化、结构设计和施工技术产生革新。本章就常见的土木工程材料进行简要的介绍。

Civil engineering material (CEM) is a generic term of all kinds of materials and products used in civil engineering (CE), which is the material basis in the whole CE buildings. CEM determines the structural forms and construction methods. Its development is closely related with the current level of science and technology, industrialization and environmental protection. In order to adapt to building industrialization, improvement of engineering quality, protection of ecological environment and social sustainable development, new materials are constantly emerging in CE field. The appearance of new materials promotes the renovation of architectural form and reformation of structural design and construction technique. The common CEMs will be introduced in this chapter.

1.1　土木工程材料分类概述

　　土木工程材料品种繁多,钢材、水泥、木材、混凝土、砖、砌块、沥青等都是常见的工程材料。土木工程材料的分类方法主要有如下3种。

1.1　Classification and overview of CEM

There are many kinds of CEM, such as steel, cement, wood, concrete, brick, building block and asphalt. Common classification methods mainly include:

① According to its usability, CEMs can be divided into structural materials (materi-

① 按使用性能分类,可以分为结构材料(受力构件或结构所用的材料,如基础、梁、板、柱等所用的材料)、墙体材料(内外及隔墙墙体所用的材料,如砌墙砖、砌块、墙板、幕墙等所用的材料)、功能材料(具有专门功能的材料,如防水材料、保温隔热材料、装饰装修材料、地面材料及屋面材料等)。

② 按用途分类,可以分为建筑结构材料、桥梁结构材料、水工结构材料、路面结构材料等。

③ 按化学成分分类,土木工程材料又可以分为无机材料、有机材料及复合材料(见表1.1)。

als used for bearing members and structures, such as foundation, beam, slab, column, etc.), wall materials (materials used for external and internal walls, such as wall brick, building block, wallboard, curtain wall), and materials with specialized functions (such as waterproof materials, thermal insulation materials, decoration materials, roofing materials, etc.).

② According to the usage of CEMs, they can be divided into building structure materials, bridge structure materials, hydraulic structure materials and pavement structure materials, etc.

③ According to chemical components of CEMs, they can be divided into inorganic materials, organic materials and composite materials, as shown in Table 1.1.

表1.1 土木工程材料按化学成分分类
Table 1.1 CEMs classified according to the chemical components

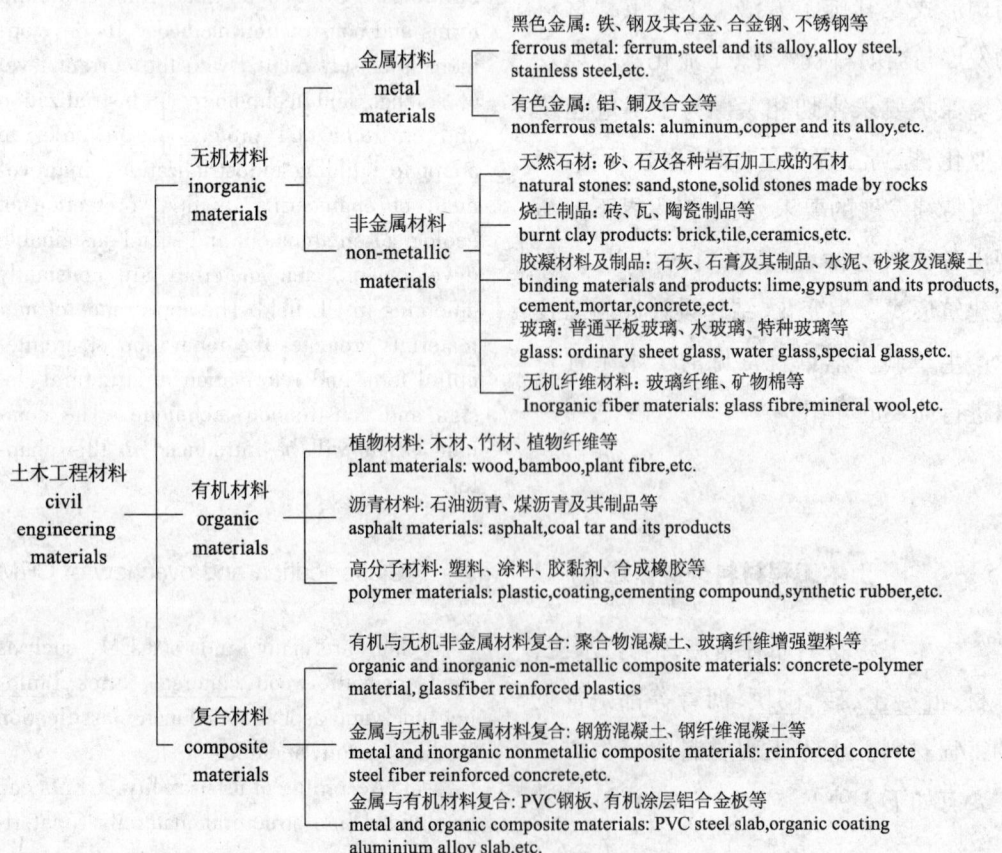

土木工程材料 civil engineering materials	无机材料 inorganic materials	金属材料 metal materials	黑色金属:铁、钢及其合金、合金钢、不锈钢等 ferrous metal: ferrum,steel and its alloy,alloy steel, stainless steel,etc.
			有色金属:铝、铜及合金等 nonferrous metals: aluminum,copper and its alloy,etc.
		非金属材料 non-metallic materials	天然石材:砂、石及各种岩石加工成的石材 natural stones: sand,stone,solid stones made by rocks
			烧土制品:砖、瓦、陶瓷制品等 burnt clay products: brick,tile,ceramics,etc.
			胶凝材料及制品:石灰、石膏及其制品、水泥、砂浆及混凝土 binding materials and products: lime,gypsum and its products, cement,mortar,concrete,ect.
			玻璃:普通平板玻璃、水玻璃、特种玻璃等 glass: ordinary sheet glass, water glass,special glass,etc.
			无机纤维材料:玻璃纤维、矿物棉等 Inorganic fiber materials: glass fibre,mineral wool,etc.
	有机材料 organic materials		植物材料:木材、竹材、植物纤维等 plant materials: wood,bamboo,plant fibre,etc.
			沥青材料:石油沥青、煤沥青及其制品等 asphalt materials: asphalt,coal tar and its products
			高分子材料:塑料、涂料、胶黏剂、合成橡胶等 polymer materials: plastic,coating,cementing compound,synthetic rubber,etc.
	复合材料 composite materials		有机与无机非金属材料复合:聚合物混凝土、玻璃纤维增强塑料等 organic and inorganic non-metallic composite materials: concrete-polymer material,glassfiber reinforced plastics
			金属与无机非金属材料复合:钢筋混凝土、钢纤维混凝土等 metal and inorganic nonmetallic composite materials: reinforced concrete steel fiber reinforced concrete,etc.
			金属与有机材料复合:PVC钢板、有机涂层铝合金板等 metal and organic composite materials: PVC steel slab,organic coating aluminium alloy slab,etc.

1.2 石材、砖、瓦和砌块

石材、砖、瓦和砌块等是最基本的建筑材料。无论是在古代还是现代的建筑领域中，石材、砖、瓦和砌块均处于不可替代的地位。

1.2.1 石材

凡采自天然岩石，经过加工或未经加工的石材，统称为天然石材（如图 1.1 所示）。一般天然石材具有强度高、硬度大、耐磨性好、装饰性及耐久性好等优点。石材的使用有着悠久的历史，古埃及的金字塔、太阳神神庙，中国隋唐时期的石窟、石塔、赵州桥等，都是具有历史代表性的石材建筑。在现代建筑中，北京人民英雄纪念碑、毛主席纪念堂、人民大会堂等，都是使用石材的典范。石材被公认为是一种优良的土木工程材料，土木工程中常用的石料根据其加工程度分为毛石、片石、料石、装饰石材和石子等（如图 1.2 所示）。

图 1.1 天然石材
Fig. 1.1 Natural stone

(1) 毛石

岩石被爆破后直接获得的形状不规划的石块称为毛石。土木工程中使用的

1.2 Stone, brick, tile and building block

Stone, brick, tile and building block are basic building materials. They are irreplaceable in CE field, not only in ancient times, but also in modern times.

1.2.1 Stone

Any stone collected from natural rock, whether processed or not, is called natural stone (Fig. 1.1). Natural stone usually has the advantages of high strength, large hardness, good abrasion resistance and durability, etc. The use of stone in CE field has a long history. Such as, ancient Egyptian Pyramids, Sun God temple, and Rock Cave, stone tower, Zhaozhou Bridge built in Sui and Tang Dynasties in China are historic stone buildings. In modern architectures, the Monument to the People's Heroes, Chairman Mao Memorial Hall, and Great Hall of the People in Beijing are typical examples buildings with stone material. Stone material is recognized as an excellent CEM, and it can be divided into rubble, flag, dressed stone, decorative stone and cobblestone according to the degree of processing (Fig. 1.2).

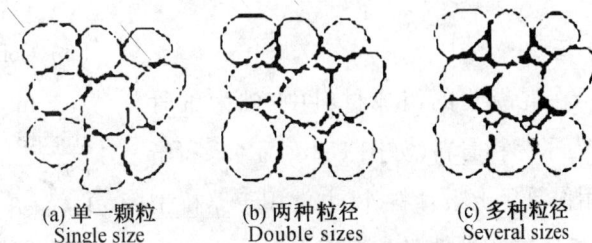

(a) 单一颗粒　　(b) 两种粒径　　(c) 多种粒径
Single size　　Double sizes　　Several sizes

图 1.2 粗骨料颗粒级配图
Fig. 1.2 Distribution diagram of coarse aggregates

(1) Rubble

Rubble is the out-of-shape stone which obtained from the blasted rocks. The average height of rubbles used in CE is not less than

毛石，一般高度应不小于150 mm。毛石可用于砌筑基础、堤坝、挡土墙等，乱毛石也可用作毛石混凝土的骨料。

（2）片石

片石也是由爆破而得的，形状不受限制，但薄片不得使用。一般片石的尺寸应不小于150 mm，主要用来砌筑圬工工程、护坡、护岸等。

（3）料石

料石是由人工或机械开采出的较规则的六面体石块，再经人工加凿而成。料石一般由致密均匀的砂岩、石灰岩、花岗岩加工而成，用于土木工程结构物的基础、勒脚、墙体等部位。

（4）饰面石材

建筑物内外墙面、柱面、地面、栏杆、台阶等处装修用的石材称为饰面石材。饰面石材一般由大理石和花岗岩制成，大理石板材可用于室内装饰，而花岗岩板材主要用于土木工程的室外饰面。

（5）石子

在混凝土的组成材料中，砂为细骨料，石子为粗骨料（如图1.2所示）。石子除用作混凝土粗骨料外，也常用于路桥工程、铁道工程的路基等。石子又分为碎石和卵石：由天然岩石或卵石经破碎、筛分而得到的粒径大于5 mm的颗粒，称为碎石或碎卵石；岩石由于自然条件作用而形成的粒径大于5 mm的颗粒，称为卵石。

150 mm. Rubble is available for laying foundation, dam, revetment, etc. Mess rubble can be also used as aggregate for rubble concrete.

(2) Flag

Flag is also obtained from the blasted rocks, but its shape is not restricted, and the slice can't be used. Generally, the size of flag should be no less than 150 mm. Flag can be used for masonry engineering, slope protection, revetment, etc.

(3) Dressed Stone

Dressed stone is gained from hexahedral stones, which is exploited by artificial or mechanical method. Generally, dressed stone is made by dense homogeneous sandstone, limestone, granite. It can be used for foundation, plinth, and wall of CE structures.

(4) Facing stone

Facing stone is the stone used for wall, cylindrical surface, ground, railing, and step surface of the building. Facing stone is usually made of marble and granite. The former is used for indoor decoration, and the latter is mainly for outside decoration.

(5) Cobblestone

In concrete ingredients, sand is fine aggregate, and cobblestone is coarse aggregate (Fig. 1.2). Besides, cobblestone also can be used for the roadbed in bridge and road engineering, railway engineering, etc. Cobblestone is divided into gravel and pebble. Gravel is gained from broken natural rock or pebble, and its diameter is greater than 5 mm. The rock, which is formed by natural conditions, is called pebble, and its diameter is also greater than 5 mm.

1.2.2 砖

砖是一种常用的砌筑材料。砖有多种分类方法：① 按生产工艺可分为两类，一类是通过焙烧工艺制成的，称为烧结砖；另一类是通过蒸压工艺制成的，称为蒸压砖，也称非烧结（免烧）砖。② 按所用原材料可分为黏土砖、页岩砖、煤矸石砖、粉煤灰砖、炉渣砖和灰砂砖等。③ 按有无孔洞砖又可分为实心砖、多孔砖和空心砖。其中，孔洞率大于等于25%，且孔的尺寸小而数量多的砖为多孔砖（如图1.3所示），常用于承重部位；孔洞率大于等于40%，且孔的尺寸大而数量少的砖为空心砖（如图1.4所示），常用于非承重部位。

1.2.2 Brick

Brick is a kind of masonry materials. Usually, bricks can be classified as follows：① Classified by manufacturing technique, brick can be divided into fired brick and autoclaved brick. The former is made by roasting process, and the latter is made by autoclaved process. ② Classified by raw material, brick can be divided into clay brick, shale brick, colliery wastes brick, fly ash brick, cinder brick, sand-lime brick, etc. ③ Classified by the hole ratio inside, brick can be divided into solid brick, perforated brick and air brick. Perforated brick is the brick with void ratio not less than 25%. The void size is small and the number of hole is large (Fig. 1.3). Therefore, it is commonly used in load-bearing areas. Air brick is the brick with void ratio not less than 40%. The void size is big but the number of hole is few (Fig. 1.4). It is commonly used in no load-bearing areas.

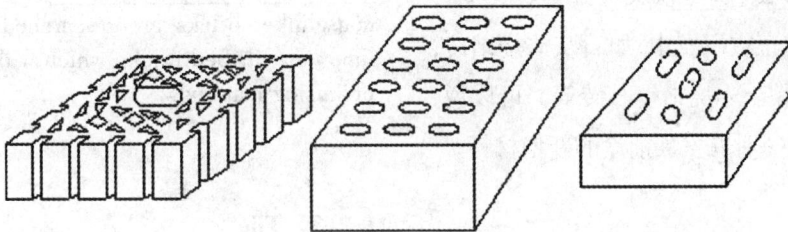

图 1.3　烧结多孔砖
Fig. 1.3　Fired perforated brick

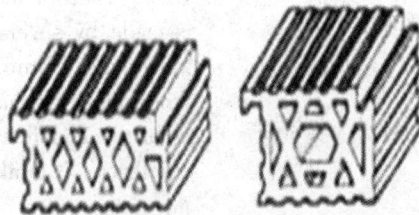

图 1.4　烧结空心砖
Fig. 1.4　Fired air brick

砖的标准尺寸为240 mm×115 mm×53 mm，通常将240 mm×115 mm的面称

The standard size of brick is 240 mm×115 mm×53 mm. The face with size of 240 mm×115 mm is commonly called big face.

为大面,240 mm × 53 mm 的面称为条面, 115 mm × 53 mm 的面称为顶面,如图 1.5 所示。

The face with size of 240 mm × 53 mm is called side face, and the face with size of 115 mm × 53 mm is called top face, as shown in Fig. 1.5.

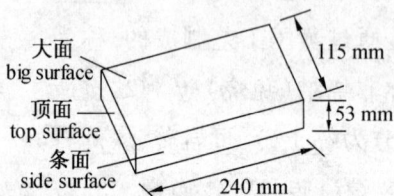

图 1.5　砖及其各部分名称
Fig. 1.5　Brick and its faces

由于生产烧结普通砖(以黏土砖为主)的过程中要大量占用耕地,能耗高、污染环境,施工生产中劳动强度高、工效低,因此我国绝大多数城市现已全面禁止使用烧结普通砖。目前应用较广的是蒸压砖,主要有蒸压灰砂砖、蒸压粉煤灰砖等。其他一些非烧结砖正在研发中,如石灰非烧结砖,水泥、石灰黏土非烧结空心砖等。可以说,非烧结砖是一种有发展前途的新型材料。

Owing to the use of a large number of cultivated land, high energy consumption, pollution, high labor intensity and low work efficiency, the application of fired common brick is prohibited in most cities in China. Autoclaved brick is widely used at present, which includes autoclaved sand-lime brick, autoclaved flyash-lime brick, etc. Many non-clinker bricks are researched, such as lime non-vitrified brick, which will become a promising material.

1.2.3　瓦

瓦,一般指黏土瓦,属于屋面材料(如图 1.6 所示)。它以黏土为主要原料,经泥料处理、成型、干燥和焙烧等环节制成。由于黏土瓦材质脆、自重大、片小,施工效率低以及破坏与污染环境等缺点,与黏土砖一样,目前已经禁止使用。

随着建筑工业的发展,新型建筑材料的涌现,目前我国生产的瓦的种类很多,按形状,可分为平瓦和波形瓦(如图 1.7 所示)两类;按所用材料,可分为陶土烧结

1.2.3　Tile

Tile generally referred to the clay tile in the past. It is a kind of roof material (Fig. 1.6). The raw material of tile is clay, and it is made by several processes such as sludge processing, forming, drying, and roasting. Since clay tile has many shortcomings, such as brittle and heavy, pollution and low work efficiency, it is also prohibited at present.

With the development of construction industry and the appearance of new materials, many kinds of tile have been produced, such as flat tile and pantile (Fig. 1.7), which are divided by shape. Classified by materials, there are clay sintered tile, con-

瓦、混凝土瓦、石棉瓦、钢丝网水泥瓦、聚氯乙烯瓦、玻璃钢瓦、沥青瓦等。

crete tile, asbestos tile, steel mesh cement tile, polyvinyl chloride tile, glass steel tile, asphalt tile, etc.

图1.6　黏土瓦
Fig. 1.6　Clay tile

图1.7　波形瓦
Fig. 1.7　Pantile

1.2.4　砌块

砌块是人造板材,外形多为直角六面体(如图1.8所示)。砌块建筑在我国始于20世纪20年代,近十年来发展较快。砌块可以充分利用地方资源和工业废渣,节省黏土资源和改善环境,实现可持续发展。且其具有生产工艺简单,原料来源广,适应性强,制作及使用方便灵活等优点。砌块除用于砌筑墙体外,还可用于砌筑挡土墙、高速公路隔音屏障及其他构筑物。

1.2.4　Building block

Building block is a kind of artificial panel, its shape is right angle hexahedron (Fig. 1.8). In China, block buildings can date back to 1920s, and have developed rapidly in the recent decade. Blocks can make use of local resources and industrial wastes, which is a good way to save clay resources and improve environments, and then to achieve sustainable development. Block has lots of advantages, such as simple production process, wide raw material source, good adaption, and flexible usage. Block can be used not only in the external wall, but also in the retaining wall, the highway noise barriers and other constructions.

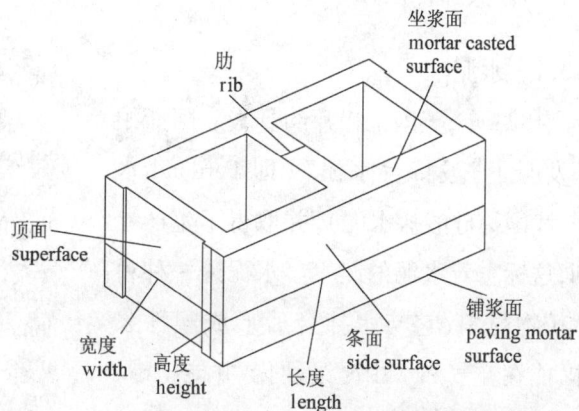

坐浆面
mortar casted surface

肋
rib

顶面
superface

宽度
width

高度
height

长度
length

条面
side surface

铺浆面
paving mortar surface

图1.8　砌块及其他部位名称
Fig. 1.8　Building block and its parts

土木工程材料　Civil Engineering Material

土木工程中,凡是经过一系列物理、化学作用,能将散粒材料(如沙子、石子等)或块状材料(如砖、石块和砌块等)黏结成具有一定强度的整体的材料,称为胶凝材料。胶凝材料的分类见表1.2。

The material, which can bond granular materials (such as sand, stone, etc.) and massive materials (such as brick, stone, block, etc.) together as a whole unit by a series of physical and chemical effects, is called cementitious(binding)material in CE. Its classification is shown in Table 1.2.

表 1.2　胶凝材料的分类
Table 1.2　Classification of cementitious material

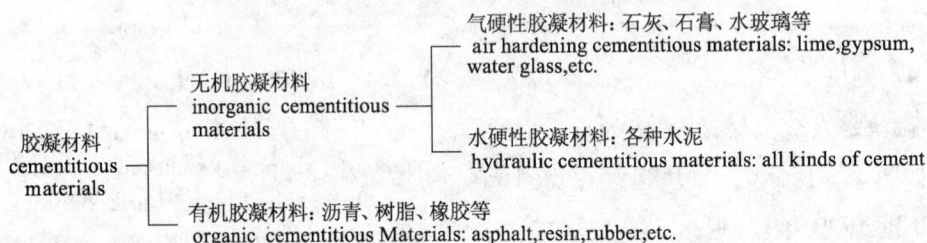

胶凝材料
cementitious
materials
- 无机胶凝材料
inorganic cementitious
materials
  - 气硬性胶凝材料:石灰、石膏、水玻璃等
  air hardening cementitious materials: lime,gypsum,
  water glass,etc.
  - 水硬性胶凝材料:各种水泥
  hydraulic cementitious materials: all kinds of cement
- 有机胶凝材料:沥青、树脂、橡胶等
organic cementitious Materials: asphalt,resin,rubber,etc.

气硬性胶凝材料在水中不能硬化,只能在空气中硬化,保持并发展其强度,因而不能用于潮湿环境和水中;水硬性胶凝材料不仅能在空气中硬化,还能更好地在水中硬化,保持并继续发展其强度,因此它既适用于地上,也适用于潮湿环境或水中。下面主要介绍土木工程中常见的胶凝材料及其拌合物。

Air hardening cementitious material can be hardened and develop its strength only in air rather than in water, so it can not be used in damp environment. Hydraulic cementitious material can be hardened not only in air but also in water, and it can be applied on the ground, damp environment and under water. Common cementitious materials and its mixtures used in CE will be introduced briefly.

1.3.1　水泥

1824 年,英国工程师约瑟夫·阿斯帕丁发明了"波特兰水泥"(即 Portland 水泥,我国称硅酸盐水泥),并取得了生产专利,这标志着水泥的诞生。水泥是一种粉状矿物材料,它与水拌合后形成塑性浆体,能在空气和水中凝结硬化,并能把砂、石等材料胶结成整体,形成坚硬石状体的水硬性胶凝材料。普通水泥的主要成分

1.3.1　Cement

Joseph Aspdin, a British engineer, discovered the "Portland cement" (which is called silicate cement in China) in 1824, and obtained production patent. This event marked the birth of cement. Cement is a kind of powdery mineral material. It becomes plastic slurry after mixed with water, and can condense and harden in air and water. This plastic slurry can combine the materials such as sand and stone into a whole, and then form a hard body like stone. The

包括硅酸三钙（3CaO · SiO$_2$）、硅酸二钙（2CaO · SiO$_2$）和铝酸三钙（3CaO · Al$_2$O$_3$）。

土木工程中应用的水泥品种众多，在我国就有上百个品种。按水泥的主要水硬化物质可分为硅酸盐系水泥、铝酸盐系水泥、硫铝酸盐系水泥、铁铝酸盐系水泥、磷酸盐系水泥、氟铝酸盐系水泥等系列；按水泥的用途和性能可分为通用水泥、专用水泥和特性水泥三大类。

1.3.2 砂浆

砂浆是由胶凝材料、细骨料、水，有时也加入适量掺和料和外加剂混合，按适当比例配制而成的土木工程材料，在工程中起黏结、衬垫和传递应力的作用。在结构工程中，砂浆可以把砖、砌块和石材等黏结为砌体；在装饰工程中，墙面、地面及混凝土梁、柱等需要用砂浆抹面，起到保护结构和装饰的作用。

按胶凝材料不同，砂浆可以分为水泥砂浆、水泥混合砂浆、石灰砂浆、石膏砂浆和聚合物砂浆等。其中，水泥混合砂浆是在水泥砂浆中加入一定量的掺和料（如石灰膏、黏土膏、电石膏等）制成的，以此来改善砂浆的和易性，降低水泥用量。

按用途不同，砂浆又可以分为砌筑砂浆、抹面砂浆和特种砂浆等。将砖、石、砌块等黏结成砌体的砂浆称为砌筑砂浆。它起着黏结砌块、传递荷载、均匀应力、协

main compositions of Portland cement include 3CaO · SiO$_2$, 2CaO · SiO$_2$ and 3CaO · Al$_2$O$_3$.

The variety of cement in CE is numerous, and there are hundreds of varieties in China. According to the main water hardening materials, cement can be divided into silicate cement (or Portland cement), aluminate cement, sulphur aluminate cement, iron aluminate cement, phosphate, fluorine aluminate cement and other series. According to its usage and performance, cement can also be divided into general cement, special cement and characteristic cement.

1.3.2 Mortar

Mortar is made up of cementitious materials, fine aggregate and water with appropriate proportion. Sometimes, the suitable amount of admixture and additive will be added. Mortar plays functions as bond, pad and transfers stress in CE. In structural engineering, mortar can bond brick, block and stone to build masonry. In decorative engineering, mortar can protect and decorate the surface of structural element, such as wall, ground, concrete beam and column, etc.

According to cementitious materials, mortar can be divided into cement mortar, mixed cement mortar, lime mortar, plaster mortar and polymer mortar, etc. Mixed cement mortar is formed by adding a certain amount of admixture into cement mortar, such as lime paste, and clay, gypsum, which can improve the workability of mortar and reduce the amount of cement.

According to its application, mortar can also be divided into masonry mortar, decorative mortar and special mortar. Masonry mortar is used to mix brick, stone and building blocks. It can bond blocks, transfer load, make stress distribution more uniform and

调变形的作用,是砌体的重要组成部分。凡粉刷于建筑物或建筑构件表面的砂浆,统称为抹面砂浆,如图1.9所示。抹面砂浆具有保护基层材料,满足使用要求和装饰的作用。特种砂浆是指具有某些特殊功能的抹面砂浆,主要有绝热砂浆、吸声砂浆、耐酸砂浆和防辐射砂浆等。

1.3.3 沥青

沥青是一种褐色或黑褐色的有机胶凝材料,在房屋建筑、道路、桥梁等工程中有着广泛的应用,采用沥青作为胶凝材料的沥青拌合料是公路路面、机场跑道面的一种主要材料(如图1.10所示)。由于沥青属于憎水材料,因此也广泛应用于水利工程以及其他防水、防渗工程中。

coordinate deformation, so it is an important part of the masonry. The mortar which is painted on the surface of the building or its component is called decorative mortar, as shown in Fig. 1. 9. Decorative mortar can protect the base materials to meet the serviceability and achieve adornment effect. Special mortar has some special functions, mainly contain sound absorption, insulation, acid-proof and radiation protection, etc.

1.3.3 Asphalt

Asphalt is a kind of brown or dark brown organic cementitious material, which has been widely used in house buildings, roads, bridges and so on. Asphalt cementitious material is mainly used in highway pavement and airport runway, as shown in Fig. 1. 10. Considering hydrophobic nature, asphalt is widely used in hydraulic projects and other waterproof, anti-seepage projects.

图1.9 抹面砂浆
Fig. 1.9 Decorative mortar

图1.10 沥青路面
Fig. 1.10 Bituminous pavement

1.4 建筑钢材

1.4.1 钢材分类

钢是由生铁冶炼而成的。在理论上凡含碳量在2.06%以下,含有害杂质较少的铁碳合金均可称为钢。钢的品种繁多,分类方法也很多(见表1.3),通常有按化学成

1.4 Construction steel

1.4.1 Classification of steel

Steel is made from pig iron. It is the iron-carbon alloy, in which the carbon content is less than 2. 06%, and few harmful impurities exist. There are many varieties of steel, which are usually classified by chemical composition, quality, application, etc., as

土木工程导论

分、质量、用途等几种分类方法。土木工程常用钢材可划分为钢结构用钢和混凝土结构用钢两大类,两者所用的钢种基本上都是碳素结构钢和低合金高强度结构钢。

shown in Table 1.3. In CE, steel can be divided into two kinds, namely, steel for steel structure and steel for the reinforcement of concrete. All steels are basically carbon steels and low-alloy high strength structural steels.

表 1.3　钢的分类
Table 1.3　Classification of steel

分类 Classification	类别 Category		特性 Character
按化学成分分类 classified by chemical composition	碳素钢 carbon steel	低碳钢 low-carbon steel	含碳量 < 0.25% carbon content <0.25%
		中碳钢 medium-carbon steel	含碳量 0.25% ~0.60% carbon content 0.25% ~0.60%
		高碳钢 high-carbon steel	含碳量 > 0.60% carbon content > 0.60%
	合金钢 alloy steel	低合金钢 low-alloy steel	合金元素总含量 < 5% total content of alloy elements < 5%
		中合金钢 medium-alloy steel	合金元素总含量 5% ~10% total content of alloy elements 5% ~10%
		高合金钢 high-alloy steel	合金元素总含量 > 10% total content of alloy elements > 10%
按脱氧程度分类 classified by deoxidization degree	沸腾钢 rimmed steel		脱氧不完全,硫、磷等杂质偏析较严重,代号为"F" deoxidization is incomplete; segregation of impurities such as sulfur and phosphorus is serious, which is code named "F"
	镇静钢 killed steel		脱氧完全,同时去硫,代号为"Z" deoxidization is complete; sulfur is abandon, which is code named "Z"
	半镇静钢 balanced steel		脱氧程度介于沸腾钢和镇静钢之间,代号为"b" deoxidization degree is between the rimmed steel and killed steel, which is code named "b"
	特殊镇静钢 special killed steel		比镇静钢脱氧程度还要充分彻底,代号为"TZ" deoxidization degree fully completely, which is code named "TZ"
按质量分类 classified by quality	普通钢 ordinary steel		含硫量≤0.055% ~0.065%,含磷量≤0.045% ~ 0.085% sulfur content ≤0.055% ~0.065%, phosphorus content ≤0.045% ~0.085%
	优质钢 high quality steel		含硫量≤0.03% ~0.045%,含磷量≤0.035% ~ 0.045% sulfur content ≤0.03% ~0.045% phosphorus content ≤0.035% ~0.045%

分类 Classification	类别 Category	特性 Character
按质量分类 classified by quality	高级优质钢 high-grade fine steel	含硫量≤0.02%～0.03%,含磷量≤0.027%～0.035% sulfur content ≤0.02%～0.03% phosphorus content ≤0.027%～0.035%
按用途分类 classified by application	结构钢 structural steel	工程结构构件用钢、机械制造用钢 engineering structural and machinery steel
	工具钢 tool steel	各种刀具、量具及模具用钢 all kinds of cutter, gauge and mould
	特殊钢 special steel	具有特殊物理、化学或机械性能的钢,如不锈钢、耐热钢、耐酸钢、耐磨钢、磁性钢等 special physical, chemical or mechanical performance, such as stainless steel, heat-resistant steel, acid-resistant steel, wear-resistant steel, magnetic steel, etc.

1.4.2　钢结构用钢材

钢结构用钢主要有型钢、钢板和钢管。型钢有热轧及冷弯成形两种;钢板有热轧(厚度为 0.35～200 mm)和冷轧(厚度为0.2～5 mm)两种;钢管有热轧无缝钢管和焊接钢管两大类。钢结构的连接方法有焊接、螺栓连接和铆接 3 种,如图 1.11 所示。

1.4.2　Steel for steel structure

In steel structure, the shape steel, steel plate and steel pipe are mainly used. Shape steel can be formed by hot rolling or cold bending. There are two kinds of steel plate: the hot rolled steel (thickness: 0.35～200 mm) and cold rolled steel (thickness: 0.2～5 mm). There are two kinds of steel pipe: the hot rolled seamless pipe and the welded pipe. Connection methods of steel structure are welding, bolt connection and riveting, as shown in Fig. 1.11.

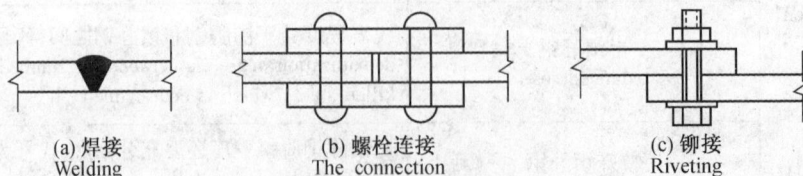

(a) 焊接　　　　　　　(b) 螺栓连接　　　　　　(c) 铆接
Welding　　　　The connection　　　　Riveting

图 1.11　钢结构连接方法
Fig. 1.11　Connection method of steel structure

(1) 型钢

型钢主要有热轧型钢和冷弯薄壁型钢两大类。其中,热轧型钢常用的有角钢(等边的和不等边的)、工字钢、槽钢、T 型钢、H 型钢、Z 型钢等,如图 1.12 所示。冷

(1) Shape steel

Shape steel mainly includes two kinds: hot rolled shape steel and cold-formed thin-walled steel. The common hot rolled shape steels contain angle steel, I-shaped steel, U-shaped steel, T-shaped steel, H-shaped

土木工程导论

弯薄壁型钢通常用1~6 mm 薄钢板冷弯或模压而成,有角钢、槽钢等开口薄壁型钢及方形、矩形等空心薄壁型钢,主要用于轻型钢结构。

steel, Z-shaped steel, as shown in Fig. 1.12. Cold-formed thin-walled steel is made by 1~6 mm thin steel plate, which includes angle steel, U-shaped steel, square or rectangular hollow thin-walled steel, etc. It mainly used in thin-walled steel structure.

(a) 角钢
Angle steel

(b) 工字钢
I-shaped steel

(c) 槽钢
U-shaped steel

(d) T 型钢
T-shaped steel

(e) H 型钢
H-shaped steel

(f) Z 型钢
Z-shaped steel

图 1.12　几种常用热轧型钢
Fig. 1.12　Common hot rolled shape steels

(2) 钢板

用光面压辊轧制而成的扁平钢材,且以平板状态供货的称钢板(如图 1.13 所示)。钢板有热轧钢板和冷轧钢板两种,热轧钢板按厚度分为厚板(厚度大于4 mm)和薄板(厚度为0.35~4 mm)两种,冷轧钢板只有薄板(厚度为0.2~4 mm)。

一般厚板用于焊接结构,薄板主要用于屋面板、墙板和楼板等。在钢结构中,单块板不能独立工作,必须用几块板通过连接组合成工字形、箱形截面等构件来承受荷载。图 1.14 所示为用钢板焊接组成的工字形截面和箱形截面。

(2) Steel plate

Steel plate is a kind of flat steel which is rolled with a smooth roller, as shown in Fig. 1. 13. There are two kinds of steel plate, namely, hot rolled plate and cold rolled plate. According to its thickness, the hot rolled plate can be divided into thick plate (thickness > 4 mm) and thin plate thickness 0. 35 mm ~ 4 mm. The thickness of cold rolled plate is 0. 2 ~ 4 mm.

Generally, thick plate is used for welding structure, while thin plate mainly for roof panel, wall panel and floor, etc. In steel structure, single plate cannot work independently. The element formed by connecting several single plates into I-section or box section could bear loads, as shown in Fig. 1. 14.

图 1.13　钢板图
Fig. 1.13　Steel plate

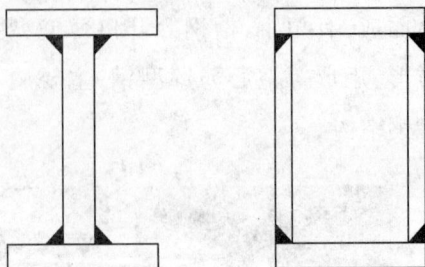

图 1.14　焊接组成截面
Fig. 1.14　Welded cross section

（3）钢管

按生产工艺不同，钢结构所用钢管（如图1.15所示）分为热轧无缝钢管和焊接钢管两大类。在土木工程中，钢管多用于制作桁架、塔桅、钢管混凝土等，广泛应用于高层建筑、厂房柱、塔柱、压力管道等工程中。

(3) Steel Pipe

According to the production process, steel pipes are divided into two categories, namely, the hot rolled seamless pipe and the welded pipe, as shown in Fig. 1.15. In CE, steel pipes are used to make truss, tower mast and concrete filled steel tubes, etc., which are widely used in high-rise buildings, columns, pillar, pressure pipe, etc.

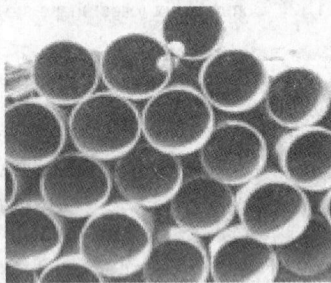

图 1.15　钢管
Fig. 1.15　Steel pipe

1.4.3　混凝土结构用钢材

混凝土结构主要包括钢筋混凝土结构和预应力混凝土结构，其所用钢材主要有普通钢筋和预应力筋。

1.4.3　Steel used in concrete structure

Concrete structure mainly includes the reinforced concrete (RC) structure and prestressed concrete (PC) structures. The steels used in concrete structure are mainly ordinary reinforcement and prestressing tendon.

（1）普通钢筋

普通钢筋指用于钢筋混凝土结构中的钢筋和预应力混凝土结构中的非预应

(1) Ordinary reinforcement

Ordinary reinforcement is the steel used in RC and the non-prestressing steel in PC. The materials include two types: plain carbon

土木工程导论

力钢筋,其材质包括普通碳素钢和普通低合金钢两大类。普通钢筋按生产工艺性能和用途的不同可分为以下几类。

① 热轧钢筋。用加热钢坯轧成的条型成品,称为热轧钢筋。按轧制外形,可分为热轧光圆钢筋和热轧带肋钢筋(如图1.16所示)两类,其中肋的形式有等高肋和月牙肋(如图1.17所示)。按热轧工艺,热轧带肋钢筋又可分为普通热轧钢筋和细晶粒热轧钢筋。

steel and ordinary low-alloy steel. According to production process and usage, ordinary reinforcement can be divided into the following categories.

① Hot rolled bars. Hot rolled bars are rolled by heating billet. According to rolled appearance, they can be divided into hot rolled plain bars (HPB) and hot rolled ribbed bars (HRB, as shown in Fig. 1. 16). There are two types of ribbed bars: contour ribs and crescent ribs, as shown in Fig. 1. 17. According to hot rolled process, HRB can be divided into the ordinary hot rolled bars and fine grain hot rolled bars.

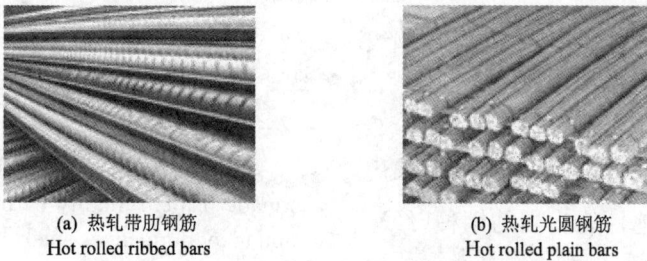

(a) 热轧带肋钢筋
Hot rolled ribbed bars

(b) 热轧光圆钢筋
Hot rolled plain bars

图 1.16 热轧钢筋
Fig. 1.16 Hot rolled bars

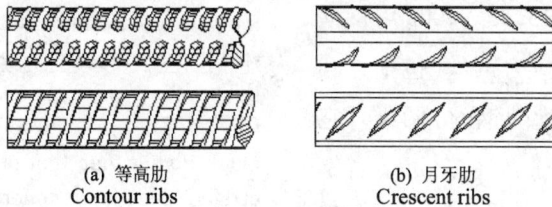

(a) 等高肋
Contour ribs

(b) 月牙肋
Crescent ribs

图 1.17 带肋钢筋
Fig. 1.17 Ribbed steel bars

② 冷拉钢筋。为了提高强度以节约钢筋,工程中常按施工规程对钢筋进行冷拉。冷拉后钢筋的强度提高,但塑性、韧性变差。因此,冷拉钢筋不宜用于受冲击或重复荷载作用的结构。

② Cold-drawn bars. In order to improve strength and save steel, cold drawing process is usually done according to the construction regulations. The strength of cold-drawn steel bars will be improved, but ductility and toughness will be deteriorated. Therefore, cold-drawn steel should not be used in structures with impact loads or repeated loads.

③ 冷轧带肋钢筋。冷轧带肋钢筋是

③ Cold rolled ribbed bars. Using the

采用普通低碳钢或低合金钢热轧的圆盘条,经冷轧在其表面形成两面或三面有肋的钢筋。

④ 热处理钢筋。热处理钢筋是用热轧螺纹钢筋经淬火和回火的调质处理而成的,公称直径主要有 6,8,10,12,14 mm 5 个规格。热处理钢筋具有高强度、高韧性和高黏结力及塑性降低少等优点,目前主要作为预应力混凝土构件的配筋。

(2) 预应力筋

预应力筋主要包括预应力钢丝、钢绞线和螺纹钢筋。

① 预应力钢丝。预应力钢丝是采用优质碳素钢或其他性能相当的钢种,经冷加工及时效处理或热处理而制得的高强度钢丝(如图 1.18a 所示)。根据《预应力混凝土用钢丝》(GB/T5223—2002),钢丝分为冷拉钢丝和消除应力钢丝(包括光圆钢丝、刻痕钢丝和螺旋勒钢丝)两类。它的强度比普通热轧钢筋高许多,可节省钢材、减少截面、节省混凝土,主要用于桥梁、吊车梁、大跨度屋架、管桩等预应力钢筋混凝土构件中。

hot rolled rods of low-carbon or low-alloy steel, the cold rolled ribbed bars with two or three ribbed sides are formed by cold rolling technique.

④ Heat-treated bars. Heat-treated bars are made with hot rolled ribbed bars by quenching and tempering treatment. The nominal diameters of this bar are 6, 8, 10, 12 and 14 mm. Heat-treated bars have some advantages such as high strength, high toughness, high bond strength and less plasticity decreases, which are mainly used as reinforcement of PC members.

(2) Prestressing tendons

Prestressing tendons mainly contain prestressing wire, strand and deformed bar.

① Prestressing wire. Prestressing wire is made of high-quality carbon steel or other similar performance steel by cold working and aging treatment or heat treatment (Fig. 1.18a). According to *Steel Wires for the Prestressing of Concrete* (GB/T5223 – 2002), the wires can be divided into cold drawn steel wire and stress relief wire (including light round wire, nicked wire and spiral wire). The strength of prestressing wire is much higher than that of HRB, so the amount of steel bars and concrete can be reduced. Prestressing wires are mainly used in many PC elements, such as bridges, crane beams, large-span roof trusses and pipe piles.

(a) 钢丝
Wires

(b) 钢绞线(7 股)
Strand(7wires)

图 1.18　钢丝和钢绞线
Fig. 1.18　Wires and strand

② 钢绞线。预应力混凝土用钢绞线由冷拔钢丝制造而成。钢绞线的规格有 2 股、3 股、7 股、19 股等,其中 7 股钢绞线(如图 1.18b 所示)由于面积较大、柔软,施工操作方便,目前已成为国内外应用最广的一种预应力钢筋。钢绞线具有强度高、柔性好、质量稳定、成盘供应无需接头等优点,适用于大型结构、薄腹梁、大跨度桥梁等负荷大、跨度大的预应力混凝土结构。

③ 预应力螺纹钢筋。预应力混凝土用螺纹钢筋又名精轧螺纹钢筋,按屈服强度分为 PSB785,PSB835,PSB930,PSB1080 4 个级别。其特点是在整根钢筋上轧有外螺纹的大直径、高精度的直条钢筋,具有连接锚固简便、张拉锚固安全可靠、黏着力强等特点,主要用于制造高强度、大跨度、钢筋用螺母方式连接的混凝土制品上,如核电站、水电站、桥梁、隧道、高速铁路等重点工程。

1.5 混凝土及其构件

1.5.1 混凝土

混凝土是指由胶凝材料、骨料(或称集料)、水按一定比例配制(也常掺入适量的外加剂和掺合料),经搅拌振捣成型,在一定条件下养护而成的人造石材。混凝土常简写为"砼(tóng)",它是现代土木工程中用途最广、用量最大的建筑材料之一。

② Strand. The strands used in PC are made with cold drawn steel wires. Specifications of strands include 2, 3, 7 and 19 steel wires and so on. Due to large area, soft and operability, the 7 steel wires strand (Fig. 1.18b) has become the most widely used prestressing strand at home and abroad. Strand has many advantages, such as high strength, good flexibility, stable quality and no joints. Therefore, it is suitable for large structures, thin abdominal beams, large span bridges Which are typical PC structure with large load and span PC structures.

③ Prestressing screw-thread bar. Deformed bar used in PC are also named as refined rolled deformed bar. According to yield strength, prestressing screw-thread bars can be divided into four levels: PSB785, PSB835, PSB930 and PSB1080. The straight bars, characterized by large diameter and high-precision, have many advantages including easy connecting anchorage, reliable tensioned anchorage and strong adhesion. It is mainly used in the manufacture of high-strength, large span RC products with nut connection, such as nuclear power plants, hydroelectric plants, bridges, tunnels, high-speed railway and other key projects.

1.5 Concrete and its components

1.5.1 Concrete

Concrete is a kind of artificial stone block made with cementitious materials, aggregate, water with certain percentage by vibrating and curing, in which some admixtures are usually added. Concrete is one of the widely-used building materials.

（1）混凝土的分类

混凝土品种众多，主要有以下几种分类方法：

① 按胶凝材料，可分为无机胶凝材料混凝土（如水泥混凝土、石膏混凝土、硅酸盐混凝土、水玻璃混凝土等）和有机胶凝材料混凝土（如沥青混凝土、聚合物混凝土、树脂混凝土等）。

② 按表观密度，可分为重混凝土（表观密度 >2 800 kg/m³）、普通混凝土（表观密度为 2 000 kg/m³ ~2 800 kg/m³，一般在 2 400 kg/m³ 左右）和轻混凝土（表观密度 <2 000 kg/m³）。

③ 按使用功能，可分为结构混凝土、保温混凝土、装饰混凝土、防水混凝土、耐火混凝土、水工混凝土、海工混凝土、道路混凝土、防辐射混凝土等。

④ 按生产和施工工艺，可分为离心混凝土、真空混凝土、灌浆混凝土、喷射混凝土、碾压混凝土、挤压混凝土、泵送混凝土等。

⑤ 按掺合料种类，可分为粉煤灰混凝土、硅灰混凝土、矿渣混凝土和纤维混凝土等。

⑥ 按混凝土抗压强度等级，可分为低强度混凝土（抗压强度 f_{cu} <30 MPa）、中强度混凝土（f_{cu} 为 30~60 MPa）、高强度混凝土（f_{cu} ≥60 MPa）、超高强混凝土（f_{cu} ≥

（1）Classification of concrete

Concrete can be classified as follows：

① Divided by cementitious materials, there are inorganic cementitious materials concrete (such as cement concrete, gypsum concrete, silicate concrete, water glass concrete, etc.) and organic cementitious materials (such as asphalt concrete, polymer concrete, resin concrete, etc.).

② Divided by apparent density, there are heavy concrete (apparent density > 2 800 kg/m³), ordinary concrete (apparent density is between 2 000 kg/m³ and 2 800 kg/m³, generally is about 2 400 kg/m³) and lightweight concrete (apparent density < 2 000 kg/m³).

③ Divided by its functions, there are structural concrete, insulation concrete, decorative concrete, waterproof concrete, fire-resistant concrete, marine concrete, road concrete and radiation concrete.

④ Divided by production and construction process, there are centrifugal concrete, vacuum concrete, grout concrete, extruded concrete and pumping concrete, etc.

⑤ Divided by the type of admixtures, there are fly ash concrete, silica fume concrete, slag concrete and fiber concrete.

⑥ Divided by compressive strength of concrete, there are low strength concrete (compressive strength f_{cu} < 30 MPa), medium strength concrete (f_{cu} is from 30 to 60 MPa), high-strength concrete (f_{cu} 60 MPa), ultra high strength concrete (f_{cu} 100 MPa).

土木工程导论

100 MPa)。

此外,随着混凝土的发展和工程的需要,还出现了膨胀混凝土,加气混凝土等各种特殊功能的混凝土。商品混凝土以及新的施工工艺给混凝土施工带来方便。

In addition, with the development of concrete and engineering needs, some special concretes, such as expansion concrete and aerated concrete are produced. Commercial concrete and new construction techniques make concrete construction more convenient.

(2) 水泥混凝土

水泥混凝土是指以水泥为胶凝材料,以砂、石为骨料,以水为稀释剂,并掺入适量的外加剂和掺合料拌制成的混凝土,也称普通混凝土。沙子和石子在混凝土中起骨架作用,故称为骨料(或称集料),沙子称为细骨料,石子(碎石或卵石)称为粗骨料;水泥和水形成水泥浆,包裹在砂粒表面并填充砂粒间的空隙而形成水泥砂浆,水泥砂浆又包裹在石子表面并填充石子间的空隙而形成混凝土,其结构如图1.19 所示。适量地掺入外加剂(如减水剂、引气剂、缓凝剂、早强剂等)和掺合料(如粉煤灰、硅灰、矿渣等)是为了改善混凝土的某些性能以及降低成本。

(2) Cement concrete

Cement concrete, also known as ordinary concrete, is made of cement, sand, stone aggregate, water, and usually incorporated with the right amount of admixtures and additives to concrete mixture Sand and gravel form the concrete skeleton, which are called aggregate. Sand is known as fine aggregate, and gravel (crushed stone or gravel) is known as coarse aggregate. Cement and water form water slurry, parcel the sand surface and fill the voids among the sand grains to form cement mortar. Cement mortar wraps the surface of stones and fills the voids among the stones and then concrete is formed (structure is shown in Fig. 1.19). In order to improve the performance of concrete and reduce costs, the right amount of additives (such as water reducer, air-entraining agent, retarder, and early strength agent) and admixtures (such as fly ash, silica fume, and slag) are added.

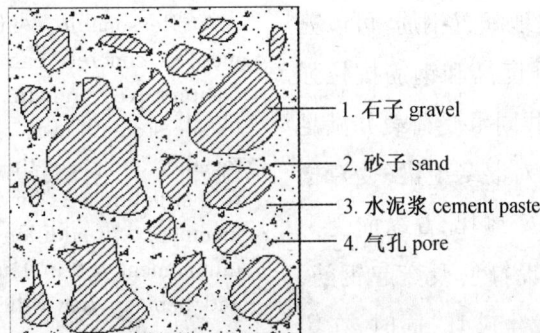

1. 石子 gravel
2. 砂子 sand
3. 水泥浆 cement paste
4. 气孔 pore

图 1.19 混凝土的结构
Fig. 1.19 The structure of concrete

砂浆与水泥混凝土的区别在于砂浆

The difference between cement mortar and concrete is that no coarse aggregate is

不含粗骨料,因此可以认为砂浆是混凝土的一种特例,也可称为细骨料混凝土。此外,在砂浆和混凝土的性能设计中,外加剂起着十分重要的作用,常见的外加剂有减水剂,缓凝剂,早强剂等。

1.5.2　钢筋混凝土

不配筋的混凝土称为素混凝土,其主要缺陷是抗拉强度很低(只有一般抗压强度的 $1/20 \sim 1/10$),也就是说混凝土受拉、受弯时易产生裂缝,并发生脆性破坏。为了克服混凝土抗拉强度低的弱点,充分利用其较高的抗压强度,一般在受拉一侧加设抗拉强度很高的(受力)钢筋,即形成钢筋混凝土(reinforced concrete, RC)。图1.20 为某简支梁破坏示意图。

added in mortar. Thus cement mortar can be considered as a special case of concrete, which is called fine aggregate concrete. Besides, the admixture will play an important role in properties of mortar and concrete. The commonly used admixtures include water reducing agent, cement retarder, early strength agent, etc.

1.5.2　Reinforced concrete

The main flaw of unreinforced concrete (referred to plain concrete) is its low tensile strength (only $1/20 \sim 1/10$ of compressive strength). That is to say, when the concrete is in tension or bending, cracks are easy to occur, which results to brittle failure. In order to overcome the shortcoming of low tensile strength and make full use of its high compressive strength, the steels with high tensile strength are set in the tensile zone of concrete, that is the reinforced concrete (RC). Fig. 1.20 is a schematic diagram of damage for simply supported beam.

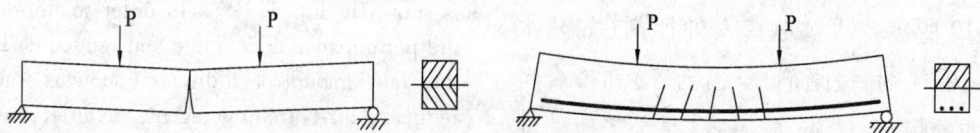

图 1.20　简支梁受力破坏图示
Fig. 1.20　Damage of simply supported beam

在混凝土中合理地配置钢筋,可以充分发挥混凝土抗压强度高和钢筋抗拉强度高的特点,使两者共同承受荷载并满足工程结构的需要。如对混凝土梁(受弯构件)来说,除了在受拉一侧配置纵向受力钢筋外,一般还要加设箍筋及弯起钢筋,以防止它沿斜裂缝发生破坏;同时,在梁的上部另加直径较小的钢筋作为架立钢筋,它与受力钢筋、箍筋和弯起钢筋一起扎结成钢筋架,如图 1.21 所示。目前,钢

Reasonable configuration of reinforcement in concrete can make use of the high compressive strength of concrete and high tensile strength of reinforcement to bear loads together and meet the needs of the engineering structure. As for concrete beam(flexural member), in addition to the longitudinal reinforcement set in tensile zone, the stirrup and bent-up bar will also be configured to prevent the diagonal crack damage in beam. At the same time, some small diameter steel bars are set in the upper portion of the beam as erection bars. All the steels in concrete beam are shown in Fig. 1.21. At present,

土木工程导论

筋混凝土是使用最多的一种结构材料。

RC is one of the most frequently used structural materials.

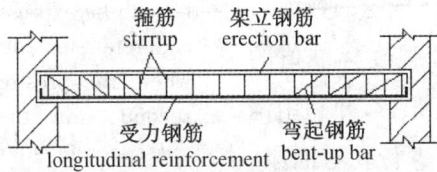

图 1.21　混凝土梁内钢筋
Fig. 1.21　Steel bars in concrete beam

1.5.3　预应力混凝土

钢筋混凝土虽然可以充分发挥混凝土抗压强度高和钢筋抗拉强度高的特性，但其在使用阶段往往是带裂缝工作的，这对某些结构如储液池等来说是不容许的。为了控制混凝土构件受荷后的应力状态，在构件受荷之前（制作阶段），人为给拉区混凝土施加预压应力，使其减小或抵消荷载（使用阶段）引起的拉应力，将构件受到的拉应力控制在较小范围内，甚至处于受压状态，即可控制构件在使用阶段不产生裂缝，这样的混凝土称为预应力混凝土（prestressed concrete，PC）。

按照施加预应力的方法（施工工艺），预应力混凝土可分为先张法预应力混凝土（简称"先张法"）和后张法预应力混凝土（简称"后张法"）两大类。先张法是先将预应力筋张拉到设计控制应力，用夹具将其临时固定在台座或钢模上，进行绑扎钢筋，支设模板，然后浇筑混凝土；待混凝土达到规定的强度后，切断预应力筋，借助于它们之间的黏结力，在预应力筋弹性回缩时，使混凝土获得预压应力，其工序

1.5.3　Prestressed concrete

RC can give full play to the compressive strength of concrete and high tensile strength of steel. However, cracks often occur in the service phase, which is not allowed for some structures such as the liquid storage tank. In order to control the stress state in concrete component, the compressive stress is applied to the concrete in tensile zone before loads are added, which can reduce or offset the tensile stress in concrete in the stage of service. Furthermore, the concrete in tensile zone may be in compressive state, which can control the appearance of cracks in the service stage. Such concrete is called prestressed concrete (PC).

According to the method of applying prestress, PC can be divided into two categories: the pre-tensioned PC (pre-tensioning method for short) and post-tensioned PC (post-tensioning method for short). The construction processes of the former can be expressed as follows: first, draw the prestressing tendons to the designed control stress and fix them in the stretching bed temporarily; second, assemble reinforcements and place concrete; finally, cut off the prestressing tendons when the strength of concrete reaches to design strength, and the compressive stress can be obtained in concrete

如图 1.22a 所示。后张法是先浇筑混凝土构件,并在预应力筋的位置预留出相应孔道,待混凝土强度达到设计规定的数值后(一般不低于混凝土设计强度标准值的75%),穿入预应力筋进行张拉,并利用锚具把预应力筋锚固,最后进行孔道灌浆,使砼产生预压应力,其工序如图 1.22b 所示。

through the adhesive force between concrete and tendon. All processes are shown in Fig. 1.24a. The construction processes of the latter can be expressed as follows: first, prepare concrete element with some prepared hole where the prestressing tendons are placed; second, draw the prestressing tendons after the concrete strength reaches to the design value (usually not less than 75% of the standard concrete design strength value) and fix the two ends with anchorage; at last, grout the hole with cement mortar. All processes are shown in Fig. 1.22b.

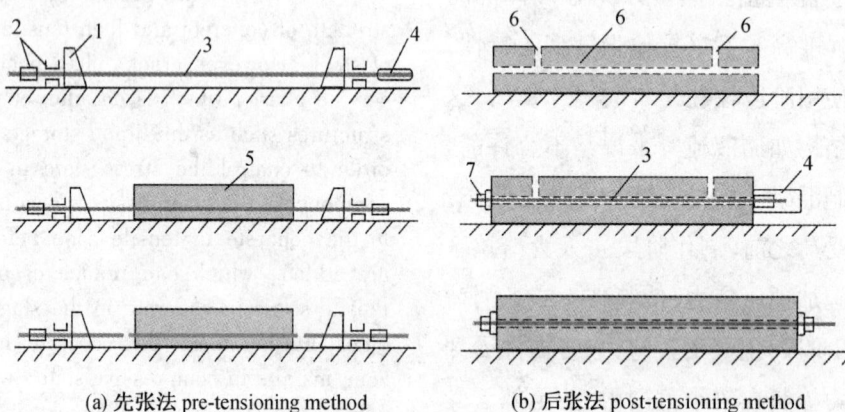

(a) 先张法 pre-tensioning method　　(b) 后张法 post-tensioning method

1—台座；2—夹具；3—预应力筋；4—千斤顶；5—构件；6—预留孔道；7—锚具
1-stretching bed; 2-grip; 3-prestressing tendon; 4-Jack; 5-element; 6-prepared hole; 7-anchorage

图 1.22　先张法和后张法施工工序
Fig. 1.22　Construction processes of pre-tensioning method and post-tensioning method

1.6　木材

木材是人类使用最早的工程材料之一。我国使用木材的历史不仅悠久,而且在技术上还有独到之处,如保存至今已达千年之久的山西佛光寺正殿、山西应县木塔等都集中反映了我国古代土木工程中应用木材的水平。

1.6.1　木材的分类

木材是由树木加工而成的,树木的种

1.6　Timber

Timber is one of the earliest engineering materials used by human beings. In China, the use of timber not only has a long history, but also has unique technologies. For examples, the main hall of Shanxi Foguang Temple and Shanxi Yingxian Wood Tower, which have existed for more than one thousand years, reflect the application level of timber in ancient CE.

1.6.1　Classification of timber

Timber is made of wood. There are

类很多,一般按树种分为针叶树和阔叶树两大类。针叶树树叶细长呈针状,树干直而高,易得大材,纹理平顺,材质均匀,木质较软而易于加工,故又称软木材。建筑上针叶树多用于承重结构构件和门窗、地面材及装饰材,常用树种有松树、杉树、柏树等。阔叶树树叶宽大呈片状,多为落叶数,树干通直部分较短,材质较硬,较难加工,故又称硬木材。建筑上阔叶树常用于制作尺寸较小的构件,常用树种有榆树、水曲柳、桦树等。

many kinds of trees, which can be generally divided into two major categories: conifer and broad-leaved tree. Conifer has long leaf, straight and tall trunk. With the features of smooth texture and uniform material quality, the wood of conifer is soft and easy to process, so it is also called soft wood. It is usually used for load-bearing member, door and window, the ground material and decorative material. Common tree species of conifer are pine, cedar, cypress and so on. Broad-leaved trees have large and flaky leaves, and their trunk is straight and short. As the material is hard and difficult to process, it is also known as hard wood. It is usually used as small size component. Common species of broad-leaved trees are elm, ashtree, birch and so on.

1.6.2 木材的主要性质

木材的构造决定着木材的性能,其宏观构造如图 1.23 所示。木材的性质包括物理性质和力学性质。物理性质主要有密度、含水率、热胀干缩等;力学性质主要有抗拉、抗压、抗弯和抗剪 4 种强度。

1.6.2 Main properties of timber

The structure of timber determines the performance of timber, and its macro-structure is shown in Fig. 1.23. The properties of timber include the physical and mechanical properties. Its physical natures include density, moisture content, thermal expansion and shrinkage. It's mechanical properties include tensile strength, compressive strength, flexural strength, and shear strength.

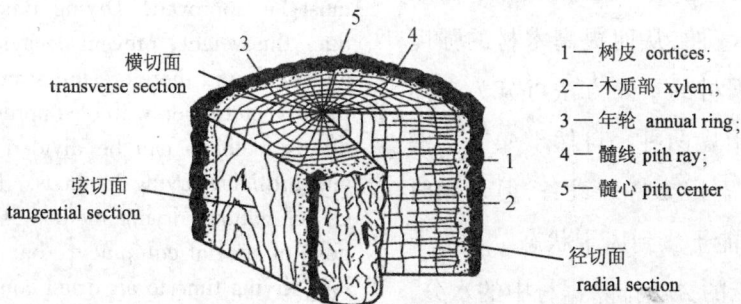

横切面 transverse section

弦切面 tangential section

1 — 树皮 cortices;
2 — 木质部 xylem;
3 — 年轮 annual ring;
4 — 髓线 pith ray;
5 — 髓心 pith center

径切面 radial section

图 1.23 木材的 3 个切面
Fig. 1.23 Three sections of timber

木材有很好的力学性质,但木材是有机各向异性材料,顺纹方向与横纹方向的

Timber has excellent mechanical properties. However, timber is an organic anisotropic material, and there are great diffe-

力学性质有很大差别,见表1.4。木材的顺纹抗拉和抗压强度均较高,但横纹抗拉和抗压强度较低。木材强度还因树种而异,并受木材缺陷、荷载作用时间、含水率及温度等因素的影响,其中以木材缺陷及荷载作用时间两者的影响最大。

rences in mechanical properties between the longitudinal direction and transverse direction, as shown in Table 1.4. The tensile and compressive strength along grain are higher. The strength of wood also varies among species, and it will be affected by timber defects, loading time, moisture content and temperature, among which the timber defects and loading time are greatest ones.

表1.4　木材各强度之间关系
Table 1.4　Relationship among strength of woods

抗压强度 Compression strength		抗拉强度 Tensile strength		抗弯强度 Flexural strength	抗剪强度 Shear strength	
顺纹 along grain	横纹 cross grain	顺纹 along grain	横纹 cross grain		顺纹 along grain	横纹 cross grain
1	1/10 ~ 1/3	2 ~ 3	1/20 ~ 1/3	3/2 ~ 2	1/7 ~ 1/3	1/2 ~ 1

注:以木材的顺纹抗压强度为1作标准。
Note:The standard compression strength along grain is 1.

1.6.3　木材的应用

在工程中,除直接使用原木外,木材一般都加工成锯材(板材、方材等)或各种人造板材使用。原木可直接用作屋架、檩、椽、木桩等。

为减小使用中发生的变形和开裂,锯材须经干燥处理。干燥能减轻自重,防止腐朽、开裂及弯曲,从而提高木材的强度和耐久性。锯材的干燥方法可分为自然干燥和人工干燥两种。自然干燥方法的优点是简单,不需要特殊设备,但干燥时间长,而且只能干燥到风干状态。人工干燥是利用人工的方法排除锯材中的水分,主要采用干燥窑法,也可用简易的烘、烤方法。使用中易腐朽的木材应事先进行防腐处理。

1.6.3　Application of timber

Besides the direct usage, timber is generally processed into lumber (sheet, etc.) or artificial plate in engineering. Logs can be directly used as roof frame, purlin, rafter, wood, etc.

In order to reduce the deformation and cracking in usage, timber drying process must be approved. Drying process can reduce the weight, prevent decaying, cracking and bending, therefore the strength and durability of timber will be improved. Timber drying methods can be divided into natural and artificial drying methods. The advantages of natural drying method is simple, no need of special equipment, but it will cost a long drying time to air dried condition. Artificial drying using artificial method to eliminate lumber moisture, mainly uses the kiln drying method, as well as the simple baking and roasting methods. Easy rotten timber should be embalmed in advance.

木材经加工成型和制作构件时,会留下大量的碎块废屑,以这些废料或含有一定纤维量的其他作物做原料,采用一般物理和化学方法加工而成的产品即为人造板材。这类板材与天然木材相比,板面宽,表面平整光洁,没有节子,不翘曲、开裂,经加工处理后还具有防水、防火、防腐、防酸等性能。常用人造板材有胶合板、纤维板、刨花板、木屑板等,如图1.24所示。

A large number of pieces of scrap will be left during the process of making models and component. The waste or other crops that contain a certain amount of fiber can be taken as raw materials. Artifical plate can be made from these raw materials through general physical methods and chemical methods. Compared with natural timber, artifical plate has the characteristic of wide and smooth surface, without knot, warping and cracking. Besides, artifical plate has the feature of waterproof, fireproof, anticorrosive, and anti-aeid performance after a series of processing. Common artifical plate are plywood fiberboard, chipboard, sawdustboard, etc., as shown in Fig. 1.24.

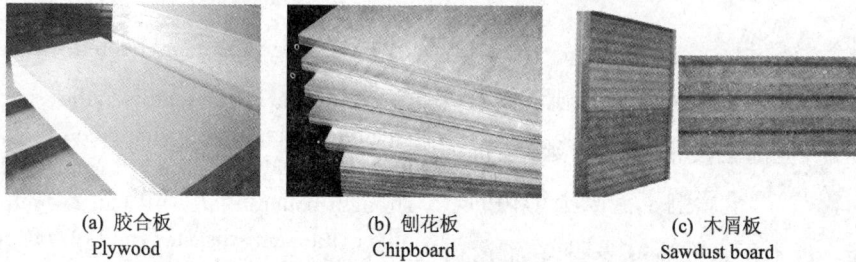

(a) 胶合板
Plywood

(b) 刨花板
Chipboard

(c) 木屑板
Sawdust board

图 1.24　人造板材
Fig. 1.24　Artificial plate

1.7　土木工程材料的发展前景

1.7　Developing prospects for CEM

土木工程材料是土木工程的重要组成部分,它和工程设计、工程施工以及工程经济之间有着密切的关系。自古以来,工程材料和工程建(构)筑物之间就存在着相互依赖、相互制约和相互推动的关系。一种新材料的出现必将推动建筑设计方法、施工程序或形式的变化,而新的结构设计和施工方法必然要求提供新的、更优良的材料。

近几十年来,随着科学技术的进步和

The CEMs are the important ingredients of civil engineering. There are close relationships among CEM, the construction and costs of buildings. Since ancient times, the engineering materials and buildings have inseparably interconnected and developed together. The appearance of a new material will improve the way of structural design and construction. Conversely, the advanced structural type and construction method will inevitably induce the innovation of excellent materials.

In recent decades, with the development of science and technology in CE, a

土木工程发展的需要,一大批新型土木工程材料应运而生,出现了仿生智能混凝土(自感知混凝土、自愈合混凝土、透光混凝土等)、高强钢材、新型建筑陶瓷和玻璃、纳米技术材料、新型复合材料(纤维增强材料、夹层材料)等。随着社会的进步,环境保护和节能减排的新形势,对土木工程材料提出了更多、更高的要求。因而,今后一段时间内,土木工程材料将向以下几个方向发展。

1.7.1 高强高性能材料

(1) 高性能混凝土

在20世纪,混凝土的强度得到了较大幅度的提高,但高强度混凝土的延性、抗火性能均较差,严重影响了混凝土结构的抗灾性能。近十年来的研究表明,在混凝土的组分中引入纳米材料、短切复合材料或有机聚合物,可以更好地改善混凝土,并不断研究开发出纳米混凝土、高延性纤维混凝土、高耐久性混凝土、良好抗疲劳和耐磨混凝土材料。此外,将混凝土材料与其他聚合物复合,可以增加混凝土材料的阻尼特性,发展高阻尼混凝土材料,提高结构的抗震性能。

(2) 高强钢材

采用高强钢材可显著减小钢结构构件的尺寸和结构重量,相应的减少焊接工作量和焊接材料用量,减少各种涂料(防锈、防火等)的用量及其施工工作量,所取

great deal of new materials has been invented, such as bionic intelligent concrete (including self-sensing concrete, self-healing concrete and light transmission concrete etc.), high strength steel, new-style building ceramics and glass, nanotechnology material, new-style composite material (including fiber reinforced material, sandwich material). It is predictable that more and more performance-advanced materials are required to meet the needs of environmental protection, energy-saving and emission-reduction of our society. Therefore, the developing prospect for CEMs will be listed as follows.

1.7.1 High strength and performance materials

(1) High performance concrete (HPC)

In the 20th century, the strength of concrete was greatly improved. However, the ductility and fire resistance of high strength concrete are still bad, which greatly affect the anti-disaster ability of concrete structure. It can be found in recent studies that the concrete performance can be greatly improved when some additives, such as nanophase material, short composite material, and organic polymer are added into concrete mixture. Besides, many high performance concretes including nanometer concrete, high ductility fiber concrete, good durability concrete, anti-fatigue and anti-abrasion concrete are developed and used. The combination of concrete materials with organic polymer can effectively improve the seismic performance of concrete structures.

(2) High strength steel

High strength steels can evidently diminish the sizes and weights of steel structural elements. Accordingly, it can reduce the welding works, the amounts of welding materials and coating materials. All related construction work will be decreased obviously.

得的经济效益可使整个工程总造价降低。同时,在建筑物使用方面,减小构件尺寸能够带来更大的使用空间。目前工程应用的钢材强度已经达到 460 MPa 以上,甚至开始推荐使用屈服强度为 500,590,620,690 MPa 的更高强度的结构钢。我国国家体育场"鸟巢"采用了 700 多吨板厚达到 110 mm 的 Q460E/Z35 高强度高性能钢材(如图 1.25 所示);中央电视台新址采用了 2 300 多吨 Q460E/Z35 高强度高性能钢材;Q460 高强度型钢已经应用于输电塔架。

欧洲已将 S460～S690 级高强度结构钢材列入规范;美国已在桥梁建设中应用屈服强度 485 MPa 级和 690 MPa 级的高性能钢材;澳大利亚在高层和大跨度建筑中成功应用了屈服强度 690 MPa 级钢材。可见,发展高强度和高性能钢材符合世界各国的技术发展规划,以及可持续发展和环境保护的基本理念,是土木工程材料的重要发展方向之一。

(3) 纤维增强复合材料

纤维增强复合材料(FRP)具有轻质、高强、耐久、高阻尼等特性,已成为一种重要的土木工程结构材料。FRP 材料主要有碳纤维(CFRP)、玻璃纤维(GFRP)、芳纶纤维(SFRP)和玄武岩纤维(BFRP)等几种。随着研究的深入,利用 FRP 复合材

As a result, the total cost of building will be reduced effectively. Besides, from the viewpoint of building's serviceability, it is advantageous to use smaller elements to get larger space. Nowadays, the strength of steel used in existed buildings has reached over 460 MPa, which is the yield strength of steel. Furthermore, some higher strength structural steels with yield strengths of 500, 590, 620 and 690 MPa begin to appear in practical engineering. For examples the National Stadium "Bird's Nest" (Fig. 1. 25) in China consumed more than 700 t high strength steels of the type of Q460E/Z35 with the thickness of 110 mm. Another unique new building is China Central Television (CCTV), which adopted the same high strength steels with the amount of 2 300 t. Besides, the Q460 steels have also been used in transmission towers.

The high level structural steels with the yield strengths of S460 to S690 have been listed in European codes. Some high performance steels with yield strengths of 485 MPa to 690 MPa have been adopted in bridge structures in America. In Australia, the steel with yield strength of 690 MPa has been successfully used in high-rise and large-span buildings. It can be clearly seen that the action to develop high-strength or high-performance steels is in accordance with sustainable development and environmental protection all over the world. High strength steel has become one of the important developing fields in CEM.

(3) Fiber reinforced composite materials

Fiber Reinforced Polymer/Plastics (FRP) has the characteristics of low-weight, high-strength, durability and high-damping, and it has become an important structural material in civil engineering. Currently, the FRP materials mainly include CFRP, GFRP, SFRP and BFRP. With the deep research on FRP, it has become a hot issue on

料替代传统的土木工程材料受到了愈来愈多的关注,如 FRP 套管替代钢管约束混凝土结构、FRP 筋代替传统的钢筋以及 FRP 索代替钢索等。图 1.26 给出了国内首座碳纤维 CFRP 索代替钢索的预应力索斜拉桥,该桥总长为 55 m,由东南大学、江苏大学以及北京特希达科技有限公司共同研究与开发,2004 年建成。实践表明,纤维增强复合材料 FRP 因其良好的力学性能和耐久性,将成为继钢材和混凝土材料之后的第三类结构材料。

how to adopt FRP materials to replace the traditional CEMs all over the world, for example, how to use FRP tube to replace the steel tube in steel-concrete composite structures, as well as how to use FRP bars to replace steel bars in concrete structures and to replace prestressing cables in prestressed structures. Here, a practical application of CFRP cable in bridge structures is shown in Fig. 1.26. This bridge is the first CFRP cable stayed bridge in China, and its total span is 55 m. It was designed and built by Southeast University, Jiangsu University and Beijing Texida Technology Co., LTD in 2004. It can be concluded from engineering practice that with the excellent mechanical properties and durability, FRP material will become the third structural material following concrete and steel.

图 1.25　中国国家体育场"鸟巢"实景图
Fig. 1.25　National Stadium "Bird's Nest"

图 1.26　中国国内首座碳纤维 CFRP 索实桥
Fig. 1.26　First CFRP suspension bridge in China

1.7.2　绿色节能材料

（1）节能减排材料

土木工程材料的生产能耗和建筑物使用能耗,一般占国家总能耗的 20% ~ 35% 左右,因此研制和生产低能耗的新型节能土木工程材料是构建节约型社会的需要。另外,充分利用工业废渣(如粉煤灰、矿渣等)、生活废渣以及建筑垃圾等生

1.7.2　Green and energy-saving material

（1）Energy-saving and emission reduction materials

Normally, the total energy consumption during the processes of producing materials and using buildings accounts for nearly 20%~35% of the national energy consumption. Therefore, it is a strong requirement for conservation-oriented society to develop and produce the energy-saving materials in CE. Besides, the emission reduction of buildings is another

产土木工程材料,将各种废渣尽可能地资源化,以保护环境、节约自然资源,是人类社会实现可持续发展的需要。生态混凝土是近几年研究开发出来的一种有利于改善生态环境和自然景观(植生绿化功能)的新型混凝土,在我国已有一定的应用(如图1.27所示),这是一种很有发展潜力的混凝土材料。

hot issue nowadays. In order to protect environment and save natural resources, it is a wise way to use industrial and construction waste to reproduce CEMs. It is also an inevitable demand to achieve sustainable development of society. Currently, the eco-concrete (or environmentally friendly concrete) is one kind of the new-style concretes, which can bring benefits to ecological system and improve environment with plant growth. It has already been used in China (Fig. 1.27) and will have a good future.

(a) 施工现场 Construction site   (b) 浇注完成 Work completion   (c) 绿化功能实现 Plant growth

图 1.27 国内某生态混凝土江堤工程
Fig. 1.27 One Eco-concrete engineering in China

(2) 绿色建材

绿色建材就是指采用清洁的生产技术,少用天然资源、大量使用工业或城市固体废弃物和农植物秸秆生产的无毒、无污染、无放射性,有利于环保与人体健康的建材。发展绿色建材,改变长期以来存在的粗放型生产方式,选择资源节约型、污染最低型、质量效益型、科技先导型的生产方式是21世纪建材工业的必然出路。当前,我国墙体材料的"绿色化"进程已取得了一定的成果。

(2) Green building materials

Green building materials are favorable to environmental protection and human's health. With clean technologies, the green building materials are usually produced by industrial and construction waste or plant straw instead of natural resources. In order to change the current extensive mode of production and create a new one with resource-saving, low pollution and leading technology, it is an inevitable way for building material industry to develop green building materials. Currently, some achievements have been obtained in green wall material's reformation in China.

1.7.3 纳米智能材料

(1) 智能混凝土

通过掺入功能相材料,使传统材料在

1.7.3 Nanophase materials and smart materials

(1) Smart concrete

By adding some functional materials,

保持原有基本力学性能不变的情况下,获得一些特殊功能,是材料科学发展的一个主要趋势。20世纪90年代,这一概念也得到了混凝土研究学者的认同,并提出了"智能混凝土"的概念。目前,混凝土的智能化主要通过以下3个途径来实现:①在混凝土内复合某些导电或半导体纳米材料,使混凝土具备自感知的功能,制备出自感知混凝土;②将混凝土材料与压电材料、磁致伸缩材料或形状记忆材料等"智能材料"相复合,制备出自集能混凝土制品;③在混凝土内埋设一些传感器或感知骨料,使混凝土具有相应的传感功能。

(2) 智能工程材料

对于智能材料的研究始于20世纪80年代的航空航天领域,目前已经在包括土木工程、机械工程、生物医学工程等各个领域中得到了广泛的研究和应用。过去的十多年里,以压电陶瓷、电/磁致伸缩材料、电磁流变液以及形状记忆材料等为代表的智能材料在土木工程领域得到了长足的发展,足尺的磁流变阻尼器已经应用于桥梁、海洋平台以及多层建筑的振动控制;形状记忆合金在古建筑的加固以及隔震座限位器等方面得到了应用;智能型压电摩擦阻尼器和磁致阻尼器也逐步走向了示范工程。进入21世纪,智能材料与智能土木工程结构是土木工程领域最具创新、最有活力的研究方向之一,也是发展高性能土木工程结构和可持续土木工程结构的重要途径。

the traditional materials will hold some special functions without changing their basic mechanical properties, which is the main trend in the development of material science. From 1990s, this idea was accepted in concrete research, and the conception of smart concrete was put forward. At present, the characteristics of smart concrete are usually been achieved through the following ways: ① Self-sensing concrete, in which some electric and semiconductor nanophase materials are added; ② Self-energy collection concrete, which is a composite material by compounding concrete and 'smart material', such as piezoelectric material, magnetostrictive material, shape memory material; ③ Sensing concrete, which has certain sensing functions by setting in some sensors or sensing aggregates in concrete.

(2) Intelligent engineering materials

Intelligent materials were firstly researched in aerospace field in 1980s. Today, it has been widely studied and applied in many fields including CE, Mechanical Engineering, Biomedical Engineering and so on. In the last decade, many representative smart materials, such as piezoelectric ceramics, electrostrictive or magnetostrictive material, magnetorheological fluid and shape memory materials, have been greatly developed and used in CE field. The intelligent materials and their related structures have become one of the most creative fields in CE field since the 21st century.

（3）纳米技术材料

　　纳米为细微的长度单位，等于十亿分之一米，一般称毫微米，记作 nm。今后建材的主导方向是绿色、环保以及高性能，而这些建材的制备主要依靠纳米技术来实现。纳米材料对颜料、陶瓷、水泥等制品的改性有很大贡献，如把纳米氧化铝加入陶瓷中，则陶瓷强度、韧性的增加将非常显著。纳米无机涂料，可以解决混凝土表面腐蚀、老化及渗水等问题。这种涂料在混凝土或水泥浆表面形成玻璃态或离子化胶态，注入微裂或孔隙中与水泥反应形成新的硅酸盐复合体，不仅可以使弯曲强度提高 2～3 倍，还可起到防水作用。但纳米建材现在还仅仅处于起步阶段，其进一步的发展应用还有很长的路要走。

　　总而言之，随着研究工作的深入和新材料的不断涌现，以及与材料有关的基础学科的日益发展，人类对材料内在规律有了进一步的了解，对各种材料的共性知识初步有了科学的抽象认识，从而诞生了"材料科学"这一新的学科领域。材料科学（更准确地说应该是材料科学与工程）是介于基础科学与应用科学之间的一门应用基础科学，其主要任务在于研究材料的组分、结构、界面与性能之间的关系及其变化规律，从而达到按使用要求设计材料、研制材料及预测使用寿命的目的。土木工程材料也属于材料科学的研究对象，随着人们逐渐将土木工程材料的研究纳入材料科学的轨道，在不久的将来土木工

(3) Nanotechnology materials

　　Nanometre（nm）is a kind of length units, which equals to 10^{-10} m. Many high performance materials with the direction of green, environmental protection, and high efficiency, can be invented by nanotechnology. Nanotechnology can bring great contributions to improve material performance, such as paint, ceramics and cement-based products. For example, the ceramic strength and toughness can be remarkably improved by adding nanophase alumina into ceramics. Nanophase inorganic coating can solve some problems to concrete with surface erosion, aging and water percolation. However, nanotechnology materials used in CE field are still in the initial stage, and further efforts are needed for their development.

　　To sum up, with the fast development of material-based scientific research and the appearance of new materials, people can get more information about materials inherent properties and make some conclusions about common knowledge of all kinds of materials. Gradually, a new discipline, Material Science, comes into being. Material Science, which can be exactly described as Material Science and Engineering, is a basic application science between basic science and application science. Its main tasks are as follows: ① Study the relationships between material's constituents, structures and interfaces with its performance. ② Study their change rules to design materials, invent new materials and predict their service lives. The materials used in CE are also the research objects in Material Science. In the future, the great development will be obtained in CEM, and the enormous changes will hap-

程材料的发展必将有重大突破,土木工程   pen in CE.

也将出现翻天覆地的变化。

注:本章图片均来源于网络。

Note:In this chapter, all pictures are from webs.

知识拓展
Learning More

相关链接   Related Links

(1) 中国建筑装饰材料网 http://www.cbh-jj.com/

(2) 中国水泥网 http://www.ccement.com/

(3) 中国工程机械工业协会钢筋及预应力机械分会 http://www.chinarm.org/

小贴士   Tips

(1) 水泥的分类、组分与材料可参见现行标准《通用硅酸盐水泥》(GB 175—2007),主要有硅酸盐水泥、普通硅酸盐水泥、矿渣硅酸盐水泥、火山灰硅酸盐水泥、粉煤灰硅酸盐水泥和复合硅酸盐水泥。

More information about the classification, ingredient and stuff of cement can be obtained from the code *Common Portland Cement* (GB 175 – 2007). There are Portland cement, ordinary Portland cement, slag portland cement, portland pozzolan cement, fly-ash portland cement and composite portland cement.

(2) 钢筋混凝土结构用的热轧带肋钢筋可参见线性标准《钢筋混凝土用钢——第2部分:热轧带肋钢筋》(GB 1499.2—2007),主要包括普通热轧钢筋(牌号:HRB335、HRB400 和 HRB500)和细晶粒热轧钢筋(牌号:HRBF335、HRBF400 和 HRBF500)。其中,HRB 是 Hot Rolled Ribbed Bars 的缩写,后面的数字是屈服强度;HRBF 是在 HRB 后面加"细"的英文(Fine)首位字母。

The information about hot rolled ribbed bars used in RC structures can be obtained from the code *Steel for the Reinforcement of Concrete—Part* 2: *Hot Rolled Ribbed Bars* (GB 1499.2 – 2007). There are two kinds of bar, ordinary hot rolled ribbed bars including HRB335, HRB400 and HRB 500, where the number is the yield strength of steel bar, and fine hot rolled ribbed bars including HRBF335, HRBF400 and HRBF 500.

思考题   Review Questions

(1) 土木工程中常用的天然石料有哪些? 它们各有何特点?

     What kinds of natural stone used in CE? What characteristics do they have?

(2) 气硬性胶凝材料与水硬性胶凝材料有何区别? 请举例说明。

     What are the differences between air hardening cementitious materials and hydraulic cementitious materials? Please show some examples.

(3) 钢材的分类方法有哪些? 土木工程中常用什么钢材?

     How many classification methods for steel? Which steels are commonly used in CE?

（4）水泥混凝土的组成材料有哪些？与砂浆有什么不同？

What are the components of cement concrete? What are the differences between mortar and concrete?

参考文献
References

［1］郑晓燕,胡白香：《新编土木工程概论》（第 2 版）,中国建材工业出版社,2012 年。

［2］柯国军,严兵,刘红宇：《土木工程材料》,北京大学出版社,2006 年。

［3］刘正武：《土木工程材料》,同济大学出版社,2005 年。

［4］柳俊哲,宋少民,赵志曼：《土木工程材料》,科学出版社,2005 年。

［5］江见鲸,叶志明：《土木工程概论》,高等教育出版社,2001 年。

［6］刘瑛：《土木工程概论》,化学工业出版社,2004 年。

［7］丁大均,将永生：《土木工程概论》,中国建筑工业出版社,2003 年。

［8］Palanicharmy M. S. Basic Civil Engineering. China Machine Press, 2005.

［9］Zuyan Shen. Introduction of Civil Engineering. China Architecture & Building Press, 2005.

［10］Shan Somayaji. Civil Engineering Materials. 2nd ed. Prentice Hall, 2001.

土木工程材料 Civil Engineering Material

第2章 岩土工程

Chapter 2 Geotechnical Engineering

岩土工程是土木工程的分支,是以岩体力学、土力学与基础工程、工程地质学为理论基础,研究和解决工程建设中与岩土有关的技术问题的一门新兴的应用科学。按照工程建设阶段,岩土工程可以分为岩土工程勘察、岩土工程设计、岩土工程治理、岩土工程监测与检测几个阶段。土木工程建设中遇到的岩土工程问题促进了该学科的发展。本章主要介绍岩土工程勘察,基础工程,基坑与地下工程,地基处理,边坡工程的相关内容,并对岩土工程的发展作了展望。

Geotechnical engineering（GE）is a branch of civil engineering（CE）, which concerns with the engineering behavior of earth materials, such as rock and soil. It is a new application discipline aiming at studying and solving the geotechnical problems in the engineering construction on the basis of the theories of rock mechanics, soil mechanics and foundation engineering, engineering geology. According to the engineering construction stage, a GE project can be divided into four stages, namely, geotechnical investigation, geotechnical design, geotechnical treatment, geotechnical inspection and monitoring. Geotechnical problems encountered in CE construction promote the development of this discipline. In this chapter, the author mainly introduces the content of geotechnical investigation, foundation engineering, foundation excavation and underground engineering, ground treatment, slope engineering, and presents the prospects for the development of GE.

2.1 岩土工程概述

人类历史上早就有将岩土材料用于防洪筑堤、水利工程、墓穴、基础建筑的实

2.1 Overview of GE

Human beings have used soil for controlling flood, irrigation, burial sites, build-

例。例如，公元前20世纪古美索不达米亚和新月沃土地带，以及印度河流域文明中心的摩亨佐达罗和哈拉帕古城，人们通过修建堤防、坝址、运河，筑堤防洪，引洪灌溉。随着城市的扩张，出现了以成型的地基支撑上部结构的基础。此外，古希腊人建造了垫板基础和筏板基础。直至18世纪，仍未出现关于岩土设计的相关理论，更多是依靠过去的经验进行建筑。

此后，与基础相关的工程问题不断出现，例如意大利的比萨斜塔，促使科学家们对岩土工程的问题进行深入研究。最早的理论进展源于挡土墙结构的土压力计算。1717年，Henri Gautier（戈蒂叶，法国工程师）提出了天然土体边坡静态休止角的概念。1773年，Charles Coulomb（库伦，物理学家、工程师、陆军上尉）最早将力学基本原理应用于土体，提出了确定军事堡垒土压力的方法。库伦理论与Christian Otto Mohr（莫尔）的二维应力状态相结合，被称为莫尔—库伦理论，至今仍然广泛应用于工程实践中。

19世纪，达西提出了达西渗透定律，以描述流体在多孔介质中的流动；布辛涅斯克提出了弹性固体中的应力分布理论，可以有效地确定地面以下不同深度的应力分布；William Rankine（郎肯，工程师和

ing foundations for a long time. First activities were linked with irrigation and flood control, as demonstrated by traces of dykes, dams and canals dating back to at least 2 000 B. C. that were found in ancient Mesopotamia and the Fertile Crescent, as well as around the early settlements of Mohenjo Daro and Harappa in the Indus valley. As the cities expanded, formalized foundations which can support the above structures appeared. For example, ancient Greeks constructed pad footings and strip-and-raft foundations. Until the 18th century, however, no theoretical basis for soil design had appeared, and the discipline was more of an art than a science, relying on past experience.

Several foundation-related engineering problems, for example, the Leaning Tower of Pisa, prompted scientists to begin taking a more scientific-based approach to examine the subsurface. The earliest advances occurred in the development of earth pressure theories for the construction of retaining walls. Henri Gautier, a French royal engineer, recognized the "natural slope" of different soils in 1717, an idea later known as the soil's angle of repose. The application of the principles of mechanics to soils was documented as early as 1773 when Charles Coulomb (a physicist, engineer, and army Captain) developed methods to determine the earth pressures against military ramparts. The theory combined Coulomb's theory and Christian Otto Mohr's 2D stress state was known as Mohr-Coulomb theory, which has been used in practice so far.

In the 19th century, Henry Darcy developed what is now known as Darcy's Law describing the flow of fluids in porous media. Joseph Boussinesq developed theories of stress distribution in elastic solids, so as to prove the estimated stresses at depth in the ground. William Rankine, an engineer

物理学家)提出了郎肯土压力理论;Albert Atterberg(阿太堡)提出了黏性土的一致性指标,至今仍被用于土的分类。

现代岩土工程始于 1925 年,以 Karl Terzaghi(太沙基,工程师和地质学家)的《土力学》(*Erdbaumechanik*) 一书为标志,因此太沙基被公认为现代土力学和岩土工程之父。他不仅提出了有效应力原理,指出土体的抗剪强度由有效应力控制,而且提出了地基承载力的理论框架及用固结理论计算黏性土层沉降的思想。

20 世纪 60 年代末至 70 年代初,岩体力学、土力学与基础工程、工程地质学三者逐渐结合为一体并应用于土木工程实际,从而形成一门新的学科——岩土工程。岩土工程运用土力学和岩石力学原理研究地质条件,确定岩土材料的物理、力学及化学性质,评价天然边坡和人造土场地的稳定性及风险分析,设计岩土工程结构与基础,设计场地条件施工监测等多个方面。岩土工程是土木工程的一个重要组成部分,在房屋、市政、能源、水利、道路、矿山、国防等各种建设中都有十分重要的意义。

2.2 岩土工程勘察

岩土工程勘察是设计和施工的基础,主要目的是查明工程地质条件,分析存在的地质问题,对建筑地区做出工程地质评价。根据勘察对象的不同,岩土工程勘察

and physicist, put forward the rankine earth pressure theory. Albert Atterberg developed the clay consistency indices which has been used so far in soil classification.

It is said that modern GE began in 1925 with the publication of *Erdbaumechanik* by Karl Terzaghi (a civil engineer and geologist). Considered as be the father of modern soil mechanics and GE, Terzaghi developed the principle of effective stress, and demonstrated that the shear strength of soil is controlled by effective stress. He also developed the framework for theories of bearing capacity of foundations, and the settlement prediction for clay layers due to consolidation.

From late of 1960s to early 1970s, a new application discipline named GE was formed based on the combination of rock mechanics, soil mechanics and foundation engineering, engineering geology. GE uses principles of soil mechanics and rock mechanics to investigate subsurface conditions and materials, determine the relevant physical/mechanical and chemical properties of these materials, evaluate stability of natural slopes and man-made soil deposits, assess risks posed by site conditions, design earthworks and structure foundations, and monitor site conditions, etc. GE is an important part of CE. It has been widely used in building municipal, energy highway, military, water conservancy, mining, etc., which has very great significance.

2.2 Geotechnical investigation

Geotechnical investigation is the basis of design and construction. The main purpose is to find out the geological conditions, analyze geological problems and assess the engineering geology of building area. According to the different objects, geotechnical

可分为水利水电工程(主要指水电站、水工构造物)、铁路工程、公路工程、港口码头、大型桥梁及工业、民用建筑等的勘察。由于水利水电工程、铁路工程、公路工程、港口码头等工程一般比较重大,投资造价及重要性高,因此国家对这几个类别的工程勘察进行了专门的分类,编制了相应的勘察规范、规程和技术标准。通常这些工程的勘察被称为工程地质勘察。

因此,通常所说的"岩土工程勘察"主要指工业、民用建筑工程的勘察,勘察对象的主体主要包括房屋楼宇、工业厂房、学校楼舍、医院建筑、市政工程、管线及架空线路、岸边工程、边坡工程、基坑工程、地基处理等。岩土工程勘察的内容主要有:工程地质调查和测绘、勘探及取土试样、原位测试、室内试验、现场检验和检测,最终根据以上几种或全部手段,对场地工程地质条件进行定性或定量分析评价,编制满足不同阶段所需的成果报告文件。

2.2.1 工程地质调查与测绘

工程地质调查与测绘是岩土工程勘察的基础工作,一般在勘察的初期进行。这一方法的本质是运用地质、工程地质理论,对地面的地质现象进行观察和描述,

investigation can be divided into different categories, such as hydraulic and hydroelectric engineering (mainly refers to the hydropower station and hydrotechnical construction), railway engineering, highway engineering, wharfs, large bridges, industrial and civil constructions, etc. Since the higher scale, investment cost and significance for hydraulic and hydroelectric engineering, railway engineering, highway engineering, port engineering, some special classifications were given in China, and the corresponding investigation standards, regulations and technical standards were also given. These engineering investigations are usually called the engineering geological investigation.

Therefore, the "geotechnical investigation" mainly refers to industrial, civil construction engineering investigation. The investigation object includes the residential buildings, industrial buildings, school buildings, hospital construction, municipal engineering, pipeline and overhead line, shore engineering, slope engineering, foundation engineering, and ground improvement. Geotechnical investigation mainly includes the engineering geological survey and mapping, exploration and sampling, in-situ test, laboratory test, field test and detection. Finally, according to several or all of the above methods, qualitative or quantitative analysis evaluations of site engineering geological conditions are developed, and report documents for the results are compiled meeting for different stages.

2.2.1 Engineering geological survey and mapping

Engineering geological survey and mapping, which is usually carried out in the early stages of geotechnical investigation, are basic work of geotechnical investigation. According to the application of geology, engineering geological theory, geological pheno-

分析其性质和规律,并以此推断地下的地质情况,为勘探、测试工作等其他勘察方法提供依据。在地形地貌和地质条件较复杂的场地,必须进行工程地质测绘。但对地形平坦、地质条件简单且较狭小的场地,则可采用调查代替工程地质测绘。通过调查和测绘,借助计算机软件或其他诸如遥感等测试技术,可以建立场地的地质地貌模型(如图2.1所示),形成地质地貌图像(如图2.2所示)。

menon on the ground can be observed and described, the nature and laws can be analyzed and followed by the underground geological conditions, which can provide a basis for exploration, testing and other survey methods. In the complex geological site conditions, the engineering geological mapping should be carried out. However, on flat terrain, simple geological conditions and relatively narrow filed space, filed survey can be adopted instead of engineering geological mapping. Through survey and mapping, a geological and geomorphic model can be built by computer software or other technologies such as remote testing (Fig. 2.1), or actual images of geology and geomorphology can be obtained (Fig. 2.2).

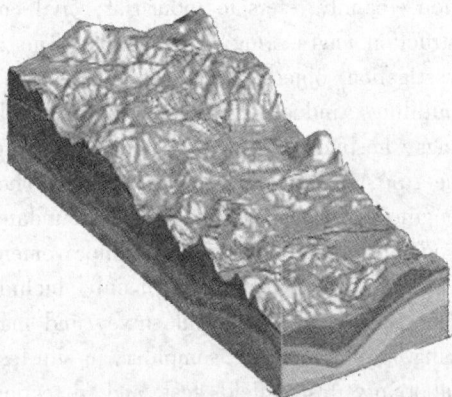

图 2.1 GeoView 软件建立的三维地质模型
Fig. 2.1 Three-dimensional geological model by GeoView

图 2.2 江苏大学遥感影像图
Fig. 2.2 Remote sensing image of Jiangsu University

2.2.2 勘探与取样

勘探工作包括物探(如图2.3所示)、钻探(如图2.4所示)、坑探(如图2.5所示)及其他方法。通过勘探,可以调查地下地质情况,并利用勘探工程取样进行原位测试和监测。在实际工程中,应根据勘察目的及岩土的特性选用不同的勘探方法。

2.2.2 Exploration and sampling

Exploration work includes geophysical survey (Fig. 2.3), drilling survey (Fig. 2.4), pitting survey (Fig. 2.5) and other methods. The underground geological conditions can be investigated through exploration and in-situ testing and monitoring by exploration engineering samplings can be carried out. Different exploration methods should be chosen carefully according to the characteristics of rock and soil, as well as geotechnical investigation purpose.

图 2.3　物探
Fig. 2.3　Geophysical survey

图 2.4　钻探
Fig. 2.4　Drilling survey

图 2.5　坑探
Fig. 2.5　Pitting survey

物探是一种间接的勘探手段,较之钻探和坑探,它的优点是轻便、经济而迅速,常与测绘工作配合使用。钻探和坑探均是直接勘探手段,能可靠地了解地下地质情况,在岩土工程勘察中是必不可少的。当钻探方法难以查明地下地质情况时,可采用坑探方法。

现场取样(如图 2.6 所示)伴随整个勘探过程,根据所取土样质量等级可划分为两种:一是原状土样,即保持原有的天然结构未受破坏的土样;二是扰动土样,即试样的天然结构已遭破坏。通过一定的取样仪器(如图 2.7 所示),可以获得原状土样。但是需要注意,土样脱离母体后,其应力状态发生了变化,这将影响到土样的结构。钻探及采样时,钻具在钻压过程中必然要对周围土体产生一定程度的扰动。此外,无论何种取土器都有一定的壁厚、长度和面积,在其压入过程中,也会使土样受到一定的扰动。

Geophysical survey is an indirect exploration method with the advantages of portability, economy and speediness than drilling survey and pitting survey. It is often used together with the mapping work. Drilling survey and pitting survey are two direct exploration means which underground geological conditions can be obtained reliably. Pitting survey method should be chosen when drilling method is too hard to ascertain the underground geological conditions.

Field sampling (Fig. 2.6) can be used in the whole exploration process. According to the quality level, soil samples can be divided into two categories, namely, undisturbed soil sample and disturbed soil sample respectively. Undisturbed soil sample keeps its unspoiled nature, and the disturbed soil structure has been destroyed. Undisturbed soil sample can be obtained by certain sampling instrument (Fig. 2.7). It should be noticed that soil stress state has been changed when soil specimens are token out form the filed soil layers, the soil structures will be affected. Drilling tool is bound to have a certain degree of disturbance on the surrounding soil in the process of drilling. It will also make the soil samples under certain disturbance in the process of penetration, no matter what kind of sampler will be used due to its different wall thickness, length and area.

图 2.6　现场取样
Fig. 2.6　Field sampling

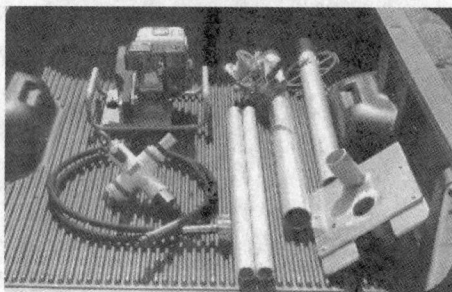

图 2.7　地表取样钻机
Fig. 2.7　Drilling equipment for sampling

2.2.3　原位测试与室内试验

原位测试与室内试验的主要目的是为岩土工程问题分析评价提供所需的技术参数,包括岩土的物理指标、强度参数、固结变形特性参数、渗透性参数和应力、应变及时间关系的参数等。原位测试一般都借助于勘探工程进行,是详细勘察阶段一种主要的勘察方法。原位测试的优点是试样不脱离原来的环境,基本上在原位应力条件下进行试验,所测定的岩土体尺寸大,能反映宏观结构对岩土性质的影响,代表性好。缺点是试验时的应力路径难以控制、边界条件也较复杂,有些试验耗费人力、物力较多,难以大量进行。原位测试的方法比较多,图 2.8 和图 2.9 分别为原位静力触探试验和原位十字板剪切试验的现场。

2.2.3　In-situ test and laboratory test

Main objectives of in-situ test and laboratory test is to provide technical parameters for the analysis and evaluation of geotechnical engineering problems. These parameters include physical properties of rock and soil, strength parameters, consolidation deformation characteristics parameter, permeability parameter and stress, strain and time related parameters, etc. In-situ test is usually carried out by exploration engineering work. It is an investigation method in the process of detailed exploration. The advantages of in-situ test are that soil specimen is not divorced from the original environment and the test is basically conducted in the in-situ stress condition. Besides, the size of rock and soil is large, which can reflect the macroscopic structure of the properties. The disadvantages of in-situ test are that stress path is difficult to control, and the boundary conditions are more complex. Meanwhile, some in-situ tests consume more manpower and material resources than that in corresponding laboratory tests. Fig. 2.8 and Fig. 2.9 show the in-situ static cone penetration test and in-situ vane shear test site, respectively.

图 2.8 原位静力触探试验
Fig. 2.8 In-situ static cone penetration test

图 2.9 原位十字板剪切试验
Fig. 2.9 In-situ vane shear test

室内土工试验是研究土的特性的试验,主要优点是试验条件比较容易控制,边界条件明确,应力、应变条件可控等。室内土工实验主要包括:① 土的物理性质试验;② 土的水理性质试验;③ 土的力学性质试验(静态)④ 土的动力性质室内试验。室内试验仪器有很多,如常规物理性质试验仪器,三轴压缩试验仪器(如图 2.10 和图 2.11 所示),固结试验仪器(如图 2.12 所示)等。

Laboratory test is used for testing soil properties. Main advantages of laboratory tests suggest that it is easy to control the test conditions; the boundary conditions are explicit; stress and strain conditions can be easily controlled, etc. The laboratory experiments mainly include: ① physical properties test for soil. ② water physical properties test for soil. ③ mechanical properties test (static) for soil. ④ dynamic properties test for soil. There are many kinds of laboratory test equipment such as conventional physical properties test instruments, triaxial compression test instrument (Fig. 2.10 and Fig. 2.11), oedometer test instrument (Fig. 2.12).

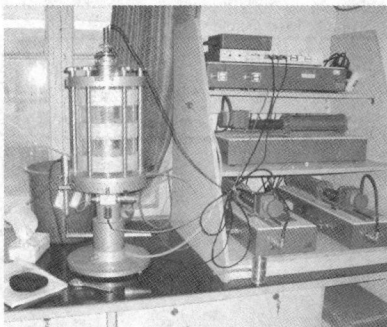

图 2.10 GDS 三轴仪
Fig. 2.10 GDS triaxial instrument

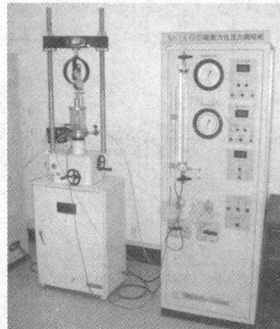

图 2.11 普通三轴仪
Fig. 2.11 Normal triaxial instrument

图 2.12 固结仪
Fig. 2.12 Oedometer

2.2.4 现场检验与监测

现场检验与监测的主要目的在于保证工程质量和安全，提高工程效益。现场检验包括施工阶段对先前岩土工程勘察成果的验证核查，以及对岩土工程的施工监理和质量控制。现场监测则主要包括施工作用和各类荷载对岩土反应性状的监测、施工和运营中的结构物监测和对环境影响的监测等。根据检验与监测所获取的资料，可以反求某些工程技术参数，并以此为依据及时修正设计，使工程在技术和经济方面得到优化。图 2.13 为地基自动化监测系统示意图。

2.2.4 Field inspection and monitoring

The main purpose of field inspection and monitoring is to ensure the project quality and safety as well as improve the engineering benefit. The field inspection includes both the verification and the validation of previous geotechnical engineering investigation in construction stage, the supervision and quality control of geotechnical engineering construction. The field monitoring mainly includes the geotechnical behavior monitoring under construction and all kinds of loads on the rock and soil, structure monitoring during construction and operation, the monitoring of influence on environment, etc. Data obtained from inspection and monitoring can recalculate some engineering parameters. A modified design could be given according to the tests data, and the optimization can be adopted in the aspects of technology and economy. Fig. 2.13 shows the schematic diagram of the foundation automation detection system.

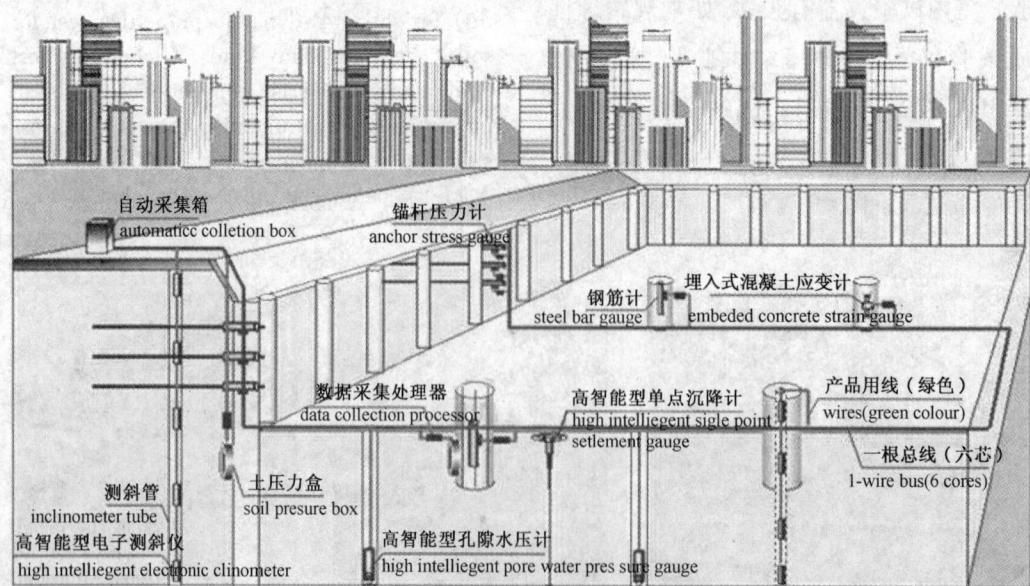

图 2.13　地基自动化监测系统示意图
Fig. 2.13　Schematic diagram of the foundation automation detection system

2.3 基础工程

建筑物向地基传递荷载的下部结构称为基础,其作用是将上部结构的荷载安全可靠地传给地基。岩土工程师设计基础时,应考虑结构的荷载特性、场地土体和/或基岩的性质。基础按埋置深度分浅基础和深基础。

① 浅基础:基础埋深不大(一般小于 5 m),只需经过挖槽、排水等普通施工程序就可建造的基础。

② 深基础:浅层土质不良,须将基础埋置于较深的良好土层,并需借助特殊施工方法建造的基础。

2.3.1 浅基础

浅基础可扩大建筑物与地基的接触面积,使上部荷载扩散。浅基础主要有独立基础(如大部分柱基)、条形基础(如墙基)、筏形基础(如水闸底板)3 种形式,分别如图 2.14、图 2.15、图 2.16 所示。

2.3 Foundation engineering

The foundation of a building transmits loads from buildings and other structures to the earth. Geotechnical engineers design foundations based on the load characteristics of the structure and the properties of the soil and/or bedrock at the site. According to buried depth, foundation can be divided into two categories, namely, shallow foundation and deep foundation, respectively.

① Shallow foundation refers to the small depth foundation (less than 5 m), which can be built only after dredging, drainage and other common construction procedures.

② Deep foundation refers to the large depth foundation for the case that the shallow soil layer is poor, and foundations should be embedded in deep soil layers by special construction methods and equipments.

2.3.1 Shallow foundation

Shallow foundations can enlarge the contact area between the building and the ground. Therefore, the upper load is diffused. Shallow foundations can be divided into independent foundation (such as the pillar foundation), strip foundation (such as the wall foundation) and raft foundation (such as the sluice plate), as shown in Fig. 2. 14, Fig. 2. 15, and Fig. 2. 16 respectively.

图 2.14　独立基础
Fig. 2.14　Independent foundation

图 2.15　条形基础
Fig. 2.15　Strip foundation

图 2.16　筏板基础
Fig. 2.16　Raft foundation

2.3.2 深基础

深基础是埋深较大,以下部坚实土层或岩层作为持力层的基础,其作用是把所承受的荷载相对集中地传递到地基的深层。当建筑场地的浅层土质不能满足建筑物对地基承载力和变形的要求,而又不适宜采用地基处理措施时,就要考虑采用深基础方案了。深基础包括桩基础(如图 2.17 所示)、墩基础(如图 2.18 所示)等类型。

图 2.17 桩基础
Fig. 2.17 Pile foundation

2.3.2 Deep foundation

Deep foundation is used to transfer the load of a structure down through the upper weak layer of topsoil to the stronger layer of subsoil below. Deep foundations are used for structures or heavy loads when shallow foundations cannot provide adequate capacity, due to the size and structural limitations. There are different types of deep foundations including piple foundation(Fig. 2.17), pier foundation (Fig. 2.18),etc.

图 2.18 墩基础
Fig. 2.18 Pier foundation

2.4 基坑与地下工程

2.4.1 基坑工程

随着基坑开挖深度和面积的增大,基坑围护结构的设计和施工也越来越复杂,需要的理论和技术也越来越高,远远超出了作为施工辅助措施的范畴,而施工单位没有足够的技术力量来解决复杂的基坑稳定、变形和环境保护问题。因此,研究和设计单位的介入解决了基坑工程的理论计算和设计问题,由此逐步形成了一门

2.4 Foundation excavation and underground engineering

2.4.1 Foundation excavation engineering

With the increase in depth and area of excavation, the design and construction of building foundation excavations have become increasingly complex. More and more advanced theories and technologies are also needed. It is not just an auxiliary construction measure now. Since the construction units do not have enough technical supports to solve the complex problems related to foundation stability, deformation and environmental protection, research and design units participate in, and the corresponding

独立的学科分支——基坑工程。基坑工程涉及岩土工程、结构工程和环境工程等众多学科领域，综合性高，影响因素多，设计计算理论目前还不成熟，在一定程度上还依赖于工程实践经验。

theoretical calculation and design problems can be solved. An independent discipline branch, namely foundation excavation engineering, is gradually formed, which covers many fields such as geotechnical engineering, structure engineering and environmental engineering. Currently, the design calculation theory is not mature due to its high comprehensiveness and large amount of influencing factors. Therefore, the design is also subjected to the experiences of engineering practice to some extent.

基坑土方开挖的施工工艺一般有两种：放坡开挖（无支护开挖）和在支护体系保护下开挖（有支护开挖），分别如图2.19和图2.20所示。前者既简单又经济，但应具备放坡开挖的条件，即基坑不太深而且基坑平面之外有足够的空间供放坡。而当不具备放坡开挖条件时，应选择有支护开挖，常见的支护开挖方法详见2.4.2节介绍。

Construction technology of soil excavation in foundation excavation engineering can be generally divided into two categories: slope excavation (without support) and excavation (with support), as shown in Fig. 2.19 and Fig. 2.20 respectively. The slope excavation is simple and economical, but excavation conditions should be guaranteed in case that the foundation is shallow and it has enough space for the slope in the outside of the foundation plane. Otherwise, excavation (with support) should be selected. Common support methods for foundation excavation are summarized in the section 2.4.2.

图2.19　无支护开挖基坑
Fig. 2.19　Slope excavation (without support)

图2.20　有支护开挖基坑
Fig. 2.20　Excavation (with support)

2.4.2　基坑支护方法

① 板桩式，主要有钢板桩、钢管桩、钢筋混凝土板桩等，适用于黏性土、砂性土

2.4.2　Foundation excavation support methods

① Panel pile type support methods mainly include steel sheet pile, steel pipe

和粒经不大于 100 mm 的沙卵石地层。由于施工时打桩噪声大,宜用于远离居民区的施工中。常见的钢板桩主要有 H 型钢(如图 2.21 所示),间距在 1.2 ~ 1.5 m 之间。钢板桩的特点是有成品制作,可反复适用;施工简便,但施工有噪音。钢管桩(如图 2.22 所示)的截面刚度大于钢板桩,在软弱土层中开挖深度可增大(在日本开挖深度达 30 m),同时需有防水措施相配合。

图 2.21　H 型钢板桩
Fig. 2.21　H type steel sheet pile

② 柱列式,又有钻孔灌注桩和挖孔灌注桩两种,分别如图 2.23 和图 2.24 所示。其特点是施工噪音小,适于城区施工,对周边地层、环境影响小;刚度大,可用于深大基坑;需与降水或止水措施配合使用。

pile, reinforced concrete sheet-pile, etc. It is suitable for cohesive soil, sandy soil and grain less than 100 mm in diameter. Construction should be away from residential areas due to the piling noise of construction. Furthermore, a common type of steel sheet pile is called H type steel with spacing between 1.2 m and 1.5 m, as shown in Fig. 2.21. It is pre-produced in the factory and can be used repeatedly in the construction. Although the construction process is simple, the construction noises can not be avoided (Fig. 2.22). The section stiffness of the steel pipe pile is larger than that of the steel sheet pile. Besides, the excavation depth of the steel pipe pice can be larger in soft soils with waterproof measures. For instance, the excavation depth of the steel pipe pile could increase to 30 m in Japan.

图 2.22　钢管桩
Fig. 2.22　Steel pipe pile

② Soldier type support methods mainly include cast-in-place bored pile and artificial drill-pouring pile, as shown in Fig. 2.23 and Fig. 2.24, respectively. Little construction noise is the main advantage for this type. Therefore, it is suitable for urban construction with small effect on the environment and the surrounding soil layers. It can also be used for deep foundation excavation engineering due to its high stiffness. However, some dewatering or watertight measures should be operated in coordination.

图 2.23　钻孔灌注桩施工
Fig. 2.23　Cast-in-place bored pile construction

图 2.24　挖孔灌注桩施工
Fig. 2.24　Artificial drill-pouring pile construction

③地下连续墙(如图 2.25 所示)可兼做永久结构,适于逆筑法、半逆筑法。其特点是施工时振动小、噪声低,墙体刚度大,对周边地层扰动小;开挖深度大,适用于所有土层;强度大,变位小,隔水性好,可兼做主体结构的一部分,但造价较高。

③ Underground diaphragm wall (Fig. 2.25) can be used as a permanent structure, and it is suitable for inverse construction method, semi-inverse construction method. Low noise is an advantage of this method due to small vibration in construction. Big wall stiffness results in small disturbance to the surrounding soil layers and big excavation depth. It is suitable for all soil layers with big strength, small deflection and good waterproof. However, the cost of this method is high.

图 2.25　地下连续墙施工
Fig. 2.25　Construction of underground diaphragm wall

④沉井(箱)法(如图 2.26 和图 2.27 所示)是在垂直方向上,将各种形状的井筒(沉井)或箱体(沉箱)边排土边沉入地

④ Sunk-well method (Fig. 2.26) and caisson method (Fig. 2.27) are two foundation excavation support methods. In vertical direction, various shapes of wellbore or box are sunk into the ground and finally fixed in

下,最后固定在地层中,形成地下建筑物或构筑物的施工方法。事先在地面上用钢筋混凝土制成的井筒形状(沉井)或箱体(沉箱)的结构,作为基坑坑壁的支撑,在井壁的保护下,用机械和人工在井内挖土,并在其自重作用下沉入土中。该方法施工占地面积小,不需要板桩维护,与大开挖相比,挖土量小,对邻近建筑的影响较小,操作简便,无须特殊的专用设备。

soil layers. Different shapes of wellbore or box are made of reinforced concrete as the support system for foundation pit wall in advance. Mechanical or artificial digging in the well or box under the protection of the borehole wall enable the well or box sink into the soil layers under self-weight. These two methods cover a small area in construction without the support of panel piles. Compared with the large excavation, it has small amount of digging and less effect on the adjacent buildings as well as easy operation without special equipments in construction.

图 2.26　沉井施工
Fig. 2.26　Sunk-well construction

图 2.27　沉箱施工
Fig. 2.27　Caisson construction

⑤ SMW 法(如图 2.28 所示),即在水泥土桩内插入 H 型钢或钢板、钢管(可拔出反复使用,经济性好)等,将承受荷载与防渗挡水结合起来,使之成为同时具有受力与抗渗两种功能支护结构的围护墙。SMW 支护结构的支护特点主要是:施工时基本无噪音,对周围环境影响小,结构强度可靠,特别适合于以黏土和粉细砂为主的松软地层;挡水防渗性能好,不必另设挡水帷幕,可以配合多道支撑应用于较深的基坑,具有较好的发展前景。

⑤ SMW method refers to soil mixing wall, as shown in Fig. 2.28. Cement-soil pile was first set by H type steel, steel sheet or steel pipes and so on, which can be pulled up and repeatedly used with lower cost. This method combines the two functions of bearing loads and seepage prevention, which has many advantages including little construction noise, small impact on the environment and enough strength reliability. Therefore, it is suitable for soft soil layers mainly composed of clay and silty sand. Blocking water curtain can not have to be set due to its excellent seepage prevention. It can be applied in the deep foundation pit cooperated with multi-layer support. SMV method has good prospects for development.

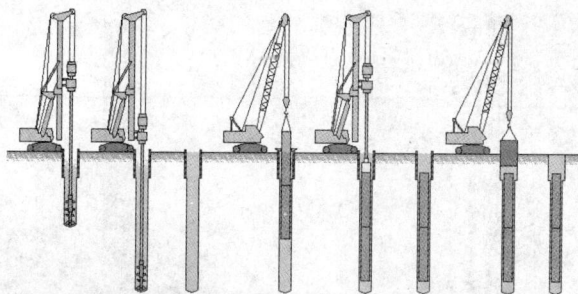

图 2.28　SMW 法示意图
Fig. 2.28　Schematic of SMW method

⑥ <u>自立式水泥挡土墙</u>,分为<u>深层搅拌桩</u>挡土墙和<u>高压旋喷桩</u>挡土墙。它利用水泥材料为固化剂,采用特殊机械,如深层搅拌机(见图 2.30)和高压旋喷机(见图 2.31),将其与<u>原状土</u>强制拌和,形成具有一定强度、整体性和稳定性的圆柱体(<u>柔性桩</u>),桩体相互搭接,组成整体结构性的水泥土墙或形成格栅状墙体,以保证基坑边坡的稳定。其优点是自立式,无需支撑,开挖方便;有良好的隔水性能;充分利用原状土,节省材料;造价低。缺点是水泥土墙体的材料强度比较低,不适于支撑作用,墙体变形和位移较大。

⑥ <u>Self-standing cement-soil retaining wall</u> (Fig. 2.29) support methods can be divided into <u>deep mixing pile</u> retaining wall and <u>high-pressure jet grouting pile</u> retaining wall, respectively. With the special machinery, such as deep mixing machine (Fig. 2.30) and high-pressure jet grouting machine (Fig. 2.31), <u>intact soil</u> is forced to mix with solidified agent of cement to form a cylinder body (<u>flexible pile</u>) possessing a certain strength, integrity and stability. These flexible piles are lapped to form cement-soil retaining wall or grille wall, which can ensure the slope stability of a foundation pit. It has many advantages including self standing, need for horizontal support, easy digging, well seepage prevention, low cost and undisturbed material. However, the deformation and displacement of the wall are relatively large since the strength of the materials are relatively low and without horizontal support.

图 2.29　水泥土墙
Fig. 2.29　Cement-soil retaining wall

图 2.30　深层搅拌机
Fig. 2.30　Machine for deep mixing pile

图 2.31　高压旋喷机
Fig. 2.31　Machine for high-pressure jet grouting pile

⑦ 土钉墙是由土钉和喷射混凝土面层(含钢筋网)及原位土体组成的基坑支护结构,如图 2.32 和图 2.33 所示。通过形成土钉复合体,显著提高基坑整体稳定性。其施工设备简单,钉长一般比锚杆的长度小得多,且不需要施加预应力;成本低,施工噪音、振动小,不影响环境;本身变形很小,对相邻建筑物影响不大。

⑦ Soil-nailed wall is one of the foundation excavation support methods, which is made up of soil nails and shotcrete surface layer (including steel mesh) together with the in-situ soil, as shown in Fig. 2.32 and Fig. 2.33. Stability of foundation pit can be improved by composite structure. Construction equipment for soil-nailed wall is simple. The length of each nail is generally much smaller than that of the anchor bolt, and it does not need to be pre-stressed. It has many advantages including small construction noise, small impact on the environment, small vibration and low cost. It also has small impact on adjacent buildings due to its small deformation.

图 2.32　土钉施工
Fig. 2.32　Soil nail construction

图 2.33　喷射混凝土面层
Fig. 2.33　Shotcrete surface layer

⑧ 锚杆支护（如图2.34所示）是一种岩土主动加固的基坑支护方法，其作用原理是将锚杆一端锚入稳定的土体中，另一端与其他形式的支护结构（如腰梁、冠梁等）连接，同时施加预应力，通过杆体受拉作用，调动深部地层潜能，从而维持基坑稳定。其具有成本低、支护效果好、操作简便、使用灵活、占用施工空间少等优点。

⑧ Anchor bolt support（Fig. 2.34）is a kind of geotechnical active reinforcement method in foundation excavation engineering. One side of the anchor bolt is driven into the soil layers, and the other end is connected to forms of supporting structure such as middle beam, and top beam. Pre-stress is put on the bolt at the same time. Through the tension effect of bolt, stability of foundation pit is maintained. It has many advantages like low cost, good supporting effect, easily operation, flexible use, and less construction space occupation, etc.

图2.34　基坑锚杆支护
Fig.2.34　Anchor support for foundation excavation

⑨ 内支撑支护（如图2.20所示）由竖向支护结构体系和水平内撑体系两部分构成。支护结构体系常用钢筋混凝土排桩墙或地下连续墙形式，内撑体系主要包括水平支撑与斜撑（钢筋混凝土或钢材）。该支护结构适用面广，适用于各种土层和不同深度的基坑。

⑨ Inner support（Fig. 2.20）is made up of vertical support retaining structure system and horizontal support system, respectively. Vertical support structure system commonly includes row of reinforced concrete pile wall which or underground continuous wall introduced in the form sections. Horizontal support system mainly includes horizontal brace and diagonal brace using reinforced concrete or steel. This support system is widely used in the foundation excavation engineering, which is applicable to different soil types and different excavation depth.

2.4.3　地下工程

城市化的高速发展，迫使人们不得不开发利用地下空间，这已经成为现代城市

2.4.3　Underground Engineering

With the rapid development of urbanization, it is urgent for people to develop and utilize the underground space, which has be-

规划和建设的重要内容之一。国际隧道协会（ITA）曾提出："大力开发地下空间，开始人类新的穴居时代"。在地面以下土层或岩体中修建各种类型的地下建筑物或结构的工程，称为地下工程。地下工程按照用途可分为不同的方面。其中，交通运输方面包括地下铁路、公路隧道、地下停车场等；国防方面包括地下指挥所、军火库等；工业与民用方面包括地下工厂、电站、车站、商场等。图2.35所示为地下空间开发利用的效果图。

地下工程施工中的开挖方法包括明挖法和暗挖法，明挖法可细分为基坑敞口开挖、基坑支挡开挖、地下连续墙、沉管法等；暗挖法可细分为矿山法、盾构机法（如图2.36所示）、掘进机（TBM）法（如图2.37所示）和顶管法（如图2.38所示）。

come an important issue in modern urban planning and construction. International tunneling association (ITA) proposed "vigorously developing the underground space, to starting a new cave era for human". Underground engineering refers to build various types of buildings or structures in the soil or rock mass below the ground. According to different uses, underground engineering can be divided into different aspects. Transportation aspects include underground railways, highways, underground parking. National defense aspects include underground command centers, shopping malls, etc. Industrial and civil aspects include underground factories, power stations, subway stations, shopping malls, etc. Fig. 2.35 shows the utilization of urban underground space.

Underground excavation construction can be divided into open excavation and concealed excavation. Open excavation method can be subdivided into open excavation of foundation pit, retaining wall of foundation pit excavation, underground continuous wall, immersed tube method, etc. Concealed excavation method can be subdivided into mining method, shield method (Fig. 2.36), tunneling boring machine (TBM) method (Fig. 2.37), and pipe jacking method (Fig. 2.38).

图 2.35　城市地下空间开发
Fig. 2.35　Utilization of urban underground space

图 2.36　盾构机
Fig. 2.36　Shield machine

图 2.37　岩石掘进机(TBM)
Fig. 2.37　Tunneling boring machine (TBM)

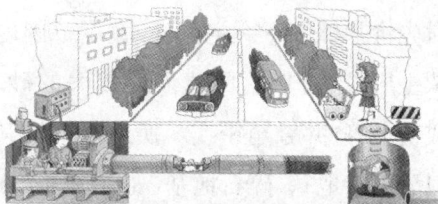

图 2.38　顶管法
Fig. 2.38　Pipe jacking method

2.5　地基处理

地基处理是一种技术,用于提高地基土的工程性质。当天然地基不能满足建(构)筑物对地基的要求时,需要对天然地基进行处理以形成人工地基。常用的地基处理方法有:换填垫层法、预压地基法、压实与夯实地基法、复合地基法和注浆加固法等。

地基处理主要分为基础工程措施和岩土加固措施。有些工程不改变地基的工程性质,而只采取基础工程措施;有些工程则需同时对地基的土和岩石进行加固,以改善其工程性质。选定适当的基础形式,不需改变地基的工程性质就可满足要求的地基称为天然地基。反之,已进行加固后的地基称为人工地基。

2.5.1　换土垫层

当建筑物基础下的持力层比较软弱,

2.5　Ground improvement

Ground improvement is a technique aiming at improving the engineering properties of the soil mass. When the natural foundation cannot meet the requirements of foundation construction, artificial foundation is needed to improve their natural foundation. Common foundation treatment methods include replacement layer of compacted fill method, preloaded ground method, compacted and tamped ground method, composite foundation method, and grouting reinforcement method, etc.

Ground improvement is mainly divided into foundation engineering measures and geotechnical reinforcement measures. For some engineering projects, foundation engineering measures are adopted without changing the properties of engineering foundation. Some projects, however, still need reinforcement of foundation soil and rocks at the same time in order to improve their engineering properties. The former is called natural foundation with selected foundation type without changing its properties. The latter is called artificial foundation due to some reinforcement measures.

2.5.1　Replacement layer of compacted fill

Replacement layer of compacted fill

岩土工程 Geotechnical Engineering

83

不能满足上部荷载对地基的要求时,常采用换土回填法来处理,如图2.39所示。施工时先将基础以下一定深度、宽度范围内的软土层挖去,然后回填强度较大的砂、石或灰土等,并夯至密实。换土回填按其材料分为砂地基、碎石地基(如图2.40所示)、灰土地基等。

method (Fig. 2.39) is often used to deal with poor ground, when the bearing soil layer under building foundation is relatively weak and cannot meet the load requirements of foundation. Soft soil layers are first dug out with a certain depth and width below the foundation in the construction. The sand, stone or lime with relatively higher strength are filled back to replace the soft soil. According to backfill materials, it can be divided into sand foundation, sand and gravel foundation (Fig. 2. 40), lime foundation, etc.

图 2.39 换土垫层
Fig. 2.39 Replacement layer of compacted fill

图 2.40 碎石垫层
Fig. 2.40 Replacement layer of gravels

2.5.2 预压地基

预压地基是指在原状土上加载,使土中水排出,以实现土的预先固结,减少建筑物地基后期沉降和提高地基承载力。按加载方法的不同,又分为堆载预压(如图2.41所示)、真空预压(如图2.42所示)、降水预压三种不同方法得到的预压地基。

2.5.2 Preloaded ground

Preloaded ground method is one of the ground improvement methods. Water can be drained from the soils by applying load on the intact soil, so as to make the soil pre-consolidate under the added load. It can reduce the settlement of building and improve the bearing capacity of foundation. According to the load source, preloaded ground method can be further divided into surcharge preloading method (Fig. 2. 41), vacuum preloading method (Fig. 2. 42), and dewatering preloading method, respectively.

土木工程导论

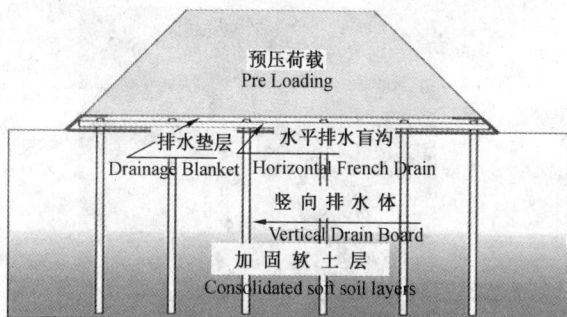

图 2.41 堆载预压
Fig. 2.41 Surcharge preloading

图 2.42 真空预压
Fig. 2.42 Vacuum preloading

排水法是采取相应措施如砂垫层、排水井、塑料多孔排水板等,使软基表层或内部形成水平或垂直排水通道,然后在土体自重或外界荷载作用下,加速土中水分的排出,使土体固结的方法。

Drainage method aims at making the soil consolidation by taking appropriate measures such as sand blanket, drainage well, and porous plastic drainage board to form drainage channels. Water in the soil layers can be discharged quickly through the horizontal or vertical drainage channel in soft ground surface or interior under its self weight or the added preload.

2.5.3 压实和夯实地基

压实法和夯实法均是有效的地基处理方法,分别如图 2.43 和图 2.44 所示。压实法是指采用重型机械将地基土压实的方法。夯实法是指用几十吨重的夯锤,从几十米高处自由落下,进行强力夯实的地基处理方法。夯锤一般重 10 ~ 40 t,落距为 6 ~ 40 m,处理深度可达 10 ~ 20 m。采用强夯法要注意振动对邻近建筑物的影响。

2.5.3 Compacted and tamped ground

Compacted ground method and tamped ground method are two effective ground treatment methods, as shown in Fig. 2.43 and Fig. 2.44, respectively. The compacted ground method refers to using heavy machinery such as road rollers to compact the soil foundation. The tamped ground method refers to improving the soil foundation by rammer with dozens of tons which drop from dozens of meters high to the ground. Rammer weight varies from 10 t to 40 t, and the drop distance varies from 6 m to 40 m. The improvement depth can be up to 10 m to 20 m. Attention should be paid to the vibration on adjacent buildings.

岩土工程 Geotechnical Engineering

图 2.43 压实法
Fig. 2.43 Compacted ground method

图 2.44 夯实法
Fig. 2.44 Tamped ground method

2.5.4 复合地基

随着地基处理技术的发展,复合地基技术得到愈来愈多的应用。复合地基是指天然地基在地基处理过程中部分土体得到增强或被置换,或在天然地基中设置加筋材料。加固区是由基体(天然地基土体)和增强体两部分组成的人工地基。复合地基中增强体和基体共同承担荷载。

复合地基一般分为水泥土搅拌桩复合地基(如图 2.45 所示)、高压喷射注浆桩复合地基、砂桩地基、振冲桩复合地基(如图 2.46 所示)、土和灰土挤密桩复合地基、水泥粉煤灰碎石桩复合地基,夯实水泥土桩复合地基等。

2.5.4 Composite foundation

Composite foundation method has been widely used with the development of the foundation treatment technology. A part of soil is enhanced or replaced in the process of the ground treatment, or reinforced material was set in natural foundation to from composite foundation. The reinforced area is composed of reinforced part and natural soil part, and this form was called artificial foundation. The strengthen body and natural soil part share the load together.

Composite foundation is generally divided into cement-soil mixing pile composite foundation (Fig. 2.45), high pressure jet grouting pile composite foundation, sand pile foundation, vibration pile composite foundation (Fig. 2.46), soil and lime compaction pile composite foundation, cement fly-ash gravel pile composite foundation, rammed soil-cement pile composite foundation, etc.

图 2.45 水泥土搅拌桩地基
Fig. 2.45   Cement-soil mixing pile foundation

图 2.46   振冲桩机
Fig. 2.46   Vibration machine

2.5.5 注浆加固

注浆加固法是通过钻孔或其他设施将浆液压送到地基孔隙或缝隙中,改善地基强度或防渗性能的工程措施,主要有固结灌浆、帷幕灌浆、接触灌浆、化学灌浆以及高压喷射灌浆法等。

（1）固结灌浆

固结灌浆是为改善基岩的力学性能而进行的灌浆操作,可以减少基础的变形和不均匀沉降,改善基础工作条件,减少基础开挖深度。其特点是灌浆面积较大、深度较浅、压力较小。

（2）帷幕灌浆

帷幕灌浆是指在基础内,平行于建筑物的轴线上钻一排或几排孔,用压力灌浆法将浆液灌入岩石的缝隙中去,形成一道防渗帷幕,截断基础渗流。其特点是深度较深、压力较大。

2.5.5 Grouting reinforcement

Grouting reinforcement aims at improving the strength of foundation and its anti-seepage performance by drilling boreholes or using other facilities to make grouting slurry pressed into pores or cracks of the foundation. It can be mainly divided into consolidation grouting method, curtain grouting method, contact grouting method, chemical grouting method, and high-pressure jet grouting method, etc.

（1）Consolidation grouting

Consolidation grouting can improve the mechanical properties of bedrock and reduce the deformation and uneven settlement. It can also improve the working conditions of foundation and reduce the depth of foundation excavation. The grouting area is large, and the depth is shallow with small grouting pressure.

（2）Curtain grouting

Curtain grouting aims at forming an impervious curtain to truncate foundation seepage by using pressure to make the grouting slurry pumped into the cracks. One row or several rows of holes are first drilled paralleling to the axis of the building in the foundation. The grouting depth is deeper due to the drilled holes, and the pressure is relatively high.

（3）接触灌浆

在建筑物和岩石接触面之间进行灌浆，以加强二者之间的结合程度和基础的整体性，提高抗滑稳定性，同时也增强岩石固结与防渗性能。

（4）化学灌浆

化学灌浆是以一种高分子有机化合物为主题材料的灌浆方法。这种浆材呈溶液状态，能灌入 0.10 mm 以下的细微管缝，浆液经过一定时间起化学作用，可将裂缝黏合起来形成凝胶，起到堵水防渗以及补强的作用。

（5）高压喷射灌浆

通过钻入土层中的灌浆管，用高压压入水泥浆液，并从钻杆下端的特殊喷嘴以高速喷射出去。在喷射的同时，钻杆以一定速度旋转，并逐渐提升；高压射流使四周一定范围内的土体结构遭受破坏，并被强制与浆液混合，凝固成具有特殊结构的圆柱体，也称旋喷桩。如采用定向喷射，可形成一段墙体，一般每个钻孔定喷后的成墙长度为 3~6 m。

2.6 边坡工程

为满足工程需要而对自然边坡进行的改造，称为边坡工程。根据边坡对工程影响的时间差别，可分为永久边坡和临时边坡两类；根据边坡与工程的关系，可分为建筑物地基边坡（必须满足稳定和有限

(3) Contact grouting

Contact grouting is performed at interface between the buildings and the rocks, which can ensure the integrity of the foundation and combination degree to improve the stability against sliding. It can also promote the rock consolidation and anti-seepage performance.

(4) Chemical grouting

It is one kind of grouting methods using macromolecular organic compounds as grouting material. The liquid grouting material can be injected into small pipe crakes less than 0.10 mm. The crakes can be bonded together to form gel after a certain time of chemical action, which can take effect of water plugging anti-seepage and reinforcement of foundation.

(5) High-pressure jet grouting

Cement grouting is injected through a special nozzle at a high speed into soil layers with high pressure by the grouting pipe drilling in the soil. Meanwhile, the drilling rod rotates at a certain speed and gradually elevated. Certain scope of soil structure around the high pressure jet grouting is damaged and forced to mix with the grouting, then it is solidified into a cylinder structure and called jet grouting pile. Wall body can be formed if use directional jet. Generally the length of wall for each jet drilling varies from 3 m to 6 m.

2.6 Slope engineering

Slope engineering refers to reform the natural slope to meet the engineering project needs. According to the difference of the time impacted on engineering, slope can be divided into two categories, namely, permanent slope and temporary slope, respectively. According to the relationship between the

变形要求）、建筑物邻近边坡（必须满足稳定要求）和对建筑物影响较小的延伸边坡（允许有一定限度的破坏）；根据边坡的材料可以分为土质边坡（如图 2.47 所示）和岩质边坡（如图 2.48 所示）。

project and the slope, slopes can be divided into three categories: building foundation slope (must meet the requirements of stability and finite deformation), nearby building slope (must meet the requirements of stability) and extension slope (allow certain destruction). According to the materials, slope can be divided into two categories, namely, soil slope (Fig. 2.47) and rock slope (Fig. 2.48), respectively.

图 2.47　土质边坡
Fig. 2.47　Soil slope

图 2.48　岩质边坡
Fig. 2.48　Rock slope

2.6.1　土质边坡稳定分析

土质边坡在重力或其他因素作用下坡体向下运动，造成土体的破坏称为滑坡或土坡破坏。土坡的稳定程度一般以安全系数的大小来衡量，表示方法有多种，通常为：① 实际的抗剪强度与维持平衡所需的抗剪强度之比；② 抗滑力矩之和与下滑力矩之和的比值；③ 总抗滑力与总下滑力的比值等。工程实践中为选用合适的安全系数，需考虑的因素有荷载组合、建筑物按重要性的类别、抗剪强度的试验条件、计算方法的选择、施工控制的可靠程度、获得的信息是否完整、工程经验、经济等。常用的安全系数值的范围是 1.1 ~

2.6.1　Soil slope stability analysis

Soil slope may move down resulting from gravity or other factors causing soil damage. The damaged soil is called landslide damage or slope damage. Slope stability is commonly measured and assessed by safety coefficient. There are many types of representation for safety coefficient, such as: ① the ratio of actual shear strength and the shear strength required for balance; ② the ratio of total anti-sliding movement and total sliding moment; ③ the ratio of total anti-sliding force and total sliding force, etc. In engineering practice, suitable safety factors should be selected according to the load combination, building importance category, test conditions of shear strength, selection of calculation methods, reliability of construction control, accuracy of information, engi-

2.0。由于土体的复杂性,在土坡分析中所涉及的因素都是随机的,即使采用大于1的安全系数,也不能认为土坡的稳定性是十分可靠的。故对重要工程,除采用较大的安全系数外,还需进行边坡稳定的概率分析。

neering experience, economy, etc. The scope of safety coefficient value commonly varies from 1.1 to 2.0. The factors involved in the analysis of soil slope are random due to the complexity of the soils. Slope stability cannot be very reliable even with the safety coefficient greater than 1. Therefore, probability analysis of slope stability should be conducted besides adopting bigger safety coefficient for important projects.

2.6.2 岩质边坡稳定分析

岩质边坡由岩石构成,由于坡度过大,坡脚受切,或作用于坡体的力发生变化,均会使边坡失稳而破坏。边坡的稳定性常以安全系数或破坏概率作为评价指标。安全系数指滑体沿滑动面所受的抗滑力和滑动力之比。破坏概率是指安全系数小于1的出现概率。破坏概率愈小,边坡稳定程度愈高。岩质边坡稳定性分析以工程地质资料为基础,内容包括:判断边坡破坏模式、确定计算参数、应用岩体力学和有限元等方法进行力学计算并做出稳定性评价。

2.6.2 Stability analysis of rock slopes

A rock slope is formed by rock, and it is unstable and can be destroyed since the slope gradient is bigger than soil slope, slope toe is sheared or the changing of external force on slope body. Safety coefficient or failure probability is often used as evaluation index for rock slope stability. Safety coefficient refers to ratio of anti-sliding force and sliding force along the sliding surface. Failure probability refers to the probability of the safety coefficient less than 1. The smaller the probability, the higher the degree of slope stability is. Rock slope stability analysis is based on the engineering geological materials including judgment of the slope failure mode, calculation of parameters, the application of the rock mechanics and finite element method to perform mechanical calculation and stability evaluation.

2.7 岩土工程发展展望

岩土工程是人类改造世界,发展生存空间,营造现代物质文明的一项重要系统工程。随着科学技术的迅猛发展,21世纪的岩土工程必将在现有先进水平的基础上获得突破性的发展。随着高层建筑、城市地下空间利用和高速公路的发展,岩土工程者的注意力较多地集中在建筑工程、

2.7 Developing prospects for GE

GE is an important engineering system aiming at reconstructing the world, developing human survival space, and creating a modern material civilization. With the rapid development of science and technology, the significant breakthroughs will be achieved based on the existing advanced level in the 21st century. Geotechnical engineers begin to focus on the engineering problems related to building engineering, municipal engineer-

土木工程导论

市政工程和交通工程建设中的岩土工程问题上。土木工程功能化、城市立体化、交通高速化，以及改善综合居住环境成为现代土木工程建设的特点。人口的增长加速了城市发展，城市化的进程促进了城市在数量和规模上的急剧发展。人们将不断拓展新的生存空间，开发地下空间，向海洋拓宽，修建跨海大桥、海底隧道和人工岛，改造沙漠，修建高速公路和高速铁路等。展望岩土工程的发展，不能离开对现代土木工程建设发展趋势的分析。

2.7.1 岩土工程可持续发展

近20年来，随着国民经济和各类土木工程的快速发展，我国岩土工程建设空前兴旺发达；但是几乎在同期，国际岩土工程学科的领域和范围发生了明显的变化。随着人们对资源和生态环境认识的深入，岩土工程已经不仅限于具体"工程"的设计施工，而是扩展到环境岩土、地质灾害，以及与生态和资源相适应的岩土工程可持续发展等大的战略问题上。例如大量开采深层地下水，将使地面普遍下降，引发地裂缝、海水入侵及地下水咸化；废弃物会污染土地、空气与地下水；大规模的水利水电工程、高速公路、高速铁路，尤其是在中国生态环境比较脆弱的西部高寒、高海拔和高纬度地区的工程，可能引发次

ing and traffic engineering with the development of high-rise buildings, urban underground space utilization and express highway. Functionalization of civil engineering, three-dimensional city, high speed transportation, and improvement of human settlement environment have become the hallmark of the modern civil engineering construction. Population growth accelerate the urban development, and the urbanization processes promote the rapid development of the number and size for big cities. People will continue to expand a new survival space, develop underground space, expand to the ocean, build sea-crossing bridges, tunnels and artificial islands, transform desert, build highways and high-speed railways, etc. When it comes to the development of geotechnical engineering, people cannot ignore the analysis of the trend of modern civil engineering construction.

2.7.1 Sustainable development of GE

With the rapid development of national economy and various types of civil engineering over the past 20 years in China, GE has undergone an unprecedented prosperity. At the same period, however, significant changes have taken place on the subject areas and scope of the international GE. With the deep understanding of resources and the ecological environment, GE is no longer limited to specific "project" design and construction, but extended to geotechnical environment, geological disasters, and sustainable development, etc. For example, a large number of exploitation of deep groundwater will make ground settlement, ground cracks, seawater intrusion, and groundwater salinization; wastes will pollute the land, air and groundwater; large-scale water conservancy and hydropower engineering, highway, high-speed railway, especially in the fragile ecological environment, cold western, high altitude and high latitude areas of China, could

生地震、崩塌、滑坡、泥石流、融陷、大面积水土流失等；在膨胀土、湿陷性黄土、多年冻土、盐渍土地区的工程一方面给工程本身造成困难，另一方面也可能存在生态环境方面的不利影响。

环境岩土工程是岩土工程与环境科学密切结合的一门新学科。它主要应用岩土工程的观点、技术和方法为治理和保护环境服务。人类生产活动和工程活动造成许多环境公害，如采矿造成采空区坍塌，过量抽取地下水引起区域性地面沉降，工业垃圾、城市生活垃圾及其他废弃物，特别是有毒有害废弃物污染环境，施工扰动对周围环境造成影响等。另外，地震、洪水、风沙、泥石流、滑坡、地裂缝等灾害也对环境造成破坏。上述环境问题的治理和预防给岩土工程师们提出了许多新的研究课题。随着城市化、工业化发展进程加快，环境岩土工程研究显得更加重要，工程师应从保持良好的生态环境和保持可持续发展的高度来认识和重视环境岩土工程研究。

2.7.2 岩土工程智能化测试技术

通过钻探取样直观地掌握岩土性状，仍将是 21 世纪岩土工程勘测的核心方法，但是钻探技术和测试技术将有长足的进

cause the secondary earthquakes, collapse, landslide, debris flow, sink, large area of soil and water loss, etc. Project in the expansive soil area, collapsible loess, permafrost, and saline soil will make its construction difficult on one hand, and there are still negative impact on ecological environment on the other hand.

Environmental geotechnical engineering is a new discipline closely combined with GE and environmental science. It serves for the improvement and environmental protection by applying GE views, technologies and methods. Human production activities and engineering activities may result in many environmental hazards. For example, mining will cause the collapse of a mined-out area; excessive pumping groundwater will cause regional land subsidence; industrial waste, municipal solid waste and other waste, especially toxic and harmful waste will pollute the environment; the construction disturbance will influence the surrounding environment. In addition, earthquake, flood, sandstorm, debris flow, landslide, ground cracks and so on will also cause damage to the environment. The prevention of the above environmental problems has posed many new research topics to geotechnical engineers. With the rapid development of urbanization and industrialization process, environmental geotechnical engineering research will become more and more important. Geotechnical engineering research should be recognized and focused from the perspective of maintaining a good ecological environment and making a sustainable development.

2.7.2 Intelligent testing technology of GE

Geotechnical properties can be obtained intuitively by sample drilling, which will also be the core method of geotechnical engineering survey in the 21st century. However, the

步,原位试验仍将是必不可少的勘测手段。随着原位试验精度的提高,不仅可使昂贵的钻探取样、室内土工试验的工作量减少到最低限度,且岩土工程勘测的效率也将大大提高。原位试验技术的发展同样以智能化为方向,而遥感遥控和数据采集处理自动化将成为普遍的技术方法。

2.7.3 城市地下空间的开发利用

随着意识转变,人们想要还城市更开阔的空间,更多的绿色景观,更浓厚的历史与人文氛围与气息,这就需要更多地利用地下空间。21 世纪人类也面临着人口增长的巨大压力,人们生活的空间是有限的,必须在保护生态环境的同时,充分开发和有效利用生存空间。一方面高层建筑仍将是解决城市人口密集区居住问题的重要手段,另一方面要充分利用既节能又不争地的地下空间。城市交通、公共设施、人居空间将大量转入地下,这也是城市防护和减灾的要求。目前,地下铁道、地下停车场、地下仓库、地下商场、越江或海底隧道等地下建筑物已经广泛地在为人们服务。

drilling technology and testing technology will be improved greatly. In-situ tests will continue to be an indispensable survey method. With the improvement of in-situ test accuracy, expensive drilling sampling and laboratory soil tests work will not only be reduced to the minimum, but also the GE exploration efficiency will be greatly improved. Intelligent will be the direction of in-situ test development. Remote sensing, remote control and automatic data acquisition process will become common technical methods.

2.7.3 Underground space development and utilization

With the changing of people's consciousness, more open space, more green landscape, more strong historical and cultural atmosphere are needed, which requires greater use of underground space. People have to face the huge pressure of population growth in the 21st century. Since the living space is limited for people, taking full advantage of human living space is needed based on the protection of ecological environment. On the one hand, high-rise buildings will still be the important method to solve the problem of the high concentration of city population. On the other hand, underground space will be fully used for both saving energy and land. Urban transportation, public facilities, residential space will transfer to underground, which meets the requirements of urban protection and disaster mitigation. Currently, subways, underground malls, underground parking lots, underground malls, river-crossing or sub-sea tunnels, etc. have been developing.

注:本章图片均来源于网络。
Note: In this chapter, all pictures are from webs.

相关链接 Related Links

(1) 岩土论坛 http://www.yantubbs.com/

(2) 国际岩土信息中心 http://www.geoengineer.org/

(3) 土木工程网 http://www.civilcn.com/

小贴士 Tips

(1) 中国注册岩土工程师

<u>注册土木工程师</u>(岩土),简称注册岩土工程师,是指取得《中华人民共和国注册土木工程师(岩土)执业资格证书》和《中华人民共和国注册土木工程师(岩土)执业资格注册证书》,从事岩土工程工作的专业技术人员。

Chinese registered geotechnical engineer

<u>Registered Civil Engineer</u> (GE), shorten as the Registered Geotechnical Engineer, refers to the professional and technical person who has been involved in the geotechnical engineering after successfully obtained both "the People's Republic of China registered civil engineer (GE) qualification certificate" and "the People's Republic of China registered civil engineer (GE) qualification certificate of registration".

(2) 岩土工程类期刊

英国:"Géotechique",美国:"Geotechnical & Environmental Engineering, ASCE",加拿大:"Canadian Geotechnical Journal",日本:"Soils and Foundations",中国:《岩土工程学报》,《岩土力学》等。

Geotechnical engineering journals

Géotechique in Britain, *Geotechnical & Environmental Engineering*, *ASCE* in USA, *Canadian Geotechnical Journal* in Canada, *Soils and Foundations* in Japan, *Chinese Journal of Geotechnical Engineering* and *Rock and Soil Mechanics*, *etc.* in China.

思考题 Review Questions

(1) 基坑支护的方法有哪些?

What are the foundation excavation support methods?

(2) 地基处理的方式有哪些?

What are the ground improvement methods?

(3) 如何实现岩土工程的可持续发展?

How can sustainable development of GE be achieved?

参考文献
References

[1] 李广信:《岩土工程50讲:岩坛漫话》(第2版),人民交通出版社,2010年。

［2］周景星,等:《基础工程》(第2版),清华大学出版社,2007年。

［3］张萌:《岩土工程勘察》,中国建筑工业出版社,2011年。

［4］龚维明,等:《地下结构工程》,东南大学出版社,2004年。

［5］高大钊:《土力学与基础工程》,中国建筑工业出版社,1998年。

［6］侔磊等:《边坡工程》,科学出版社,2010年。

［7］刘永红:《地基处理》,科学出版社,2005年。

［8］龚晓南:《21世纪岩土工程发展展望》,《岩土工程学报》,2000年第2期。

［9］钱家欢,殷宗泽:《土工原理与计算》(第2版),中国水利水电出版社,1996年。

［10］Mitchell, J. K. , Soga, K. Fundamentals of Soil Behavor. 3rd ed. Jone Wiley & Sons,2005.

［11］Klaus Kirsch, Alan Bell. Ground Improvement. 3rd ed. CRC Press Inc,2012.

［12］Braja M. Das. Principles of Foundation Engineering. Cengage Learning, 2010.

［13］Robin Chowdhury, Phil Flentje, Gautam Bhattacharya. Geotechnical Slope Analysis. CRC Press Inc,2009.

第3章　房屋建筑工程

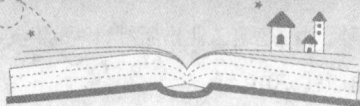

Chapter 3　Building Engineering

建筑工程是运用数学、力学、材料学等知识研究各种建筑物设计和建造方法的一门学科，是土木工程最具代表性的分支。本章首先介绍建筑工程中的荷载，接着介绍建筑工程中的基本构件及其应用，在此基础上，进一步说明由这些基本构件组合成的各种建筑结构形式和结构体系。

Building engineering is a discipline about how to design and construct buildings by using mathematics, mechanics, material science etc. It is the representative branch of civil engineering. In this chapter we will firstly introduce the loads in the building engineering, and then introduce the basic structural members and their applications, on which a variety of building structures composed of basic structural members are further presented.

3.1　荷载

荷载是指主动作用于建筑物上的外力，如建筑物的自重，工业厂房上的吊车荷载，行驶在桥梁上的车辆荷载以及作用在水工结构上的水压力和土压力等。另外，温度变化、支座沉陷、制造误差等因素也可使建筑结构产生位移或内力，这些因素可视为广义荷载，与荷载一起统称为作用。作用对建筑结构产生的效应，称为作用效应，如内力、变形、应力、应变和位移。

3.1　Loads

Loads refer to the forces acting on the structures actively, such as the gravity of the structure, the crane load, vehicular load, water pressure and soil pressure. The change of temperature, settlement of supports, manufacturing errors and other factors which can be regarded as generalized loads can also make the structure generate displacements and internal forces. For convenience, loads and generalized loads are collectively called action. Effects of the actions on the structure mainly include internal forces, deformation, stress, strain and displacements.

根据荷载在建筑结构上出现时间的变异性和可能性，可分为永久荷载、可变荷载和偶然荷载；根据荷载作用的性质，可分为静力荷载和动力荷载。各专业方向的荷载可查阅相关的规范，如《建筑结构荷载规范》（GB 50009—2012）提供了工业与民用建筑中常用的永久荷载和可变荷载取值方法；《建筑抗震设计规范》（GB50011—2010）给出了建筑结构的地震作用确定方法。

3.2 基本构件

结构是建筑物的骨架，用以承受建筑上的各种荷载和作用。结构失效将会造成建筑物的倒塌，带来生命和财产的巨大损失，因此保证结构的安全是建筑物的基本要求。结构是由基本构件组成的，建筑结构的基本构件主要有梁、柱、板、墙、拱等。

3.2.1 梁

梁是建筑结构中的受弯构件（如图3.1所示），在建筑工程中有着广泛的应用。它的长度一般大于截面高度和截面宽度。梁通常水平放置，有时也斜向放置以满足使用要求，如楼梯梁（如图3.2所示）。

According to the variability and possibility of loads on the structure, it can be categorized into permanent loads, variable loads and accidental loads. The loads can also be classified as static loads and dynamic loads by its features. The load magnitude and calculation method can be found in the relevant Codes. For examples, *Load Code for the Design of Building Structures* (GB50009-2012) provides the determination methods of permanent loads and variable loads in industrial and civilian structures, and *Code for seismic design of buildings* (GB50011-2010) provides the calculation methods of the earthquake action.

3.2 Basic structural members

The structure is the skeleton of architectures, which is used to hold various loads and actions. The structural failure will result in the collapse of the architecture which brings huge losses of life and property, so it is the basic requirements to ensure safety of the structure. Structures are formed by basic structural members, which mainly include beam, column, wall, plate, arch and so on.

3.2.1 Beam

Beam is the flexural member (Fig. 3.1). In engineering practice, beam has a variety of application, whose length is usually larger than its cross-sectional dimensions. The beam is almost horizontal, but sometimes it is placed obliquely to meet requirements, such as a stair beam (Fig. 3.2).

房屋建筑工程 Building Engineering

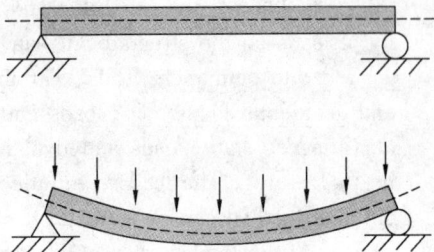

图 3.1　梁
Fig. 3.1　Beam

图 3.2　楼梯梁
Fig. 3.2　Stair beam

梁按截面形式分矩形梁、T 形梁、倒 T 形梁、花篮梁、I 形梁、Z 形梁、槽形梁、箱型梁、叠合梁等；按所用材料可分为钢梁、钢筋混凝土梁、预应力混凝土梁、木梁以及组合梁等（如图 3.3、图 3.4 所示）。

According to the cross section forms, beams can be categorized into rectangular beam, T-beam, inverted T-beam, ledger beam, I-beam, Z-beam, channel beam, box beam, composite beam, etc. From the aspect of used materials, beams can also be classified into steel beam, concrete beam, pre-stressed concrete beam, wooden beam and combination beam and etc (Fig. 3.3 and Fig. 3.4).

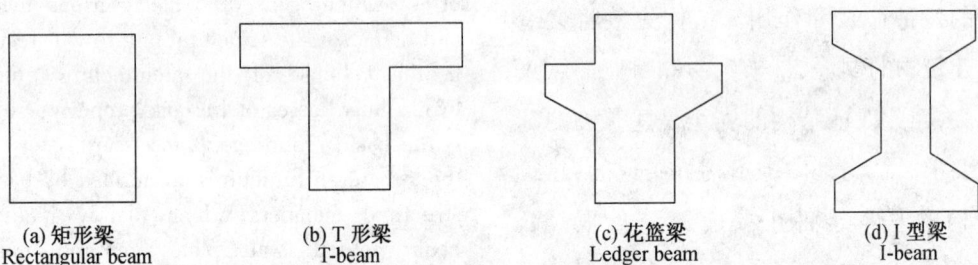

(a) 矩形梁
Rectangular beam

(b) T 形梁
T-beam

(c) 花篮梁
Ledger beam

(d) I 型梁
I-beam

图 3.3　钢筋混凝土梁的截面类型
Fig. 3.3　Cross-section types of reinforced concrete beam

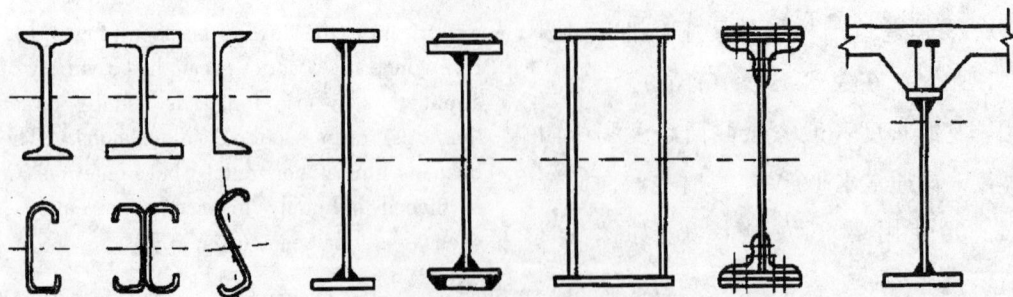

图 3.4　钢梁的截面类型
Fig. 3.4　Cross-section types of steel beam

根据梁跨数的不同，可分为单跨梁和多跨梁。常用的单跨梁有简支梁、悬臂

According to the number of spans, beams can be categorized into single span beam and multispan beam. The commonly used single

梁、伸臂梁(如图3.5所示)。桥梁工程中应用较多的是多跨梁(如图3.6所示)。

span beams include simply supported beam, cantilever beam and overhanging beam (Fig. 3.5). Multıspan beams are widely used in bridge engineering (Fig. 3.6).

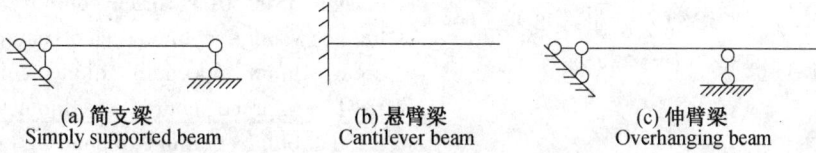

(a) 简支梁
Simply supported beam

(b) 悬臂梁
Cantilever beam

(c) 伸臂梁
Overhanging beam

图3.5　单跨梁
Fig. 3.5　Single span beam

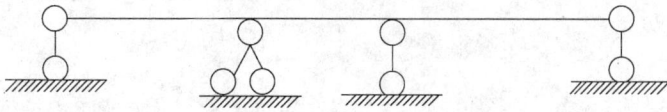

图3.6　多跨梁
Fig. 3.6　Multispan beam

根据梁在结构中的位置和使用功能,又可分为主梁、次梁、过梁、圈梁、联系梁等。图3.7是某框架结构的局部梁格布置。

According to their position and function, beams can be categorized as girders, stringers, lintels, periphery beams and tie beams etc. Fig. 3.7 shows a floor framing system of a frame structure.

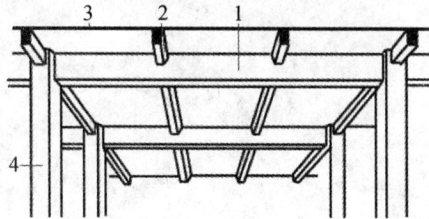

1—主梁 girder;
2—次梁 stringer;
3—楼板 floor;
4—柱 column

图3.7　某框架结构的梁格布置
Fig. 3.7　Floor framing system

3.2.2　柱

柱在建筑结构中是主要承受压力,有时也同时承受弯矩的竖向构件。

柱按截面形式可分为矩形柱、方柱、

3.2.2　Column

Columns, which are vertical members in structures, are mainly used to withstand compressive loads (sometimes along with bending moments).

According to cross section forms, the

圆柱、工字形柱、L形柱、十字形柱、格构柱（如图 3.8 所示）等；按所用材料，可分为石柱、砖柱、砌块柱、木柱、钢柱、钢筋混凝土柱、钢与混凝土组合柱（如劲性钢筋混凝土柱、钢管混凝土柱）等（如图 3.9 所示）。

columns can be categorized as rectangular column, square column, circular column, I-column, L-beam, cruciform column, lattice column (Fig. 3.8) etc. Columns can also be classified as pillar, brick column, block column, wooden column, steel column, concrete column and combination column (i. e. steel reinforced concrete column, concrete filled tubular column) depending on the materials (Fig. 3.9).

图 3.8　格构柱
Fig. 3.8　Lattice column

(a) 石柱
Pillar

(b) 砖柱
Brick column

(c) 木柱
Wooden column

(d) 钢柱
Steel column

(e) 钢筋混凝土柱
Concrete column

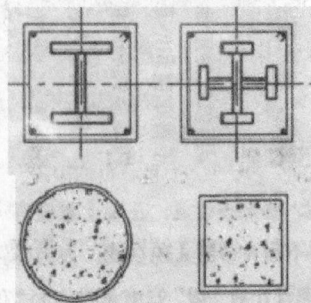

(f) 组合柱
Composite column

图 3.9　柱的种类
Fig. 3.9　Column types

3.2.3 板

板是指平面尺寸远大于厚度的受弯构件,通常水平放置,但有时也斜向放置(如楼梯板)。板在建筑中一般应用于楼板、屋面板、基础板等。

板按截面形式可分为实心板、空心板(如图 3.10a 所示)、槽形板(如图 3.10b 所示);按所用材料可分为木板、钢板、钢筋混凝土板、预应力板等。

3.2.3 Plate

Plate is a flexural member whose length and width are much larger than its thickness. Plates are almost horizontal, sometimes they are also placed obliquely (i. e. stair plates). Plates are usually used as floor slabs, roof boardings and base plates in the construction.

According to cross section forms, the plate can be categorized as solid slab, hollow-core slab (Fig. 3. 10a), channel slab (Fig. 3. 10b). The plate can also be classified as plank, steel plate, concrete slab and prestressed slab etc. from the aspect of materials.

(a) 空心板
Hollow-core slab

(b) 槽形板
Channel slab

图 3.10 空心板与槽形板
Fig. 3.10 Hollow-core slab and channel slab

3.2.4 墙

墙是指竖向尺寸的高与宽较大,而厚度相对较小的构件。

根据承重情况和使用功能,墙可以分为承重墙、非承重墙。一般来说,承重墙主要承受平行于墙面的荷载,而在多、高层钢筋混凝土结构的剪力墙结构体系中,墙体则承受全部水平作用和竖向荷载。

根据墙所处的位置,可以分为横墙、纵墙、山墙、外墙和内墙等。

3.2.4 Wall

Wall is a component whose vertical depth and width are larger than its thickness.

According to their stress condition and functions, walls can be categorized as bearing wall and non-bearing wall. Generally, bearing walls main bear loads which parallel to the wall. In shear wall structure systems, the wall will withstand all horizontal and vertical loads.

Walls can be categorized as cross wall, longitudinal walls, head wall, external and internal wall.

房屋建筑工程 Building Engineering

根据不同施工工艺,墙可以分为预制墙、现浇墙和砌筑墙。

According to their construction technologies, walls can be categorized as prefabricated wall, cast-in-situ wall and masonry wall.

3.2.5 拱

拱是曲线结构,主要承受轴向压力。拱结构广泛应用于桥梁工程和建筑工程中(如图3.11所示)。拱在竖向荷载作用下,支座不仅产生竖向反力,还产生水平推力,这使得拱的内弯矩比梁小,截面上的应力分布比较均匀,因而更能发挥材料的作用。拱形结构主要受压力作用,便于使用抗压性能好而抗拉性能差的材料,如砖、石、混凝土等。

3.2.5 Arch

An arch is a curved structure whose dominated internal force is axial compression. Arches are widely used in bridge structures and building structures (Fig. 3.11). Under the action of vertical loads, the arch gives rise to both vertical reactions and horizontal thrusts, as a result, the bending moments in arch are smaller than those in beams under the same situation. Since the stress distribution on the sections of arches is more uniform than that of beams, the capabilities of constructional materials can be effectively utilized in arch structures. Moreover, arches mainly under compressive stresses when subjected to external loads, so the constructional materials which are stronger in compression but weaker in tension such as bricks, stone and concrete are suitable to build arch-typed structures.

(a) 赵州桥
Zhaozhou Bridge

(b) 凯旋门
Triomphal Arch

(c) 圣路易斯拱门
Gateway Arch

(d) 南京奥林匹克体育场
Nanjing Olympic Stadium

图 3.11 拱的应用
Fig. 3.11 Application of arch

3.3 建筑物分类

由基本构件可以组成各种形式的建筑物,这些建筑物通常按以下几种方法进行分类。

3.3.1 按使用功能分类

建筑物根据其使用功能的不同,可分为民用建筑、工业建筑和农业建筑。民用建筑主要包括住宅建筑(如宿舍、公寓等)和公共建筑(如教学楼、体育馆等)。工业建筑主要指各类厂房和生产辅助用房(如厂房、储仓等)。农业建筑主要包括饲养牲畜、存放农具和农产品的用房(如蔬菜大棚、养鸡场等)。

3.3.2 按建筑材料分类

建筑物根据其所用建筑材料的不同,可分为木结构建筑、钢筋混凝土结构建筑、钢结构建筑、钢 – 混凝土组合结构建筑、砖混结构建筑和其他混合结构建筑(如图 3.12 所示)。

3.3 Classification of building

All types of buildings are formed by basic structural members, and these building structures are usually classified in the following manners.

3.3.1 Classification by function

According to their functions, buildings can be classified as civil structures, industrial structures and agricultural structures. Civil structures mainly include residential buildings (e. g. dormitories, hotels) and public buildings (e. g. teaching buildings, gyms). Industrial structures mainly refer to various types of production and auxiliary buildings (e. g. factory buildings, storage bunkers). Agricultural structures mainly refer to the buildings which are used for raising livestock and the storage of agricultural products and implements (e. g. vegetable green houses, poultry yards).

3.3.2 Classification by material

According to their different materials, buildings can be classified as wooden structure building, reinforced concrete structure building, steel structure building, steel-concrete structure building, brick-concrete structure building etc. as shown in Fig. 3. 12.

房屋建筑工程 **Building Engineering**

(a) 木结构
Wooden structure

(b) 钢筋混凝土结构
Reinforced concrete structure

(c) 钢－混凝土组合结构
Steel-concrete structure

(d) 砖混结构
Brick-concrete structure

(e) 钢结构
Steel structure

图 3.12　不同材料建成的结构
Fig. 3.12　Building structures constructed by various materials

3.3.3　按建筑层数分类

建筑物按其层数可分为单层建筑、多层建筑、高层建筑等。下面将主要按该分类方式对建筑工程的常用结构形式进行介绍。

3.4　单层建筑

单层建筑通常指层数为一层的建筑，可分为一般单层建筑和大跨度建筑。

3.4.1　一般单层建筑

一般单层建筑按使用目的可分为民用单层建筑和单层工业厂房。

3.3.3　Classification by storey number

According to the storey number, buildings can be classified as single-storey building, multi-storey building, tall building etc. In the following sections, we will introduce the structure forms generally used in building engineering by the classification of storey number.

3.4　Single-storey building

Single-storey building usually refers to the building with one storey. It can be classified as general single-storey building and long-span structure building.

3.4.1　General single-storey building

Based on the purpose of its application, general single-storey building can be classified as civil single-storey building and indus-

民用单层建筑一般采用砖混结构（墙体采用砖或砖砌体，梁和板采用钢筋混凝土），多用于单层住宅、公共建筑等（如图3.13所示）。

Civil single-storey buildings are mostly used in residence and public buildings (Fig. 3.13) The brick-concrete structure is the frequently used structural form in civil single-storey buildings, whose wall material is brick or brick mansonry while the beam and plate material is concrete.

图 3.13　民用单层建筑
Fig. 3.13　Civil single-storey buildings

单层工业厂房通常由柱、屋架、吊车梁、天窗架和柱间支撑等构件组成，如图3.14 所示。

As shown in Fig. 3.14, industrial single-storey buildings are generally composed of columns, roof truss, crane girder, skylight frame and column brace etc.

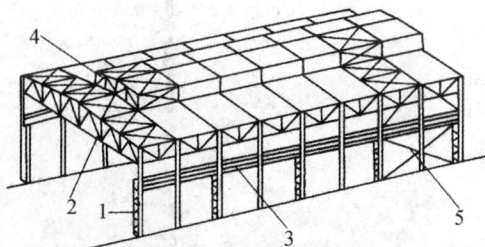

1—柱 column; 2—屋架 roof truss; 3—吊车架 crane girder;
4—天窗架 skylight frame; 5—柱间支撑 column brace

图 3.14　单层工业厂房
Fig. 3.14　Industrial single-storey buildings

单层工业厂房按结构形式可分为排架结构和刚架结构。排架结构中柱与基础刚接，屋架与柱顶铰接；刚架结构的梁或屋架与柱的连接均为刚性连接。

Industrial single-storey buildings can be categorized as bent structure building and rigid-framed structure building. In bent structure buildings, the columns are rigid connected with the base, while the roof truss is hinge connected with the column upper

房屋建筑工程
Building Engineering

3.4.2 大跨度建筑

大跨度建筑通常是指跨度在 60 m 以上的建筑,在民用建筑中主要应用于影剧院、体育场馆、展览馆和航空港等,如图 3.15 所示;在工业建筑中则主要应用于飞机装配车间、飞机库和其他大跨度厂房,如图 3.16 所示。大跨度建筑结构体系有很多种,如桁架结构、网架结构、网壳结构、悬索结构、膜结构、薄壳结构等。

3.4.2 Long-span structures

end. In the rigid-framed structure, the beams or roof truss are rigidly connected with the columns.

The long-span structure refers to the one whose span is over 60 m. Long-span structures are widely used in civil structures (such as cinema, stadium, exhibition hall, aviation port, as shown in Fig. 3.15) and industrial structures (such as aircraft assembly workshops, hangars, other long-span factories, as shown in Fig. 3.16). The long-span structures can be classified into many types, such as truss, space grid structure, latticed shell structure, suspended structure, membrane structure, thin-shell structure, etc.

(a) 中国国家体育场
China National Stadium

(b) 中国国家歌剧院
China National Opera House

(c) 香港国际会议展览中心
Hong Kong convention and exhibition center

(d) 北京首都机场 T3 航站楼
Beijing capital airport T3 terminal building

图 3.15 民用大跨度建筑
Fig. 3.15 Civilian long-span structures

(a) 飞机仓库
Hangar

(b) 储煤仓库
Coal storage bunker

图 3.16 工业大跨度建筑
Fig. 3.16 Industrial long-span structures

（1）桁架结构

桁架结构是由直杆铰接而成的平面或空间结构。桁架通常由<u>上弦杆</u>、<u>下弦杆</u>和<u>腹杆</u>（竖腹杆和斜腹杆）组成（如图 3.17 所示）。在节点荷载作用下，桁架杆件主要承受轴向拉力或压力，从而能充分利用材料，因此，桁架结构广泛地应用于大跨度建筑，如图 3.18 所示。

（1）Truss

Truss is one kind of planar or spatial structure composed of straight bars pinned at the connections. The truss is usually constituted by <u>upper chord</u>, <u>lower chord</u> and <u>web members</u> (vertical web members and diagonal web members), as shown in Fig. 3.17. The truss can take full advantage of the material for its members mainly generating tension or precession under joint loads. Truss is widely used in long-span structures (Fig. 3.18).

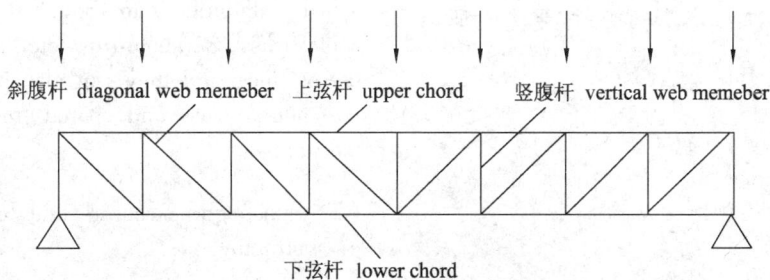

斜腹杆 diagonal web memeber　　上弦杆 upper chord　　竖腹杆 vertical web memeber

下弦杆 lower chord

图 3.17 桁架的组成
Fig. 3.17 Composition of truss

(a) 北京理工大学体育馆
Gymnasium of Beijing Institute of Technology

(b) 深圳机场候机楼
Shenzhen airport terminal

(c) 南京火车站
Nanjing Railway Station

(d) 汉中收费站
Hanzhong highway toll-gate

图 3.18 桁架结构的应用
Fig. 3.18 Application of truss

桁架结构根据所用材料可分为木桁架、钢桁架、钢筋混凝土桁架以及组合桁架(如钢-木桁架等);按外形可分为三角形、梯形、抛物线形桁架等;按受力性质可分为平面桁架和空间桁架。

(2) 网架结构和网壳结构

网架结构是由许多杆件通过节点,按照一定规律组成的网状空间杆系结构,分为平板网架(如图 3.19a 所示)和曲面网架(也称网壳,如图 3.19b 所示)。网架结构根据材料可分为木网架、钢网架、钢筋混凝土网架等,其中以钢网架应用最多。

Trusses can be classified into wooden truss, steel truss, reinforced concrete truss and composite truss (e. g. Steel-wooden truss) according to the material, and it can also be classified into triangle truss, trapezoidal truss, parabola truss etc. according to the outline, and they can also be classified into planar truss and spatial truss by force mechanism.

(2) Space grid structure and latticed shell structure

The grid structure is a reticular spatial structure connected by many rods in certain law. The grid structure can be categorized into flat grid (also called space grid structure, Fig. 3.19a) and surface grid (also called latticed shell structure, Fig. 3.19b). According to the material curved, the grid structure can be classified into wooden grid structure, reinforced concrete grid structure

网架结构在节点处通常采用焊接球(如图3.20a 所示)或螺栓球(如图 3.20b 所示)进行连接。

and steel grid structure which is most widely used. Rods of grid structure are usually connected by welded ball joints (Fig. 3.20a) or bolt ball joints (Fig. 3.20b).

(a) 平板网架
Flat grid

(b) 曲面网架（网壳）
Surface grid (latticed shell structure)

图 3.19　网架结构
Fig. 3.19　The grid structure

(a) 焊接球连接节点
Welded ball joint

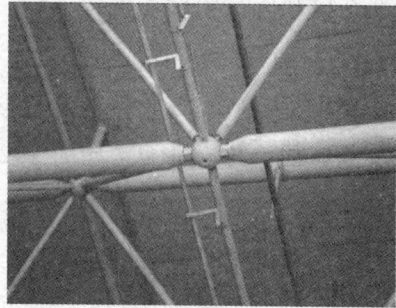

(b) 螺栓球连接节点
Bolt ball joint

图 3.20　网架结构连接节点
Fig. 3.20　Joints of grid structure

（3）悬索结构

悬索结构是以索作为主要受力构件抵抗外荷载作用的一种结构体系（如图 3.21 所示）。悬索结构有很多优点,如自重轻、跨度大、材料省且易于实现丰富多彩的建筑造型。索的材料通常采用受拉性能良好的钢丝束、钢丝绳、钢绞线等线材。

（3）Suspended structure

The suspended structure takes cables as the primary member to resist external loads, and it has many advantages, such as light weight, long span, material savings and easy to implement a variety of architectural styles (Fig. 3.21). The cables used in suspended structures are usually steel tendon, steel wire rope and steel strand which are excellent in tension performance.

图 3.21 悬索结构(代代木体育馆)
Fig. 3.21 Suspended structure (Yoyogi Stadium)

(4) 膜结构

膜结构是由高强薄膜材料(PVC 或 Teflon)及加强构件(钢架、钢柱或钢索)形成的一种空间结构形式。膜结构分为张拉膜结构和充气膜结构两大类。张拉膜结构通过柱及刚架支撑或索张拉成型(如图 3.22a 所示);充气膜结构通过向膜内充气以形成刚度和形状(如图 3.22b 所示)。膜结构具有良好的抗震性能,且制作方便、施工快速、造价经济。

(4) Membrane structure

The membrane structure is a spatial structure which is made up by high-strength thin-film materials (PVC or Teflon) and reinforcement members (steel frame, steel columns or cables). The membrane structure is usually categorized as tensioned membrane structure and air-supported membrane structure. The former is formed by column and rigid frame or cable tensioning (Fig. 3.22a); the latter is formed by pumping air into the membrane in order to get the desired stiffness and shape (Fig. 3.22b). The membrane structure has the characteristics of good seismic performance, convenient manufacturing, short construction progress, and low cost.

(a) 张拉膜结构
Tensioned membrane structure

(b) 充气膜结构
Air-supported membrane structure

图 3.22 膜结构
Fig. 3.22 Membrane structure

土木工程导论

(5) 张弦梁结构

张弦梁结构是一种由刚性上弦、柔性拉索、中间连以撑杆形成的混合结构体系（如图 3.23 所示）。在跨度较小时，刚性上弦一般为实腹式梁；跨度较大时，刚性上弦可采用桁架结构，即张弦桁架。

(5) Beam string structure

Beam string structure is a mixed structure system with the combination of rigid upper chord, flexible cable and middle strut (Fig. 3.23). The rigid upper chord generally uses solid-web beams when the span is not large. In the case of large span, it is appropriate to use truss as the upper chord of beam string structure which is also called the truss string structure.

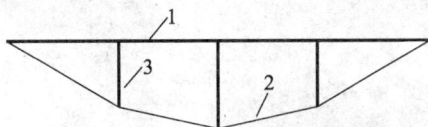

1— 刚性上弦 rigid upper chord；2— 索 cable；3— 撑杆 strut

图 3.23 张弦梁结构的组成
Fig. 3.23 Composition of beam string structure

张弦梁结构受力明确、施工方便、充分发挥了刚柔两种材料的优势，因此具有良好的应用前景（如图 3.24 所示）。张弦梁结构按受力特点可以分为平面张弦梁结构和空间张弦梁结构。

The beam string structure has good prospects since it has the advantage of clear force distribution, convenient construction and taking full advantage of both rigid and flexible materials (Fig. 3.24). According to the force-bearing feature, the beam string structure can be categorized as plane beam string structure and space beam string structure.

(a) 上海浦东国际机场
Shanghai Pudong International Airport

(b) 天津火车站站台
Tianjing Railway Station

(c) 南京国际博览中心
Nanjing International Expo Center

图 3.24 张弦梁结构的应用
Fig. 3.24 Applications of beam string structure

(6) 薄壳结构

薄壳结构是曲面的薄壁结构（如图

(6) Thin-shell structure

The thin-shell structure is a curved thin-wall structure (Fig. 3.25), and it has

3.25 所示),其优点是受力均匀,能充分利用材料强度,同时又能将承重与围护两种功能融为一体。薄壳结构按曲面的形式分为圆柱筒壳、双曲抛物面壳、球面薄壳、折板等。

the advantage of even force distribution, making full use of material and integrating the functions of load-bearing and building envelope. The thin-shell structure can be classified into cylindrical shell, hyperbolic paraboloid shell, spherical shell and folded plate etc.

图 3.25   薄壳结构
Fig. 3.25   Thin-shell structures

3.5   多层和高层建筑

多层和高层建筑主要应用于居民住宅、商场、办公楼、旅馆等建筑。随着经济的发展和房地产业的兴起,越来越多的高层和多层建筑涌现在中国大地上。

对于多层和高层建筑的区分界限,各国不一。我国对于多层和高层建筑的划分见表 3-1[依据为《民用建筑设计通则》(GB 50352—2005)]。

3.5   Multi-storey buildings and tall buildings

Multi-storey and tall structures are mainly used in residential buildings, shopping malls, office buildings, hotels and other buildings. With the development of economy and the rise of the real estate industry, more and more multi-storey and tall buildings have sprung up in China.

The boundary between multi-storey and tall buildings are different from one country to another. The division of multi-storey and tall buildings in China is shown in Table 3.1 [according to *Code for Design of Civil Buildings* (GB 50352 – 2005)].

建筑类型　Building types		划分规定　Classification rules
住宅 residential buildings	多层 multi-storey buildings	多于三层且少于十层 more than 3 storeys and less than 10 storeys
	高层 tall buildings	十层及以上 more than 10 storeys
住宅以外的 民用建筑 civil engineering except residential buildings	多层 multi-storey buildings	高于 10 m 且低于 24 m more than 10 m and less than 24 m
	高层 tall buildings	24m 及以上(不包括建筑高度大于 24 m 的单层公共建筑) more than 24 m (except the single storey buildings which are more than 24 m)

注:建筑高度超过 100 m 的为超高层建筑。
Note：The buildings with more than 100 meters high are called super tall buildings.

3.5.1　多层建筑

多层建筑常用于住宅、教学楼、旅馆等使用要求较简单的民用建筑中(如图 3.26 所示),其结构形式主要有框架结构和混合结构。框架结构是由梁和柱以刚接或者铰接连接而形成的承重结构(如图 3.27 所示),它具有空间布置灵活,构件易于标准化和施工周期短等优点。混合结构指用不同的材料建造的房屋,如常用的砖混结构(墙体采用砖砌体,屋面和楼板采用钢筋混凝土结构)。

3.5.1　Multi-storey buildings

Multi-storey buildings are usually used in civil buildings with simple requirements such as residential buildings, teaching buildings, and hotels(Fig. 3. 26). The structure forms mainly include the frame structure and mixed structure. The frame structure is composed of beams and columns which are connected by rigid or hinged connections (Fig. 3. 27). The frame structure has many advantages, such as flexible spatial arrangement, standardized members and short construction period. The mixed structure building is constructed by different kinds of materials, such as brick-concrete structure in which wall material is brick while roof and floor material is concrete.

图 3.26　多层建筑
Fig. 3.26　Multi-storey buildings

房屋建筑工程 Building Engineering

1—楼盖
floor;

2—梁
beam;

3—柱
column;

4—墙板
wall;

5—支撑
brace

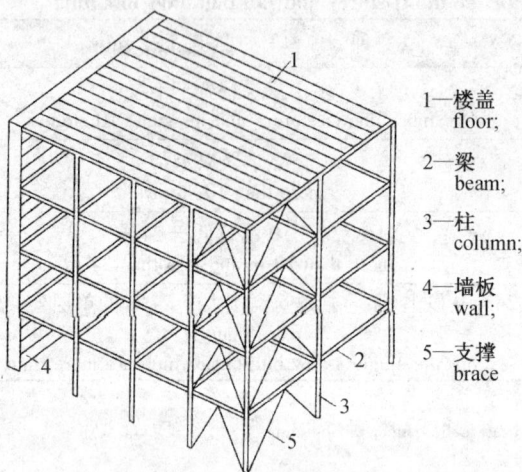

(a) 框架结构基本组成
Components of frame structure

(b) 钢筋混凝土
Reinforced concrete frame

(c) 木结构框架
Wooden frame

图 3.27　框架结构
Fig. 3.27　Frame structures

3.5.2　高层建筑

高层建筑在国外已经有一百多年的发展历史。1885 年,美国第一座根据现代钢框架结构原理建造起来的 11 层芝加哥家庭保险公司大厦(如图 3.28a 所示)是近代高层建筑的开端。1931 年,纽约建造了著名的帝国大厦(如图 3.28b 所示),地上建筑高 381 m,102 层。

20 世纪 50 年代后,轻质高强材料的应用、新的抗风抗震结构体系的发展、电子计算机的推广以及新的施工方法的出现,使得高层建筑得到了迅速发展。1972 年,纽约建造了 110 层、高 402 m 的世界贸易中心大楼(如图 3.28c 所示),该大楼由两座塔式摩天楼组成,可惜的是该大楼在 2001 年"9·11"恐怖袭击中被毁。1973 年,在芝加哥又建造了当时世界上最高的

3.5.2　Tall buildings

The development of the tall building has more than one hundred years in foreign countries. In 1885, Home Insurance Building in Chicago was built with 11 storeys according to the principle of the modern steel frame structure in American (Fig. 3.28a), and it is the beginning of modern tall buildings. In 1931, the famous Empire State Building (Fig. 3.28b), which is 381 meters above the ground and has 102 storeys, was built in New York.

After the 1950s, the tall building developed rapidly because of the application of lightweight and high strength materials, the development of new wind and earthquake-resistant structural system, the extension of the electronic computer and the emergence of the new construction method. World Trade Center Twin Tower (Fig. 3.28c) in New York was built in 1972, which is 402 meters in height and has 110 storeys above the ground, and this building consists of two tower skyscraper. Unfortunately, the building has been destroyed in the "9 11" terror-

希尔斯大厦（如图3.28d 所示），高443 m（若包括两个线塔则高达520 m），地上110 层，地下还有3 层，截至2010 年底该楼为世界第六、美国第一高楼。

目前世界上最高的建筑为哈利法塔（原名迪拜塔，如图3.28e 所示），它位于迪拜，高828 m，共162 层，2010 年建成。中国内地最高的建筑为上海的环球金融中心（如图3.28f 所示），高492 m，共104 层，建成于2008 年，其高度世界排名第四。

ist attacks in 2001. The Sears Tower (Fig. 3.28d) in Chicago which is the tallest building in the world at that time was built in 1973. It is 443 meters in height (520 m if include the two pylons) and has 110 storeys above the ground and 3 storeys underground. Sears Tower was the sixth-tallest building in the world, and the tallest building in the America at the end of 2010.

At present, the Burj Khalifa in Dubai is the world's tallest building (formerly known as the Burj Dubai, Fig. 3.28e) which was built in 2010, and it is 828 meters in height and has 162 storeys above the ground. The tallest building in mainland China and the forth-tallest building in the world is Shanghai World Financial Center which was built in 2008, and it is 492 meters in height and has 104 storeys (Fig. 3.28f).

(a) 芝加哥家庭保险公司大厦
Home Insurance Building in Chicago

(b) 纽约帝国大厦
Empire State Building

(c) 纽约世界贸易中心大楼
World Trade Center Twin Tower

(d) 芝加哥西尔斯大厦
Sears Tower in Chicago

(e) 哈利法塔
Burj Khalifa Tower

(f) 环球金融中心
Shanghai World Financial Center

图3.28　世界高层建筑
Fig. 3.28　Tall buildings in the world

高层建筑的结构形式主要有框架结构、框架-剪力墙结构、剪力墙结构和筒体结构等。

The structure forms of tall building mainly include frame structure, frame-shear wall structure, shear wall structure and tube structure, etc.

（1）剪力墙结构

剪力墙结构采用钢筋混凝土墙板来承受水平和竖向荷载，因钢筋混凝土墙板的抗剪能力很强，能有效地抵抗水平荷载，故称剪力墙。剪力墙结构的整体性强，抗侧刚度大，水平荷载作用下的侧向变形小。剪力墙结构在高层建筑中应用较多，如朝鲜的柳京饭店（如图 3.29 所示）。但剪力墙结构对平面和空间布置要求较高，难以满足大空间建筑功能的要求。

(1) Shear wall structure

Reinforced concrete wallboard has been adopted in shear wall structure to bear horizontal and vertical loads. Due to its strong shear capacity, reinforced concrete wall panel can effectively resist horizontal loads. Therefore we call it "shear wall". The characteristics of shear wall structure are strong integrity, large lateral stiffness and small lateral deformation under the action of horizontal loads. Shear wall structure has been widely applied in the tall buildings, such as Ryugyong Hotel in North Korea (Fig. 3.29). However, shear wall structure has a high demand on the plane and space layout, so it is difficult to meet the requirements of large space buildings.

（2）框架-剪力墙结构

框架-剪力墙结构是由框架和剪力墙两种不同结构体系组成的结构形式，即在框架结构中布置一定数量的剪力墙（如图 3.30 所示）。框架-剪力墙结构既有框架结构空间布置灵活的优点，又有剪力墙结构的抗侧刚度大的特点。

(2) Frame-shear wall structure

Frame-shear wall structure is the structure style composed of frame and shear wall, which means a certain number of shear walls are arranged in frame structure (Fig. 3.30). Frame-shear wall structure has the advantages of flexible space layout of the frame structure and the characteristics of the large lateral stiffness of shear wall structure.

图 3.29　柳京饭店
Fig. 3.29　Ryugyong Hotel

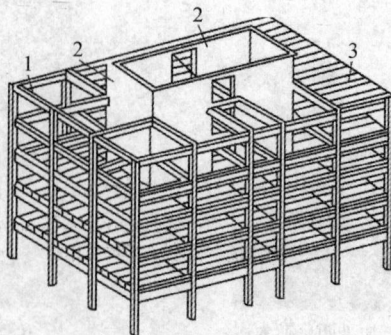

1—框架 frame; 2—剪力墙 shear wall; 3—板 floor
图 3.30　框架-剪力墙结构
Fig. 3.30　Frame-shear wall structure

（3）筒体结构

　　将剪力墙集中到房屋的内部或外部形成封闭的筒体，由筒体来承受水平和竖向荷载的结构称为筒体结构。筒体可以是剪力墙薄壁筒，也可以是密柱框筒。

　　筒体结构可分为框筒结构、筒中筒结构、框架-核心筒、多重筒体和束筒结构（如图 3.31 所示）。

（3）Tube structure

　　Tube structures bear horizontal and vertical loads with a closed tube formed by putting shear walls to the internal or external of the building. The tube can be a thin-shell tube or a dense column tube.

　　Tube structure can be classified into frame-tube structure, tube-in-tube structure, frame-core tube, multiple tube structure and beam tube structure (Fig. 3.31).

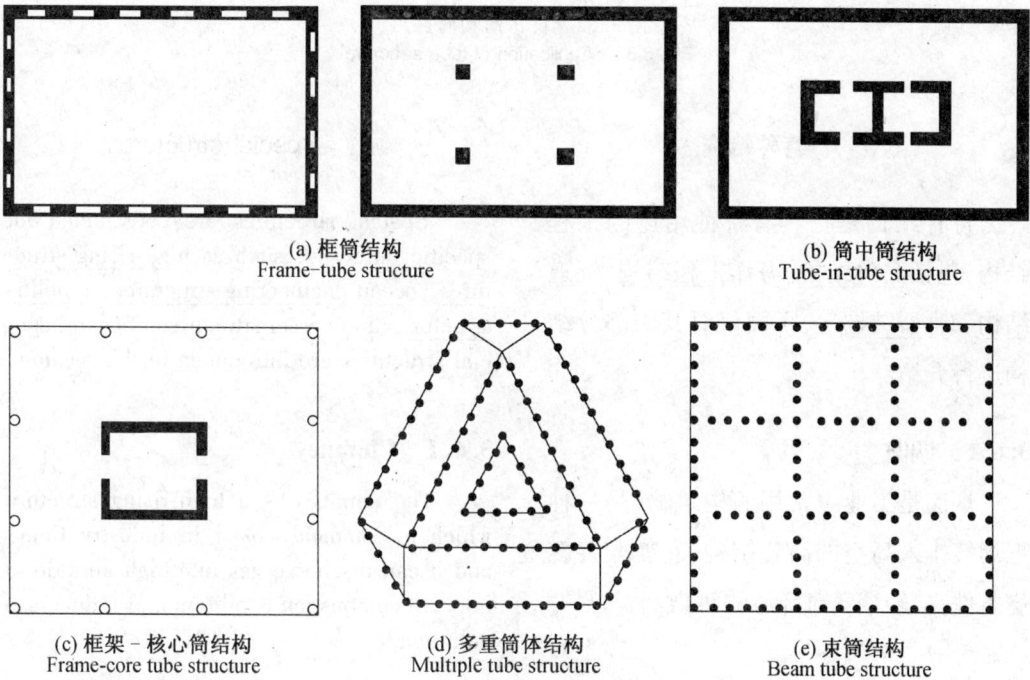

(a) 框筒结构
Frame-tube structure

(b) 筒中筒结构
Tube-in-tube structure

(c) 框架-核心筒结构
Frame-core tube structure

(d) 多重筒体结构
Multiple tube structure

(e) 束筒结构
Beam tube structure

图 3.31　筒体结构分类
Fig. 3.31　Classification of tube structure

　　筒体结构的空间刚度极大、抗扭性能也很好，广泛应用于超高层建筑（如图 3.32 所示）。

　　The tube structure is widely used in super tall buildings because of its high space-stiffness and good torsional behavior (Fig. 3.32).

房屋建筑工程 Building Engineering

(a) 西尔斯大厦（束筒结构）
Sears Tower(beam tube structure)

(b) 合肥海顿国际广场（框筒结构）
Hefei Haydn International Plaza
(frame-tube structure)

(c) 广州国际会展中心（筒中筒结构）
Guangzhou International Convention Centre
(tube-in-tube structure)

图 3.32　筒体结构的应用
Fig. 3.32　Application of tube structure

3.6　特种结构

特种结构是指具有特种用途的工程结构,包括高耸结构、海洋工程结构、管道结构和容器结构等,下面仅对其中部分结构进行介绍。

3.6.1　烟囱

烟囱是工业中常用的构筑物,是一种把烟气排入高空的高耸结构,它能改善燃烧条件,减轻烟气对环境的污染。

烟囱一般为圆锥体,截面直径自下向上逐渐减小。烟囱按材料分类,一般有砖烟囱、钢筋混凝土烟囱和钢烟囱三类;按结构形式可分为自立式烟囱和拉线式烟囱(如图 3.33 所示)。

3.6　Special structures

Special structures are constructed for specific purposes, such as high-rising structures, ocean engineering structures, pipeline structures and vessel structures. Several special structures are introduced in this section.

3.6.1　Chimney

The chimney is a high-rising structure which is commonly used in industry field, and it can discharge gas into high altitude to improve combustion condition and reduce gas pollution.

The chimney usually has a conical body whose diameter gradually diminishes from the bottom to the top. According to the material, chimney can be classified into brick chimney, concrete chimney and steel stack. The chimney can be classified into self-supporting chimney and guyed chimney according to the different structure forms (Fig. 3.33).

(a) 砖烟囱
Brick chimney

(b) 钢筋混凝土烟囱
Concrete chimney

(c) 钢烟囱
Steel stack

(d) 拉线式烟囱
Guyed chimney

图 3.33　烟囱种类
Fig. 3.33　Chimney types

3.6.2　水塔

水塔是一种用于储水和配水的高耸结构，是给水工程中常用的构筑物。水塔由水箱、塔身和基础三部分组成。

水塔按建筑材料分为钢筋混凝土水塔、钢水塔、砖混组合水塔。水箱的形式有圆柱壳式、倒锥壳式、球形、箱形等（如图 3.34 所示），其中前两种形式在我国应用最多。塔身通常有支筒和支架两种形式，支筒一般用钢筋混凝土或砖石做成圆筒形，支架多数为钢筋混凝土钢架或钢构架。

3.6.2　Water tower

The water tower is a high-rising structure which is used for storage and distribution of water, and it is widely used in water supply engineering. The water tower is usually composed by water tank, tower body and foundation.

The water tower is classified into concrete water tower, steel water tower and brick-concrete water tower by the material. The shapes of water tanks are various, such as cylinder, turbination, sphere and box (Fig. 3.34), and the cylindrical water tank and turbinate water tank are widely used in China. The structural forms of the tower body are generally tube or bracket, and the tube is always constructed by reinforced concrete or brick, while the bracket usually adopts reinforced concrete frame or steel frame.

(a) 圆柱壳式 Cylinder

(b) 倒锥壳式 Turbination

(c) 球形 Sphere

(d) 箱形 Box

图 3.34　水箱的形式
Fig. 3.34　Types of water tank

3.6.3 筒仓

筒仓是贮存粒状和粉状松散物体(如谷物、面粉、水泥、碎煤、精矿粉等)的立式容器。筒仓的平面形状有正方形、矩形、多边形和圆形等,圆形筒仓因仓壁受力合理,在实际工程中应用最多。

筒仓根据所用材料可分为钢筋混凝土筒仓、钢筒仓、砖砌筒仓和塑料筒仓等(如图3.35所示),其中钢筋混凝土筒仓由于造价低、耐久性好、施工方便以及抗冲击性能良好等优点在我国应用最广。

3.6.3 Silo

Silo is an upright container which is used to store granular and loose powder substance, such as cereal, flour, cement, crushed coal, and ore powder. The flat-shapes of silo are usually square, rectangle, polygon and circle, and the circular silo is mostly used for its reasonable force distribution.

According to the material used, the silo can be classified as reinforced concrete silo, steel silo, brick silo and plastic silo (Fig. 3.35) etc. Reinforced concrete silos are most widely used in China for its advantages of low cost, good durability, convenient construction and good performance on impact resistance.

(a) 钢筋混凝土筒仓
Reinfored concrete silo

(b) 钢筒仓
Steel silo

(c) 砖砌筒仓
Brick silo

图 3.35 筒仓类型
Fig. 3.35 Types of silo classified by materials

按照筒仓的贮料高度与直径或宽度的比例关系,可将筒仓划分为浅仓和深仓两类。浅仓主要用于短期贮料,深仓主要用于长期贮料。

Silo can be classified into bunker and deep bin according to the relationship between the height of storage in the silo and its diameter or width. The bunker is mainly used for short-term storage, while the deep bin is mainly used for long-term storage.

3.6.4 电视塔

电视塔是用于广播电视发射传播的建筑。电视发射天线越高,其信号的传播范围就越大,这就促使电视塔愈建愈高。

3.6.4 Television tower(TV tower)

The TV tower is a structure used to transmit radio and television signals. The higher the TV transmitter antenna is, the larger the transmission range of signal is. This theory prompts the TV tower to be construc-

现在,电视塔的功能不单是发射电视信号,还能观光旅游,有些电视塔上面还设有餐厅,成为一种多用途的塔。

目前,世界上最高的电视塔是 2012 年建成的高 634 m 的东京晴空塔(如图 3.36a 所示),我国最高的电视塔是广州塔(如图 3.36b 所示),高 600 m,建成于 2009 年,是世界第三高电视塔(第二高的电视塔是美国的 KVLY 电视塔,高 628 m,桅杆结构,如图 3.36c 所示),同时它也是第二高自立式电视塔。

ted higher and higher. Now, the TV tower is a multi-use tower which can not only broad TV but also be used for tour and sightseeing. Especially, there are even restaurants in some TV towers.

The Tokyo sky tower is the highest TV tower in the world so far (Fig. 3.36a) which was built in 2012, and its height is 634 m. The highest TV tower in China is the Guangzhou tower (Fig. 3.36b) which was constructed in 2009. The height of Guangzhou tower is 600 m which rank third in the world (The second tallest building is the KVLY tower in America which is 628 m in height and mast structured, as shown in Fig. 3.36c), and it also rank second among the self-standing towers.

| (a) 东京晴空塔
Tokyo sky tower | (b) 广州塔
Guangzhou tower | (c) KVLY 塔
KVLY tower |

图 3.36 电视塔
Fig. 3.36 Television towers

3.6.5 立体停车库

立体停车库是用来存取、储放车辆的机械或机械设备系统(如图 3.37 所示),主要由钢构架、回转台、输送车或升降电梯、监控操作台及辅助设备(消防、配电、防盗机构)六大部分组成。最早的立体车库,是建于 1918 年的美国芝加哥市的一家宾馆停车库。目前,立体停车库在我国很

3.6.5 Stereo garage

Stereo garage is a machine or mechanical system used for access and storage of vehicles (Fig. 3.37), and it is composed by steel frame, turret, transfer car or elevator, control console and accessory equipments (such as fire protection equipment, electrical distribution plant and security facilities). The earliest stereo garage is a hotel parking garage in Chicago which was built in 1918. Currently, the stereo garage are extensively

多城市得到开发和推广,相关技术已经日趋成熟。

developed and promoted in many cities in China in order to ease the traffic pressure, and the relevant technique is increasingly mature.

图 3.37　立体停车库
Fig. 3.37　Stereo garages

3.7　建筑结构设计

为了能使建筑物建造起来,必须进行相关的结构分析设计。我国的《建筑结构可靠度统一设计标准》(GB50068—2001))对结构设计作了以下几方面的要求:

① 在正常施工和正常使用时,能承受可能出现的各种作用;

② 在正常使用时,具有良好的工作性能;

③ 在正常维护下,具有足够的耐久性能;

④ 在设计规定的偶然事件发生时及发生后,仍能保持必需的整体稳定性。

建筑工程的结构设计步骤一般为:方案设计→结构分析→构件设计→施工图绘制。

3.7　Design for building structures

We should make a relevant design and analysis for building structures by referring to the relevant Codes before construction. According to the *Unified standard for Reliability Design of Building Structures* (GB50068-2001) in China, the structure design should meet the following requirements.

① The designed structure can withstand all kinds of actions in the condition of normal construction and regular service.

② The designed structure should have good performance under regular service conditions.

③ The designed structure should have adequate durability under regular maintenance.

④ The designed structure should be able to maintain the required stability during and after the accidental events mentioned in the Codes.

The design procedure of building structures can be generally divided into: schematic design → structure design → member design → construction drawing.

3.7.1 方案设计

方案设计又称初步设计，主要包括结构选型、结构布置和主要构件截面尺寸估算。

① 结构选型主要是根据建筑物的功能要求和工程地质条件，通过对不同结构体系方案和技术经济指标进行比较，选择最优的结构方案。

② 结构布置主要是根据使用功能要求和结构体系来确定构件的位置和方式。

③ 主要构件尺寸估算主要是根据初步方案和结构布置，对一些主要构件的大小进行预估，并对其合理性进行判断。

3.7.2 结构分析

结构分析是结构设计的重要内容，主要包括以下几方面内容：

① 结构模型的建立。在确定结构模型时，需要对实际结构进行简化假定。简化后的模型应能反映结构的实际受力特性且尽可能简单，以便计算分析。

② 荷载的计算和施加。根据建筑物服役期间可能遭受的荷载种类，准确计算荷载作用的大小，并施加在结构模型中的相应位置。

③ 计算分析。在对结构模型进行计算分析之前，首先需要根据结构特点确定计算理论，如线弹性理论、非线性理论等；然后再采用相应的计算分析手段（如分析软件）对结构模型进行计算分析。

3.7.1 Schematic design

Schematic design（also called preliminary design）mainly includes structure selection, structural arrangement and section estimation of the primary members.

① Based on the functional requirements of the building and engineering geological conditions, the structure selection is to choose the optimal structure form by comparing different structural systems, technical and economic index.

② Structural arrangement mainly includes the determination of the place and the way to place the structural members according to the functional requirements and structural form.

③ The primary member section estimation is to prospect the members' sizes and judge their rationality according to the preliminary design and the structural arrangement.

3.7.2 Structure analysis

Structure analysis is an important part of structure design which mainly including the following aspects：

① Construction of structural model. when determining the structural model, we need to make necessary assumptions of the actual structure. The simplified structural model should reflect the force characteristics of actual structure, and be as simple as possible in order to facilitate the analysis.

② Calculation and application of loads. The designer should first determine the load types during the service lives of structure and calculate them accurately, and then apply these loads on the right places of structural model.

③ Analysis. Before analyzing the structural model, the computing theory such as linear-elastic theory, nonlinear theory, should first be determined according to characteristics of the structure; then the designer chooses the

3.7.3 构件设计

通过对结构模型进行计算分析,可以得到模型中每根杆件的内力。然后根据内力和相应的设计规范,对构件进行设计。构件设计包括杆件截面设计和节点设计。

3.7.4 施工图绘制

设计的最后一个阶段是绘制施工图,要求图纸表达正确、规范、简明和美观。

3.8 未来展望

在今后很长一段时间内,钢筋混凝土结构以及钢结构仍然是主要的建筑结构形式,同时型钢混凝土和钢管混凝土等组合结构的应用也将越来越广泛,但随着科学技术的进步和土木工程发展的需要,建筑工程中将不断涌现出新的课题。总体来说,未来建筑工程将主要朝以下几个方向发展:

① 建立更加准确的分析方法和更加完善的设计理论,能够准确地分析模拟建筑结构在各类荷载作用下的受力状态,并对其进行合理的设计。

② 形成新的结构体系,满足现代建筑日新月异的使用要求,建立更高的超高层建筑和更大跨度的空间结构,同时建筑物的造型也将更加丰富多彩。

③ 结合机械工程和信息工程等领域

relevant means (e. g. analysis softwares) to analyze the structural model.

3.7.3 Member design

Members' internal forces can be obtained by analyzing the structural model. On the base of the internal forces and Codes, the members can be designed. The context of member design includes section design and joint design.

3.7.4 Construction drawing

The final stage of structural design is to make the construction drawing which is required to be correct, canonical, concise and beautiful.

3.8 Future prospects

The concrete structure and steel structure will still be the main structural form for a long time in the future, and the mixed structure will also be widely used. On the other hand, with the scientific and technological progress and development of civil engineering, new issues will continue emerging in building engineering. In general, building engineering will develop mainly in the following directions:

① More accurate analysis methods and more perfect design theories will be developed so as to simulate and analyze the building structures accurately and design reasonably.

② In order to satisfy the increasing complex formation requirements of modern architectures, a lot of new structure systems will be formed. Many higher skyscrapers and larger long-span structures will be established, and the shapes of the architectures will be more colorful.

③ The structural theory of intelligent building will be further developed combined

的知识,发展智能建筑中的相关结构工程
理论和分析设计方法。

with the mechanical and information knowledge.

注:本章除图3.23外,其余图片均来源于网络。
Note: In this chapter, the pictures are from webs except Fig. 3.23.

知识拓展
Learning More

相关链接　Related Links

(1) 土木工程在线论坛 http://bbs.co188.com/

(2) 中华钢结构论坛 http://www.okok.org/

(3) 仿真科技论坛 bbs.simwe.com

小贴士　Tips

(1) 除本章3.1节中列出的结构形式外,大跨度建筑的结构形式还包括折板结构、蒙皮结构等;特
种结构还包括海洋钻井平台、压力容器、核电站等,有兴趣的同学可查阅相关的书籍和网站进行了解。

Besides the structure forms introduced in section 3.1, there are also many other structure forms, such as
the long-span structure also includes folded plate structure and stressed skin structure, and the special structure
also includes ocean drilling platform, pressure vessel, nuclear power station. If you have interest in these subjects, you can refer to the relevant books and websites.

(2) 对工程结构进行分析设计时,需要用到相关的软件,如 PKPM,SAP2000,3D3S,ANSYS,
ABAQUS 等,有兴趣的同学可查阅相关的书籍和网站了解与学习。

When we analyzing or designing a building structure, we may use the relevant softwares, such as PKPM,
SAP2000, 3D3S, ANSYS, ABAQUS. If you want to study these softwares, you can refer to the relevant books
and websites.

(3) 目前,与结构工程专业相关的注册工程师种类主要有注册结构工程师、注册监理工程师和注册
建造师,相关内容和要求可参考国家城乡建设部网站 http://www.mohurd.gov.cn/。

At present, registered engineers of structural engineering mainly include registered structural engineer,
certified supervision engineer and certified architect. If you want to get more information, you can refer to the
web of ministry of housing and urban-rural of China: http://www.mohurd.gov.cn/.

(4) 结构工程设计过程中,需要参照相应的国家规范或规程,如《建筑结构荷载规范》、《混凝土结
构设计规范》、《钢结构设计规范》等,有兴趣的同学可查阅相关的资料进行学习。

When constructing a building, we must abide the national Codes, such as *Load Code for the Design of
Building Structures*, *Code for Design of Concrete Structures*, *Code for Design of Steel Structures*. If you have interest in the context of the Codes, you can refer to relevant books or websites.

思考题 Review Questions

（1）尝试以一个实际建筑为例，寻找其中的基本构件。

Try to find the basic structural members in an actual building.

（2）尝试按照建筑结构所用材料对建筑结构进行分类，描述每种结构类型的优缺点及其应用范围。

Try to classify the building structure by construction materials, and point out the characteristics of each structural type and the range, of its application.

（3）你对哪种结构形式最感兴趣？你周围的建筑物都有哪些结构形式？

Which kind of structure are you most interested in? What kinds of structures do the buildings around you have?

参考文献
References

［1］丁大均，蒋永生：《土木工程概论》（第 2 版），中国建筑工业出版社，2010 年。

［2］叶志明：《土木工程概论》（第 3 版），高等教育出版社，2009 年。

［3］Palanichamy. M. S. Basic Civil Engineering. 3rd ed. Mc GrawHill, 2005.

［4］江见鲸，叶志明：《土木工程概论》，高等教育出版社，2001 年。

［5］郑晓燕，胡白香：《新编土木工程概论》（第 2 版），中国建材工业出版社，2012 年。

［6］吕志涛：《新世纪我国土木工程活动与预应力技术的展望》，《东南大学（自然科学版）》，2002 年第 3 期。

［7］董羡，黄林青：《土木工程概论》，水利水电出版社，2011 年。

［8］Narayanan R. S., Beeby. A. W. Introduction to Design for Civil Engineers. Spon Press, 2001.

［9］Scott. J. S. Dictionary of Civil Engineering. 4th ed. Penguin Books Publisher, 1991.

［10］曹双寅：《工程结构设计原理》（第 3 版），东南大学出版社，2012 年。

第4章 交通与水利工程

Chapter 4 Traffic Engineering and Hydraulic Engineering

陆上交通分为公路和铁路,本章第一节和第二节分别介绍道路工程和铁路工程中的结构组成。机场是为航空运输服务的一系列建筑体系的组合,本章第三节从机场的功能区划和结构组成来介绍机场工程。第四节阐述水利工程的内涵并介绍常见的水工建筑物。

The land transportation includes road and railway. The road engineering and railway engineering are discussed mainly from the aspect of their structural composition in the first and second sections of this chapter respectively. The airport is a composite construction system in the service of aviation transportation, which will be introduced in the third section with the focus on functional distribution and structural composition. Hydraulic engineering is an important part of structural engineering. The connotation of the hydraulic engineering and common hydraulic structures are introduced in the fourth section.

4.1 道路工程

4.1 Road engineering

道路是一种带状的三维空间人工构造物(如图 4.1、图 4.2 所示),它包括路基、路面、桥梁、涵洞、隧道等结构实体。道路的设计一般从几何和结构两大方面进行考虑。道路的结构设计一般要求用最小的投资,使其在自然力及车辆荷载的共同作用下,在使用期限内保持良好状

The road is a banded three-dimensional space of artificial structure, which includes subgrade, pavement, bridge, tunnels and other structural entities (Fig. 4. 1 and Fig. 4. 2). Usually, the design of the road proceeds in two points, namely, geometry and structure. Based on the requirement of good state for road operation under the combined action of natural force and vehicle loading during service life, the investment of the

态,满足适用要求。

structural design for road should be minimized.

图 4.1　高速公路
Fig. 4.1　Expressway

图 4.2　市政道路
Fig. 4.2　Municipal road

4.1.1　公路路基

路基是行车部分的基础,它由土、石按照一定尺寸、结构要求建筑成带状的土拱结构物。公路路基的横断面一般有 3 种形式:路堤、路堑和半填半挖,如图 4.3 所示。路基的几何尺寸由高度、宽度和边坡组成。其中,路基高度由路线纵断面设计确定,路基宽度根据设计交通量和公路等级而定,路基边坡根据路基整体稳定性的要求而定。

4.1.1　Highway subgrade

Subgrade is a zonal soil arch structure which consists of soil and stone according to certain size and structural requirements. Subgrade is the foundation of driving. There are three kinds of forms about the cross-section of subgrade: embankment, cutting and cut-and-fill, as shown in Fig. 4.3. The geometric parameters of subgrade include height, width and slope. The height is determined by the route longitudinal profile design, the width is determined by the design of traffic volume and road grade, and the slope is determined by the requirements of embankment stability.

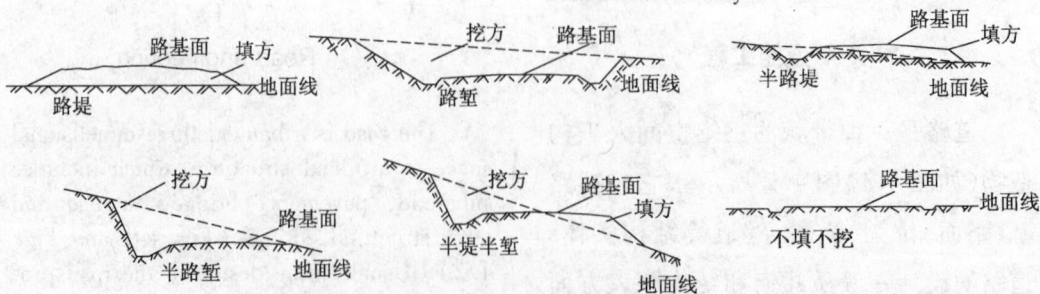

1. 路堤:embankment; 2. 路基面:subgrade surface; 3. 填方:fill; 4. 地面线:ground line;
5. 挖方:excavation; 6. 路堑:cutting; 7. 半路堤:half of the embankment; 8. 半路堑:half of the cutting;
9. 半堤半堑:half cutting and half filling; 10. 不填不挖:unfilling and unexcavation

图 4.3　路基的断面形式
Fig. 4.3　Cross section of subgrade

128

4.1.2 公路路面

公路路面是用各种坚硬材料分层铺筑而成的路基顶面的结构物,以供汽车安全、迅速和舒适地行驶。

路面一般按其力学性能主要分为柔性路面和刚性路面两大类。还有一类是半刚性路面,由无机结合料(水泥、石灰)、水硬性材料(稳定土、砂、砾石)和工业废料(如粉煤灰、矿渣等)组成,此类材料后期强度增长较大,最终强度比柔性路面强度高,但比刚性路面低。

近年来发展起来的沥青玛蹄脂碎石(Stone mastic asphalt,SMA)路面和彩色路面也越来越引起关注(如图4.4,图4.5所示)。

SMA 是一种新型沥青混合料,由沥青、纤维稳定剂、矿粉和少量细集料组成的沥青玛蹄脂填充间断级配粗集料而成。这种结构能全面提高沥青混合料和沥青路面的使用性能,从而减少维修养护费用,延长使用寿命。彩色路面具有美观,与景观搭配,且车流引导效果好等优点。

4.1.2 Highway pavement

Highway pavement is a structure at the surface of the subgrade which is composed of various hard paving layer materials for vehicle safety, rapidly and comfortability.

According to the mechanical properties, pavement is generally divided into two categories, namely, flexible pavement and rigid pavement. There is another type named as semi-rigid pavement, which mainly includes inorganic binder (cement, lime), hydraulic material (stabilized soil, sand, gravel) and industrial waste (such as fly ash, slag). The strength of this pavement grows larger in the late period, and its ultimate strength is stronger than that of flexible pavement, but lower than that of rigid pavement。

SMA(stone mastic asphalt, called SMA for short) pavement and colored pavement, which have developed in recent years (Fig. 4.4 and Fig. 4.5), have attracted more and more attention.

SMA is a new kind of asphalt mastic filled with novel asphalt mixture gradation of coarse aggregate, which is composed of asphalt, fiber stabilizer, powder and a small amount of fine aggregate. This structure improves the performance of asphalt mixture and asphalt pavement, and reduce the maintenance cost with the extension of the service life. Colored pavement has the advantages of beauty, landscape collocation and efficient guidance of traffic flow.

图4.4 沥青玛蹄脂碎石
Fig. 4.4 Stone mastic asphalt

图4.5 彩色路面结构
Fig. 4.5 Colored pavement structure

4.1.3 公路排水结构

为了确保路基稳定,免受地面水和地下水的侵害,公路还应修建专门的排水设施。地面水的排除系统按其排水方向不同,分为纵向排水和横向排水。纵向排水有边沟、截水沟和排水沟等结构物;横向排水有桥梁、涵洞、路拱、过水路面、透水路堤和渡水槽等结构物,如图4.6,图4.7所示。

4.1.3 Highway drainage structure

In order to ensure the stability of the subgrade and protect it against the surface water and groundwater, drainage facilities should be built specifically. According to the different drainage direction, surface water exclusion system is divided into transverse drainage and longitudinal drainage. Longitudinal drainage includes the side ditch, the intercepting ditch, the drainage ditch, etc., while the transverse drainage consists of bridges, culverts, roadway crowns, overflow pavement, permeable embankment, flume, etc., as shown in Fig. 4.6 and Fig. 4.7.

图 4.6 道路边沟
Fig. 4.6 Roadway side ditch

图 4.7 涵洞
Fig. 4.7 Culvert

4.1.4 公路特殊结构物和附属结构

公路的特殊结构物有隧道、高架桥、悬出路台、防石廊、挡土墙和防护工程等,如图4.8所示。

4.1.4 Special structure of highway and accessories

Special structures of highway include tunnel, viaduct, hanging out embankment, anti-stone veranda, retaining wall and protection engineering etc, as shown in Fig. 4.8.

(a) 公路隧道
Vehicular tunnel

(b) 高架桥
Viaduct

图 4.8 公路特殊结构物
Fig. 4.8 Special structures of highway

一般在公路上,除了上述各种基本结构外,为了保证行车安全、便捷还需要设置交通管理设施、服务设施和环境美化设施等附属结构。

In order to ensure the traffic safety and convenience, other accessories like traffic management facilities(Fig. 4.9), service facilities, landscaping facilities are required, besides the basic structures mentioned above.

(a) 标识立柱
Identifications of the colum

(b) 路面标线
Pavement marking

图 4.9 交通管理设施结构
Fig. 4.9 Construction of traffic management facilities

交通安全设施是为了保证行车安全和发挥公路作用,在各级公路的急弯、陡坡等路段设置的安全设施,如护栏、护网、护柱,如图 4.10 所示。

Traffic safety facilities (Fig. 4. 10), such as guardrail, protective screen and guard post, play important roles in the driving safety of highway. They are usually set up at the sharp turn and steep slope.

(a) 护栏
Guardrail

(b) 护网
Protective screen

(c) 护柱
Guard post

图 4.10 交通安全设施结构
Fig. 4.10 Construction of traffic safety facilities

4.1.5 高速公路

高速公路的建设情况反映着一个国家和地区的交通发达程度,乃至经济发展的整体水平,如图 4.11 所示。我国拥有世界上路程最长的高速公路,总里程为

4.1.5 Expressway

The development of expressway reflects the traffic development of a region or country, and even the overall level of economic development(Fig. 4. 11). The amount of expressway in China is the largest all over the world, which reached 95 600 km in 2012.

95 600 km（2012 年）。

(a) 高速公路（中国）　　　(b) 不限时高速公路（德国）　　　(c) 高速公路立交（美国）
Expressway(China)　　　Unlimited expressway(Germany)　　　Expressway interchange (America)

图 4.11　世界各地的高速公路
Fig. 4.11　The expressway around the world

目前国际上对高速公路还没有一个公认的定义。我国对高速公路的基本定义为：一般能适应 120 公里/小时或者更高的速度,路面有 4 个以上车道的宽度;中间设置分隔带,采用沥青混凝土或水泥混凝土高级路面,设有齐全的标志、标线、信号及照明装置;禁止行人和非机动车在路上行走,与其他线路采用立体交叉、行人跨线桥或地道的形式交汇。

At present, expressway has not been defined officially. In China, expressway is defined as follows: the surface of the expressway uses asphalt and cement concrete which will be adapted to the speed of 120 km/h or more, the road has more than four lanes with the separation zone; indicating device are set up on the road such as mark line, signal and lighting devices; pedestrians and non-motor vehicles are forbidden to walk on the road; people and cars use grade separation pedestrian flyover and tunnel to cross it at intersections.

4.2　铁路工程

铁路是供火车等交通工具行驶的轨道。铁路运输是一种陆上运输方式,以机车牵引列车在两条平行的铁轨上行走。

广义的铁路运输工具包括磁悬浮列车、城市轻轨和地下铁路等,或称轨道交通。铁路运输的最大优点是运量大、安全可靠、速度快、成本低、对环境污染小,基

4.2　Railway engineering

The railway is a track for the smoothly running of trains and other transportation means. Railway transportation is one of modes of land transportation, dragging the trains to travel on two parallel tracks by the engine.

Generally, railway transportation, also called rail transit, includes maglev, urban light rail and underground railway etc. The railway transportation has the advantages of abundant transportation volume, high security, high speed, low cost, less environmental

本不受气候影响,能耗远远低于航空和公路运输,是现代运输体系中的主干力量。

铁路按照牵引动力分为电力牵引、内燃牵引及蒸汽牵引3种;按照轨距分为标准轨距铁路、宽轨铁路和窄轨铁路3种;按任务和运量各国铁路一般分为若干等级,有些国家的铁路分为干线、支线和山区线,中国国家铁路划分为Ⅰ级、Ⅱ级、Ⅲ级及地方铁路。

轨距、牵引动力种类和铁路等级不但体现铁路的性能,而且也决定铁路上各种建筑物的标准、总的工程投资和运营支出。

4.2.1 基本结构组成

铁路是由线路、路基、线上结构三部分构成的。此外,属于铁路工程的还有桥梁、涵洞、隧道、车站设施、机务设备、电力供应等结构设施。

我国铁路的等级和主要技术指标见表4.1。

pollution, little weather effect, with lower energy consumption than aviation and highway transportation. It is the main force in the transportation system nowadays.

Railway can be divided into three types according to dragging power, namely, electric power drag, combustion drag and steam drag. It can also be divided into three types according to the rail distance, namely, standard-rail-distance railway, wide-rail-distance railway and narrow-rail-distance railway. Railway can be classified into various classes according to the mission and transportation volume, such as main line, branch line and mountain line in some countries, as well as Ⅰ-class-railway Ⅱ-class-railway Ⅲ-class-railway and regional railway.

The rail distance, the kind of dragging power and the rank of railway not only affect the performance of the railway, but also determine the standard of buildings besides the railway, along with the total investment and service cost.

4.2.1 Basic components of the structure

The railway consists of route, roadbed and structures on the line. Besides, structures and facilities such as bridges, culverts, tunnels, station facilities, maintenance equipment, power supplying system also belong to the railway construction.

The grades and the main technical indexes of railway are shown in the Table 4.1.

表4.1 我国铁路等级和主要技术指标
Table 4.1 The railway grades and major technical indexes

等级 Grade	Ⅰ	Ⅱ	Ⅲ
路网中的作用 function in the route net	骨干 backbone	骨干、联络、辅助 backbone,contact or assistance	地区性 regional
远期年客货运量/万吨 forward in the traffic	≥1 500	750 ~ 1 500	≤750

交通与水利工程 Traffic Engineering and Hydraulic Engineering

等级 Grade	I	II	III
最高行车速度/(km/h) the highest speed/(km/h)	120	100	80
最大坡度/(‰) the biggest gradient(‰)	6(12)	12(15)	15(20)
最小半径/m the shortest radius /m	1 000(400~350)	800(350~300)	600(300~250)

注:括号外数值为一般地段,括号内数值为困难地段。

Note: The value of common areas is outside the brackets, and the parentheses values are for difficult areas.

4.2.2 路基及断面形式

铁路路基是一种土石结构,承受并传递轨道重力及列车动态作用,是轨道的基础,也是保证列车运行的重要建筑物。路基处于各种地形地貌、地质、水文和气候环境中,有时还遭受各种灾害,如洪水、泥石流、崩塌、地震等。

(1) 铁路路基的断面形式

铁路路基断面形式有路堤、半路堤、路堑、半路堑、半填半挖等,与公路路基断面形式基本相同,如图 4.12 所示。

4.2.2 Subgrade and the forms of fracture surface

Railway subgrade is a kind of soil-rock structure, enduring and passing orbital gravity and dynamic effect from trains, which is the basis of the orbit, as well as the important structure to ensure the normal operation of trains. Subgrade is subject to various kinds of topography, geology, hydrology and climate environment, sometimes also suffering from all kinds of disasters, such as floods, landslides, collapse, and earthquake.

(1) The forms of fracture surface in railway

The forms of fracture surface in railway include embankment, half embankment, cutting, half cutting, half filling and half digging, etc., and like the forms of fracture surface in highway, as shown in the Fig. 4.12.

(a) 路堤
Embankment

(b) 路堑
Cutting

(c) 半路堑
Half cutting

图 4.12　铁路路基的断面形式
Fig. 4.12　The forms of fracture surface in railway

（2）路基的稳定性

路基的稳定性是指路基抵抗列车动荷载及各种自然力影响所出现的道砟陷槽、翻浆冒泥和路基剪切滑动与挤起等作用的能力。道路（铁路）设计中必须对路基的稳定性进行验算。

4.2.3 线上结构

（1）轨枕

轨枕又称枕木，是铁路配件的一种。轨枕既要支承钢轨，又要保持钢轨的位置，还要把钢轨传递来的巨大压力传递给道床，因此它必须具备一定的柔韧性和弹性，列车经过时，轨枕可以适当变形以缓冲压力。钢筋混凝土轨枕和预应力轨枕如图4.13，图4.14所示。

图4.13 钢筋混凝土轨枕
Fig. 4.13 Reinforced concrete rail sleeper

（2）钢轨

钢轨的作用在于引导机车车辆的车轮前进，承受车轮压力，并将其传递到轨枕上，如图4.15所示。在电气化铁道或自动闭塞区段，钢轨还可兼做轨道电路之用。

钢轨的断面形状采用具有最佳抗弯性能的工字形断面，有轨头、轨腰以及轨

（2）The stability of roadbed

The stability of roadbed refers to its resistance ability of the collapse of road, mud pumping and shear sliding and squeezing of roadbed affected by dynamic load from trains and all kinds of natural force. The stability of roadbed is supposed to be calculated in the design.

4.2.3 Structures on the line

（1）Rail sleeper

Rail sleeper is a kind of railway accessories. Rail sleeper not only supports and fixes the rail, but also transfers the great pressure from the steel rail to the bed of the rail, so it must have certain flexibility and elasticity to provide some appropriate deformation to buffer the pressure while the train is passing. Reinforced concrete rail sleeper is shown in Fig. 4.13, and prestressing force rail sleeper is shown in Fig. 4.14.

图4.14 预应力轨枕
Fig. 4.14 Prestressing force rail sleeper

（2）Steel rail

Steel rail can guide the wheels, bear the pressure from the wheel and transfer the pressure to the rail sleeper(Fig. 4.15). Besides, it can be used as the track circuit in electrified railway or automatic block section.

I-beam is adopted in the shape of the steel rail section, which consists of waist rail head, rail body and rail base. I-beam has the

底三部分组成。

我国铁路上使用的钢轨有 75,60,50,43,38 kg/m 等几种(如图 4.16 所示)。钢轨的标准长度为 12.5 m 和 25.0 m 两种,特重型、重型轨采用 25.0 m 的标准长度钢轨。近年来大力发展的时速不小于 250 km/h 客运专线的钢轨标准长度为 100 m。

optimal bending capacity.

The types of steel rail applied in China are 75,60,50,43,38 kg/m, etc. The standard length of steel rail is 12.5 m and 25.0 m (Fig. 4.16). Extra-heavy-mode and heavy-mode adopt the 25.0 m standard steel rail. In recent years, the 100 m standard steel rail is applied in the passenger transport line whose speed is over 250 km/h.

图 4.15 钢轨
Fig. 4.15 Steel rail

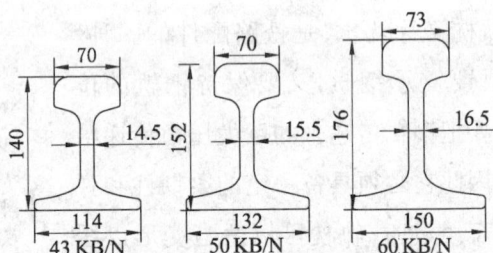

图 4.16 43,50,60 kg 钢轨截面图(单位:mm)
Fig. 4.16 Sectional view of steel rail of 43,50 and 60 kg

(3) 道床

道床通常指的是铁路轨枕下面,路基面上铺设的石碴(道碴)垫层。其主要作用是支撑轨枕,把轨枕上部的巨大压力均匀地传递给路基面,并固定轨枕的位置,阻止轨枕纵向或横向移动,大大减少路基变形的同时缓和机车车轮对钢轨的冲击,便于排水。

道床分为普通有碴道床、沥青道床和混凝土整体道床。有碴道床通常由具有一定粒径、级配和强度的硬质碎石堆集而成。沥青道床是为了改善普通石碴道床的散体特性而加入乳化沥青或沥青砂浆的结构形式。整体道床多为现浇钢筋混

(3) Roadbed

Roadbed is the ballast cushion layer laying on the surface of the road foundation and below the railway sleeper. The main functions of roadbed are supporting the rail sleeper, transferring the great pressure from the rail sleeper to the surface of foundation of road, fixing position of the rail sleeper to prevent sleeper from the longitudinal or lateral movement, greatly reducing the transformation of roadbed, as well as relieving the shock to the stell rail caused by the vehicles at the same time. Furthermore, it can provide convenience for dewatering.

Roadbed is divided into ordinary ballast bed, asphalt bed and concrete ensemble bed. Ballast bed usually consists of hard rocks which have certain particle size, grading and strength. Asphalt bed is the ballast bed mixed with emulsified asphalt or asphalt mortar in order to improve the character of direct prose style of ordinary ballast bed.

土木工程导论

凝土结构,常用于高铁、不易变形的隧道内或桥梁上(如图4.17所示)。

Ensemble bed often uses the structure of cast-in-place reinforced concrete, and it is usually used in high-speed rails and tunnel or bridges which are not easy to generate deformation(Fig. 4.17).

(a) 有碴道床
Ballast bed

(b) 混凝土整体道床
Concrete ensemble bed

图 4.17　铁路道床
Fig. 4.17　Roadbed of railway

高速铁路采用的无碴轨道是整体道床的一种,它是以混凝土或沥青砂浆取代散粒道碴道床而组成的轨道结构形式(如图4.18、图4.19所示),具有轨道稳定性高,刚度均匀性好,结构耐久性强和维修工作量显著减少等特点,相对于高速铁路较传统的有碴轨道有更好的适应性。

Ballastless roadbed, used in most high-speed railways, is a kind of concrete ensemble bed, consisting of concrete or asphalt mortar instead of particulate ballast bed (Fig. 4. 18 and Fig. 4. 19). It has the advantages of high stability, perfect uniformity of rigidity, high durability of structure, great reduction of maintenance and better adaptability over traditional ballast bed in high-speed railways.

图 4.18　无碴轨道
Fig. 4.18　Ballastless roadbed

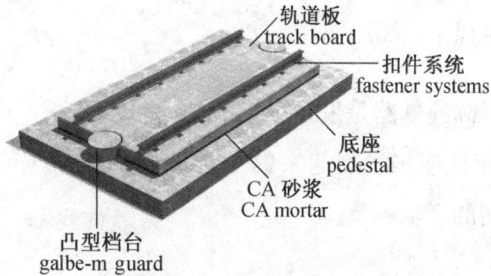

轨道板
track board

扣件系统
fastener systems

底座
pedestal

CA 砂浆
CA mortar

凸型档台
galbe-m guard

图 4.19　无碴轨道道床结构图
Fig. 4.19　Structure chart of the ballastless roadbed

(4) 道岔

道岔是一种使机车车辆从一股道转入另一股道的线路连接设备,通常在车站、编组站大量铺设(如图4.20所示)。

(4) Turnout

Turnout is a kind of wiring devices transferring the vehicle from one track to another, which is usually paved at the station and the marshalling station(Fig. 4. 20). It

有了道岔,可以充分发挥线路的通过能力,即使是单线铁路,铺设道岔,修筑一段大于列车长度的叉线,也可以对开列车。

can give full play to the capacity of passing trains with turnout, even if it is a single-track railway. As long as a turnout is built, whose distance is more than that of the train, when turnout is paved, this railway can have the capacity of passing trains.

(a) 普通铁路
Common railway

(b) 高速铁路
High-speed railway

图 4.20　铁路道岔
Fig. 4.20　Turnout of railway

4.2.4　高速铁路

铁路现代化的一个重要标志是大幅度地提高列车运行速度。一般来讲,铁路的速度分级为:① 常速:时速 100～120 公里;② 中速:时速 120～160 公里;③ 准高速和快速:时速 160～200 公里;④ 高速:时速 200～400 公里;⑤ 特高速:时速 400 公里以上。

高速铁路是发达国家于 20 世纪 60—70 年代逐步发展起来的一种城市与城市之间的运输工具。

高速铁路具有载客量高、输送能力大、速度快、安全性好、正点率高、舒适方便、能源消耗低、环境影响轻、经济效益好等优点。中国是世界上高速铁路发展最快、系统技术最全、集成能力最强、运营里程最长、运营速度最高、在建规模最大的

4.2.4　High-speed railway

An important symbol of the modernization of railway is the great improvement of train speed. Generally, the speed of railway can be divided into: ① normal speed:100～120 km/h;② medium speed:120～160 km/h;③ rapid speed:160～200 km/h;④ high speed: 200～400 km/h; ⑤ special high speed:more than 400 km/h.

High-speeds railway is a transportation tool developed in the developed countries during 1960 s to 1970 s.

High-speed railway has the advantages of high volume, high capacity of transportation, high speed, good safety, great convenience, lower energy consumption, small environmental effect and good economic benefit. In the field of high-speed railway, China has the most abundant technology, the longest running mileage, the largest scope of construction in the

国家(如图 4.21 所示)。截至 2012 年,中国铁路营业里程已达11 万公里以上,其中时速 200~350 公里的高速铁路将达 1.3 万公里。随着京港高铁、徐兰高铁、京沪高铁、沪昆客运专线等建成通车,中国将形成"四纵四横"为主骨架的高速、快速铁路网。

world(Fig. 4.21). At the end of 2012, the running mileage of high speed railway in China has been more than 110 000 kilometers, among which 13 000 kilometers can ensure the trains run at 200 to 350 km/h. With the accomplishment of the Beijing-HongKong High speed Railway, Xuzhou-Lanzhou High speed Railway, Beijing-Shanghai High-speed Railway and Shanghai-Kunming High Speed Railway, China will form the rail net with high speed, which is based on the framework of "four verticals and four horizontals".

图 4.21　中国的高速铁路
Fig. 4.21　High speed railway in China

4.2.5　磁悬浮铁路

磁悬浮铁路是一种新型的交通运输系统,利用电磁系统产生的排斥力将车辆托起,使整个列车悬浮在导轨上,利用电磁力进行导向,利用直线电机将电能直接转换成动力推动列车前进(如图 4.22 所示)。

与传统铁路相比,磁悬浮铁路由于消除了轮轨之间的接触,因而无摩擦力,线路垂直荷载小,适合高速运行。目前,磁悬浮铁路的最高试验时速为 552 km/h。磁悬浮列车及其工作原理如图 4.23 所示。

4.2.5　Magnetic levitation railway

Magnetic levitation railway is a new type of transportation using electromagnetic repulsion system to suspend the entire train on rails, pulling the train with electromagnetic force, and driving the train by electrical energy converted from motor(Fig. 4.22).

Compared with the traditional railway, magnetic levitation railway is frictionless because of the elimination of contact between the wheel and the rail. The vertical load acting on the line is small, which is suitable for high-speed operation. The current maximum test speed is 552 km/h. The working principle of maglev train is shown in Fig. 4.23.

图 4.22 磁悬浮列车和轨道(上海)
Fig. 4.22 Maglev train and track(Shanghai)

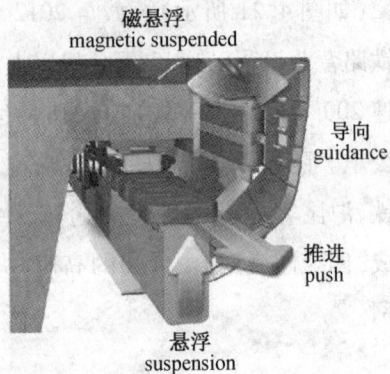

图 4.23 磁悬浮列车工作原理
Fig. 4.23 Working principle of maglev train

磁悬浮铁路系统由线路高架、导轨和轨道系统等结构体系和磁悬浮列车组成。

高架线路结构主要有钢或钢筋混凝土梁,车站一般为钢结构形式,如图 4.24、图 4.25 所示。导轨和轨道系统主要由导轨梁、车轮支撑轨、推进线圈、悬浮和导引线圈等结构组成,如图 4.26 和图 4.27 所示。

Magnetic levitation railway system is composed of its structure system, such as the line elevated track, rail and maglev train.

The elevated line is in form of steel or reinforced concrete, and the station is generally in form of steel structure(Fig. 4.24 and Fig. 4.25). The rail system is mainly supported by rail beams, wheel-rails, propulsion coils, levitation and guidance coil structures, etc. (Fig. 4.26 and Fig. 4.27).

图 4.24 高架线路(钢筋混凝土)
Fig. 4.24 Elevated lines (reinforced concrete)

图 4.25 车站结构
Fig. 4.25 The station structure

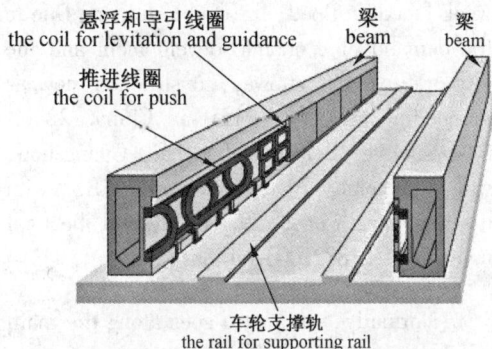

图 4.26 轨道系统结构组成
Fig. 4.26 Track system structure

图 4.27 磁悬浮导轨
Fig. 4.27 Magnetic tract

4.2.6 城市轻轨与地下铁道

地下铁道简称地铁,狭义上专指以地下运行为主的城市铁路系统或捷运系统(如图 4.28 所示)。广义上,由于许多此类的系统为了配合修筑的环境,可能也会有地面化的路段存在,因此通常涵盖了各种地下与地面上的轨道交通运输系统。地铁和轻轨不单纯是走地下和走地上的区别的,一般区分这两个概念要根据铁路的运输能力、车辆大小来判断。

4.2.6 City light rail transit and subway

The underground railway (also named as subway) specifically refers to the operation of the main city underground railway system or rapid transit system (Fig. 4.28). Generally, it also includes the up-ground line for compromising with the construction environment. Therefore, it covers rail transportation system of underground and on the ground. The subway and the light rail are not only classified by the locations, but also the railway transport capacity and the size of the vehicle.

(a) 上海地铁 (3 号线)
The Shanghai subway (line 3)

(b) 重庆轻轨 (3 号线)
Chongqing light rail transit (line 3)

图 4.28 城市轨道交通系统
Fig. 4.28 City rail transit system

地铁和轻轨作为城市轨道交通,具有噪音和干扰少、节约能源和土地、污染少、运量大、速度快、准时、促进城市发展等优点。然而,它也有投资大、建设周期长,对水灾、火灾和恐怖主义等抵御能力弱等不

Metro and light rail, which advantages less noisy and interrupting, high speed and large capacity, can save energy, reduce pollution, and promote the development of the city. At the same time, it needs large investment and has a long construction period, and it is

足。由于经济发展和城市交通的需要,目前我国的城市轨道交通建设规模庞大,预计至 2015 年前后,北京、上海、广州等 22 个城市将将建设 79 条轨道交通线路,总长 2 259.84 公里,总投资 8 820.03 亿元。

地下铁道一般沿城市主要街道布置,在市区和郊区修建,因而施工方案的选取应充分考虑地铁对城市交通、建筑物拆迁以及沿线管线的影响,即从技术、经济等方面综合考虑。地下铁道的修建方法很多,概括起来主要有两大类,即明挖法和暗挖法。

(1) 明挖法

明挖法也叫基坑法,是先将隧道部位的岩(土)体全部挖除,然后修建地下结构,再进行回填的施工方法,适用于浅埋轨道交通(如图 4.29 所示)。明挖法具有施工简单、快捷、经济、安全的优点,其缺点是对周围环境的影响较大。城市地铁工程发展初期都把它作为首选的开挖技术,目前常作为地铁车站的施工方法。

weak faced on flood, fire and terrorism. Due to the demand of economic development and the city traffic, the subway construction now is proceeding in several cities in China. 22 cities, such as Beijing, Shanghai, Guangzhou, will build totally 79 rail lines until 2015, with the total length of 2 259.84 km, and the total investment is 882.003 billions yuan.

Normally, the metro goes along the main street in the city, extending from the urban to the suburban areas, so the effect on city traffic, building demolition and utilities along the line should be fully considered in construction plan, which means making a comprehensive consideration from the point of technology, economic, and other aspects. There are several methods applied into metro construction, which can be generally divided into open-cut method and undermining method.

(1) Open-cut method

The open-cut method is also called foundation pit excavation method. At first, digging out the soil then constructing the underground structure, and backfilling at last (Fig. 4.29). It is suitable for shallow subway, which has the advantages of simple construction, fast, economic, safety and the disadvantages of large impact on surrounding environment. In the early stage of construction development, it is deemed as a prior construction method. At present, it is usually applies in construction of metro station.

(a) 明挖法基坑开挖
The excavation with open-cut method

(b) 明挖法地铁结构施工
The metro construction with open-cut method

图 4.29　明挖法地下结构
Fig. 4.29　The underground structure of open-cut method

在城市繁忙地带修建地铁车站时，往往占用道路，影响交通。当地铁车站设在主干道上，而交通不能中断，且需要确保一定交通流量要求时，可选用盖挖法。盖挖法是由地面向下开挖至一定深度后，将顶部封闭，其余的下部工程在封闭的顶盖下进行施工（如图 4.30 所示）。根据工程实际情况盖挖法又可分为盖挖顺作法、盖挖逆作法和盖挖半逆作法。

The construction of subway station in busy areas often cause the traffic problems. When the subway station is located in the main street, and the traffic cannot be interrupted, the cut and cover excavation method can be more suitable. Cut and cover excavation method is a simple construction method for shallow tunnels where a trench is excavated and roofed over with an overhead support system, the rest structure is to be built below the roof. Three basic forms of cut-and-cover tunnel are available, namely, cover-excavation method, covered top-down excavation and cover excavation semi-inverse method.

(a) 盖挖法施工图
Cut and cover method

(b) 盖挖法（支撑柱和覆盖板）
Cut and method (support column and the cover plate)

图 4.30　盖挖法地下结构
Fig. 4.30　The underground structure in covered excavation

（2）暗挖法

当城市轨道交通工程埋深超过一定限度后，明挖法不再适用，而要改用暗挖法，即不挖开地面，采用在地下挖洞的方式施工。矿山法和盾构法等均属暗挖法。

①矿山法是目前暗挖法中最常用的一种方法，是指主要用钻眼爆破方法开挖断面来修筑隧道及地下工程的施工方法，因借鉴矿山开拓巷道的方法，故名矿山法。但地铁施工多在浅部松软土层中进行，因此在传统矿山法和新奥法（如图4.31b 所示）的基础上发展出了"浅埋暗挖法"（如图 4.31a 所示）。

（2）Undermining method

When the depth of metro construction exceeds a certain limit, the open-cut is no longer applicable. It is more suitable to use tunneling method. Both mining method and shield method belong to this one.

Mining method is one of the most commonly used methods in underground engineering and tunnel construction. It uses drilling and blasting method to extract geological materials and construct tunnel subsequently. It derives from the method used in mine roadway in mine engineering. While the subway construction is normally carried out in shallow soft soil, the "shallow tunneling method" is developed based on the traditional mining method(Fig. 4.31a) and new aus-

(a) 台阶开挖（浅埋暗挖法）
The step excavation (shallow tunneling method)

(b) 光面爆破（新奥法）
The smooth blasting (NATM)

图 4.31　暗挖法
Fig. 4.31　Undermining method

采用矿山法进行暗挖施工时，隧道结构一般分为初期支护和二次衬砌。开挖之后立即进行的支护形式称之为初期支护（如图 4.32a 所示），初期支护一般有喷射混凝土、喷射混凝土加锚杆、喷射混凝土锚杆与钢架联合支护等形式。二次衬砌（如图 4.32b 所示）是指在隧道已经进行初期支护的条件下，用混凝土等材料修建的内层衬砌，以达到加固支护，优化路线防排水系统，美化外观，方便设置通信、照明、监测等设施的作用。

The tunnel structure is generally divided into underlined support and secondary lining in mining tunnel. Initial support which consists of shotcrete, rock bolt, steel combined supporting（Fig. 4. 32a）will be carried out immediately after soil extraction. The internal structure will be constructed by concrete after the initial support, so as to support reinforcement, optimize the drainage system, smooth appearance, and give facility to set communication, lighting, and monitoring facilities（Fig. 4. 32b）.

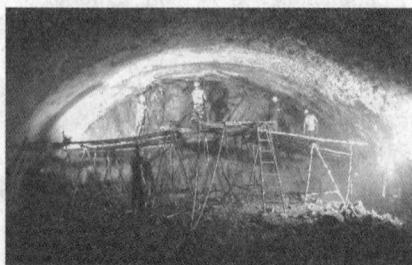

(a) 初期支护施工
The intital support construction

(b) 二次衬砌施工
The secondary lining construction

图 4.32　衬砌结构
Fig. 4.32　Lining structure

② 盾构法是暗挖法施工中的一种全机械化施工方法，它是将盾构机械在地中

A tunneling shield is a protective structure used in the excavation of tunnels through soil that is too soft or fluid to remain

推进,通过盾构外壳和管片支承四周围岩以防止发生隧道内的坍塌,同时在开挖面前方用切削装置进行土体开挖,通过出土机械运出洞外,靠千斤顶在后部加压顶进,并拼装预制混凝土管片,形成隧道结构的一种机械化施工方法。盾构法修建如图4.33所示。

盾构法施工具有施工速度快、洞体质量稳定、对周围建筑物影响较小等特点,适合在软土地基段施工。常见的盾构机有泥水加压盾构、土压平衡盾构、异型盾构等(如图4.34所示)。泥水加压盾构结构如图4.35所示。

stable during the time it takes to line the tunnel with a support structure of concrete. In effect, the shield serves as a temporary support structure for the tunnel while it is being excavated. At the front of the shield a rotating cutting wheel is located. Behind the cutting wheel there is a chamber. Depending on the type of the TBM, the excavated soil is either mixed with slurry (so-called slurry TBM) or left as-is (earth pressure balance or EPB shield). The choice for a certain type of TBM depends on the soil conditions. Systems for removal of the soil (or the soil mixed with slurry) are also present.

Behind the chamber there is a set of hydraulic jacks supported by the finished part of the tunnel which is used to push the TBM forward. Once a certain distance is excavated (roughly 1.5 ~ 2 meters), a new tunnel ring is built by the erector. The erector is a rotating system which picks up precast concrete segments and places them in the desired position.

Behind the shield, inside the finished part of the tunnel, several support mechanisms which are part of the TBM can be found: dirt removal, slurry pipelines if applicable, control rooms, rails for transport of the precast segments, etc.

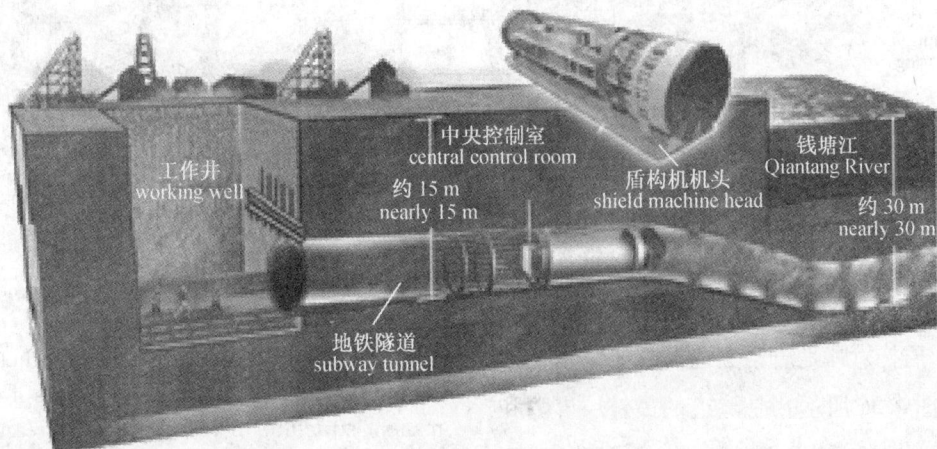

图 4.33 盾构法修建地铁隧道(杭州地铁1号线过钱塘江)
Fig. 4.33 Subway construction shield (Hangzhou Metro Line 1 over the Qiantang River)

(a) 土压平衡盾构
The earth pressure balance shield

Shield tunneling has the characteristics of quick construction, stable excavation face, minimal impact on the surrounding buildings. Therefore, it is suitable for soft layer. The common types of TMB are slurry shield, earth pressure balance shield and other special-shaped shield (Fig. 4.34). The slurry shield structure diagram is shown in Fig. 4.35.

(b) 双圆盾构
Double circle shield

(c) 矩形盾构
Rectangular shield

图 4.34 盾构机
Fig. 4.34 The shield machine

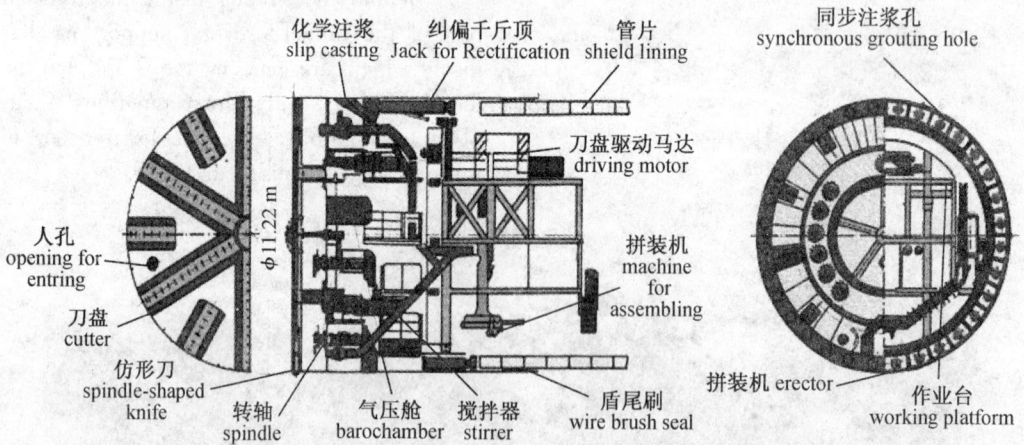

图 4.35 泥水加压盾构结构图（上海地铁 8 号线）
Fig. 4.35 Slurry shield structure diagram (Shanghai Metro Line 8)

盾构管片是盾构隧道的主要结构构件（如图 4.36 所示），是隧道的最外层屏障和永久衬砌结构，承担着抵抗土层压力、地下水压力以及一些特殊荷载的作用。管片质

The shield lining is the main structural component of shield tunnel, which is a permanent structure to resist the soil pressure, water pressure and other certain load (Fig. 4.36). The overall tunnel quality, safety, waterproofing performance and durability of

量直接关系到隧道的整体质量和安全,影响隧道的防水性能及耐久性能。

tunnel depend on the lining quality.

(a) 盾构管片
Shield segment

(b) 盾构隧道结构（外）
Shield tunnel structure (external)

(c) 盾构隧道结构（内）
Shield tunnel structure (internal)

图 4.36　盾构隧道管片结构
Fig. 4.36　The structure of shield tunnel segment

4.3 机场工程

民航运输和经济发展是相辅相成的,当代的国际中心城市,如纽约、上海、巴黎、东京,无一不是重要的国际航空枢纽中心。近年来,我国的机场建设逐渐提速,"十二五"末机场的数量将达到 230 个左右,80% 以上的人口在直线距离 100 公里以内都能够享受到航空服务(如图 4.37 所示)。

4.3 Airport project

The construction of civil aviation transportation and the development of economy are complementary to each other. International center cities such as New York, Shanghai, Paris, Tokyo are all major international aviation hub. In recent years, airport construction in China has accelerated gradually, and at the end of "Twelfth Five-Year Plan", the number of airport will reach about 230. More than 80% of the population will be able to enjoy the air service in a straight line distance of 100 kilometers (Fig. 4.37).

(a) 北京首都国际机场
Beijing Capital International Airport

(b) 上海浦东国际机场
Shanghai Pudong International Airport

图 4.37　民航机场
Fig. 4.37　Civil aviation airport

4.3.1 机场的分类和组成

(1) 机场的分类

机场,亦称飞机场、空港,较正式的名称是航空站,是专供飞机起降活动的飞行场。机场大小各不相同,除了跑道之外,机场通常还设有塔台、停机坪、航空客运站、维修厂等设施,并提供机场管制服务、空中交通管制等其他服务。机场一般分为军用和民用两大类(这里主要介绍民用机场),我国把大型民用机场称为空港,把小型机场称为航站。

按机场规模和旅客流量可将机场分为 3 种类型:国际机场、干线机场、支线机场。

(2) 民航机场的组成

机场主要由飞行区、地面运输区和候机楼区 3 个部分组成。飞行区是飞机活动的区域;地面运输区是车辆和旅客活动的区域;候机楼区是旅客登记的区域,是飞行区和地面运输区的接合部位。

① 飞行区分空中部分和地面部分。空中部分指机场的空域,包括进场和离场的航路;地面部分包括跑道(如图 4.38 所示)、滑行道、停机坪和登机门,以及一些为维修和空中交通管制服务的设施和场地,如机库、塔台、救援中心等。

② 候机楼区包括候机楼建筑本身(如图 4.39 所示)以及候机楼外的登机坪和旅

4.3.1 The classification and composition of airport

(1) The classification of airport

Airport, also known as airdrome or aerodrome, or officially the air station, is special for aircraft taking off and landing. Beside the runway, the airport also includes tower, apron, aviation passenger station, repair factory and other facilities. It provides airport services, air traffic control and other services. The airport is generally divided into two types, military and civilian. The civil airport are mainly introduced in this chapter. The large civil airport is called aerodrome, and the small airport is called terminal in China.

According to the size and the passenger flow, airport can be divided into international airport, trunk airport and regional airport.

(2) The composition of airport

The airport is mainly composed of flight area, ground transportation area and waiting area. The flight area is for aircraft. The ground transportation area is for vehicle and passengers, and the waiting area is the connection area of flight area and ground transportation area with the function of check-in.

① Flight area is divided into aerial part and ground part. The aerial part refers to the airport airspace, including the approach and departure route; the ground part includes runway(Fig. 4. 38), taxiway, apron, boarding gate, and some facilities and venues for repair and the air traffic control service, such as the hangar, tower, and rescue center.

② Waiting area(Fig. 4. 39) includes the terminal building as well as the terminal, boarding ramp and the passenger lanes. It is

客出入车道。它是地面交通和空中交通的结合部，是机场对旅客服务的中心地区。

③ 地面运输区是城市进出空港的通道（如图 4.40 所示），大型城市为了保证机场交通的通畅修建了从市区到机场的专用高速公路，甚至还开通地铁和轻轨交通，方便旅客出行。

a combination of ground traffic and air traffic, as well as the central area of the airport passenger service.

③ Ground transportation area is the channel of the city for import and export (Fig. 4.40). In order to ensure the smooth traffic of the airport, most of large cities have built special highway from the urban area to the airport, and even built the subway and light rail traffic for travel convenience.

图 4.38　跑道
Fig. 4.38　Runway

图 4.39　候机楼
Fig. 4.39　Waiting area

图 4.40　进出机场的磁悬浮铁路
Fig. 4.40　Maglev train for enter and out of the airport

4.3.2　跑道结构

跑道结构是机场飞行区的主体，直接提供飞机起飞滑跑和着陆滑跑的路径。跑道材质可以是沥青或混凝土，简易的机场跑道还可以是平整的草、土或碎石地面。跑道要有一定的长度、宽度、坡度、平坦度，以及结构强度和摩擦力等。

4.3.3　机坪与机场净空区

机场的机坪主要有等待坪和掉头坪。前者供飞机等待起飞或让路而临时停放之用，通常设在跑道端附近的平行滑行道旁边；后者则供飞机掉头使用，当飞行区不设平行滑行道时，应在跑道端头部设掉头坪。

4.3.2　Track structure

Track structure is the main airport zone, directly providing service for aircraft take-off and landing. The track line can be made of asphalt or concrete, and simple airport runway can also be surfaced grass, soil or gravel ground. The runway must have a certain length, width, slope, flatness, structural strength and friction, etc.

4.3.3　The apron and the airfield clearance zone

The airport apron includes a waiting area and a U-turn area. The former is for the plane waiting for take-off or out of the way and temporary parking, and it is usually located in the parallel taxiway next to the end of runway. The latter one is for U-turn of the plane. When the flight zone is not provided with a parallel taxiway, the U-turn area is needed at the end of runway.

交通与水利工程　Traffic Engineering and Hydraulic Engineering

机场净空区是指飞机起飞着陆涉及的范围,保证在起飞和降落的低高度飞行时不能有地面的障碍物妨碍导航和飞行(如图4.41所示)。

The airfield clearance zone is the place where aircraft takes off and land on, ensuring that no obstacle will hinder the ground navigation and flight, while the plane is taking off and landing at low altitude as shown in Fig. 4.41.

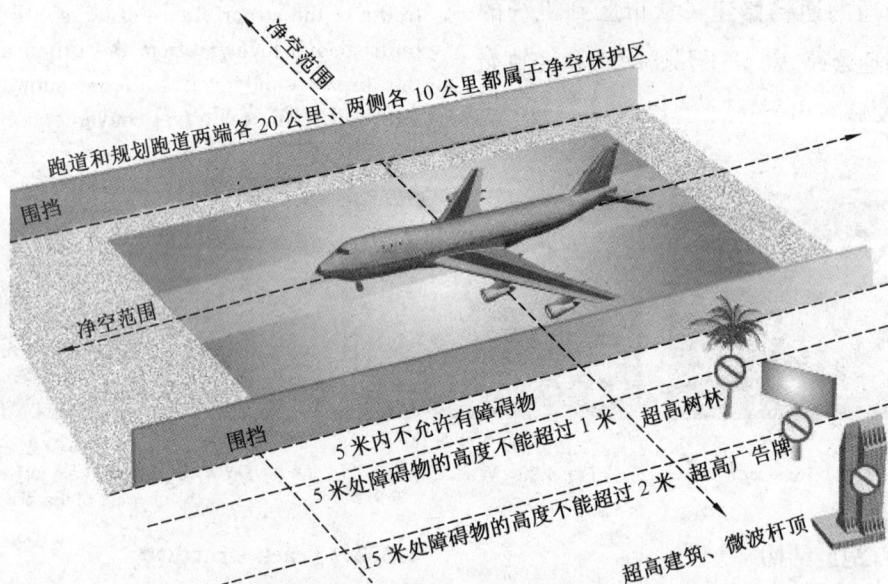

① 跑道和规划跑道两端各20公里、两侧各10公里都属于净空保护区:
20 kilometers on both ends of runway and 10 kilometers on both sides of runway are airfield clearance zone;
② 5米内不允许有障碍物: Within 5 meters is not allowed to have obstacles;
③ 5米处障碍物的高度不超过1m: The height of 5 meters obstacle is less than 1m;
④ 15米处障碍物高度不超过1m: The height of 15 meters obstacle is less than 1 m;
⑤ 超高树木: trees;
⑥ 超高广告牌: sky sign;
⑦ 超高建筑: super-high construction

图4.41 机场净空区示意图
Fig. 4.41 Sketch map of airport clearance area

4.3.4 航站区

航站区主要由候机楼、站坪及停车场组成。航站楼的结构设计涉及位置、形式、建筑面积等要素。

4.3.4 Terminal area

The terminal area is mainly composed of the air-terminal, the station site and the parking lot. The structure design of the terminal involves location, form, construction area, etc.

（1）航站楼

候机楼是航站区的主要建筑物(如图4.42所示)。航站楼的设计,不仅要考虑

（1）Air-terminal

The air-terminal is the main structure of the terminal area, as shown in Fig. 4. 42.

其功能,还要考虑其环境、艺术氛围及民族(或地方)风格等。航站楼一侧连着机坪,另一侧与地面交通系统相联系。其基本功能是安排好旅客、行李的流程,为其改变运输方式提供各种设施和服务,使航空运输安全有序。

旅客航站楼的基本结构设施应包括:① 车道;② 公共大厅;③ 安全检查设施;④ 政府机构;⑤ 候机大厅;⑥行李处理设施(行李分检系统和行李提取系统);⑦ 机械化代步设施(人行步道,自动扶梯);⑧ 登机梯;⑨ 旅客信息服务设施等。

The design of air-terminal needs to consider its feature, the environment, the artistic and national (or local) style. The air-terminal connects the apron with the ground transportation system. Its fundamental function is to arrange passengers, baggage, and facility their transportation to make sure the safety and order of the air transportation.

The basic structural facilities of the passenger terminal building include: driveway, public hall, safety inspection facilities, federal government agencies, waiting hall, baggage handling facilities (baggage sorting system and baggage reclaim system), mechanized transport facilities (pedestrian space, escalator), airstairs, passenger information service facilities, etc.

(a) 昆明长水国际机场航站楼
The terminal building of Changshui
International Airport in Kunming

(b) 上海浦东国际机场航站楼
The terminal building of Pudong
International Airport in Shanghai

图 4.42 航站楼
Fig. 4.42 Terminal building

(2) 站坪、机场停车场与货运区

站坪或称客机坪,是设在航站楼前的机坪,供客机停放、上下旅客、完成起飞前的准备和到达后的各项作业使用(如图4.43 所示)。

机场停车场设在机场航站楼附近,如图4.44 所示。停车场建筑面积主要根据高峰小时车流量、停车比例等确定。

(2) Standing platform, airport parking lot and freight

Standing platform, also named as the passenger platform is located in front of the terminal building. It's a platform for aircraft parking, passengers' alighting and boarding, the departure preparation and miscellaneous actions after arriving(Fig. 4.43).

Airport parking lot is near the airport terminal, as shown in Fig. 4.44. The construction area is mainly based on peak hour traffic flow, traffic parking ratio, etc.

机场货运区供货运办理手续、飞机装卸货、临时存储等使用,主要包括业务楼、货运库、装卸场及停车场(如图 4.45 所示)。货运区应离开旅客航站区及其他建筑物适当距离,以便将来发展。

Airport freight area, which provides space for cargo handling procedures, loading or unloading cargo, temporary storage service, etc. (Fig. 4.45), includes the freight business building, loading and unloading of field and parking lot. The freight area should keep an appropriate distance from the passenger terminal area and other buildings for its future development.

图 4.43　站坪
Fig. 4.43　Standing platform

图 4.44　机场停车场
Fig. 4.44　Airport parking lot

图 4.45　机场货运区
Fig. 4.45　Airport freight area

4.4　水利工程

4.4.1　水资源特点

作为一种独特的自然资源,水资源有优势也存在缺点。由于水利工程的本质目的是发挥水资源优势并克服水资源利用时的缺点,所以在学习水利工程之前需要了解水资源的特点(如图 4.46 所示)。

4.4　Hydraulic engineering

4.4.1　The characteristics of water resources

As a kind of unique natural resources, water resources have advantages and disadvantages. The essential purpose of hydraulic engineering is enhancing the advantages of water resources and overcoming the disadvantages of water resources utilization. Therefore, we need to comprehend the characteristics of water resources before learning hydraulic engineering (Fig. 4.46).

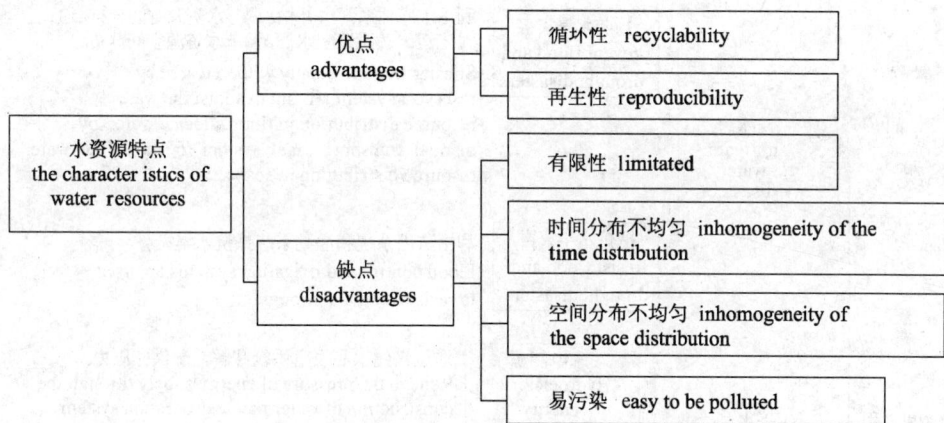

图 4.46　水资源特点
Fig. 4.46　The characteristics of water resources

4.4.2　水利工程的内涵

（1）水利工程的概念

　　水利工程是通过对自然界的水资源进行控制和调配,以实现除水害、兴水利、护水源的目的而兴建的工程（如图 4.47 所示）。

4.4.2　Connotation of the hydraulic engineering

（1）The conception of hydraulic engineering

　　Projects, which are used to control and redeploy water resources for mitigating water disasters, promoting water conservation and protection of water resources, protecting the environment of water resources, are called hydraulic engineering（Fig. 4.47）.

水利工程
hydraulic
engineering

防水害
mitigate
water
disasters

兴水利
promote water
conservancy

护水源
protect the
environment
of water
resources

预防水旱灾害
prevent flood and
drought disasters

利用水库体系蓄洪补枯，调节水资源时间分布；通过
人工运渠体系调配水源，调节水资源空间分布。
Storing flood to supply withered area by
reservoir system,aiming to adjust the water
resource distribution in time; allocate water by
manual transport canal system to adjust the water
resource distribution in space.

抵抗水旱灾害
Resist flood and
drought disasters

利用防洪系统和灌溉系统抵抗水旱灾害。
Flood control and irrigation systems are used
to resist flood and drought .

水电能源
Hydroelectric
energy

通过兴建水利发电系统缓解能源供应压力。
Relieve the pressure of energy supply through the
construction of water power generation system.

水利交通
Transportation of
hydro-engineering

通过拓展、疏浚和兴建河道、码头等水利交通系统缓
解陆路交通压力。
Ease the land transportation by the system of
hydro-engineering transportation such as
expanding, dredging or building the river and
wharf.

水资源保护和改善
protection and
improvement of
water resources

通过兴建水资源质量检测、保护和改善工程，纠正
以往无序开发留下的隐患，维持水资源的可持续发展。
Through the construction of quality detecting,
protecting water resources, it could correct
and remove the left hidden dangers because
of disorderly development before, atthe same
time maintain sustainable development
of water resources.

水资源的节约
economization of
water resources

为提高水资源的利用效率而兴建的相关水利工程。
Construct related hydraulic engineering in
order to improve the utilization efficiency of
water resources.

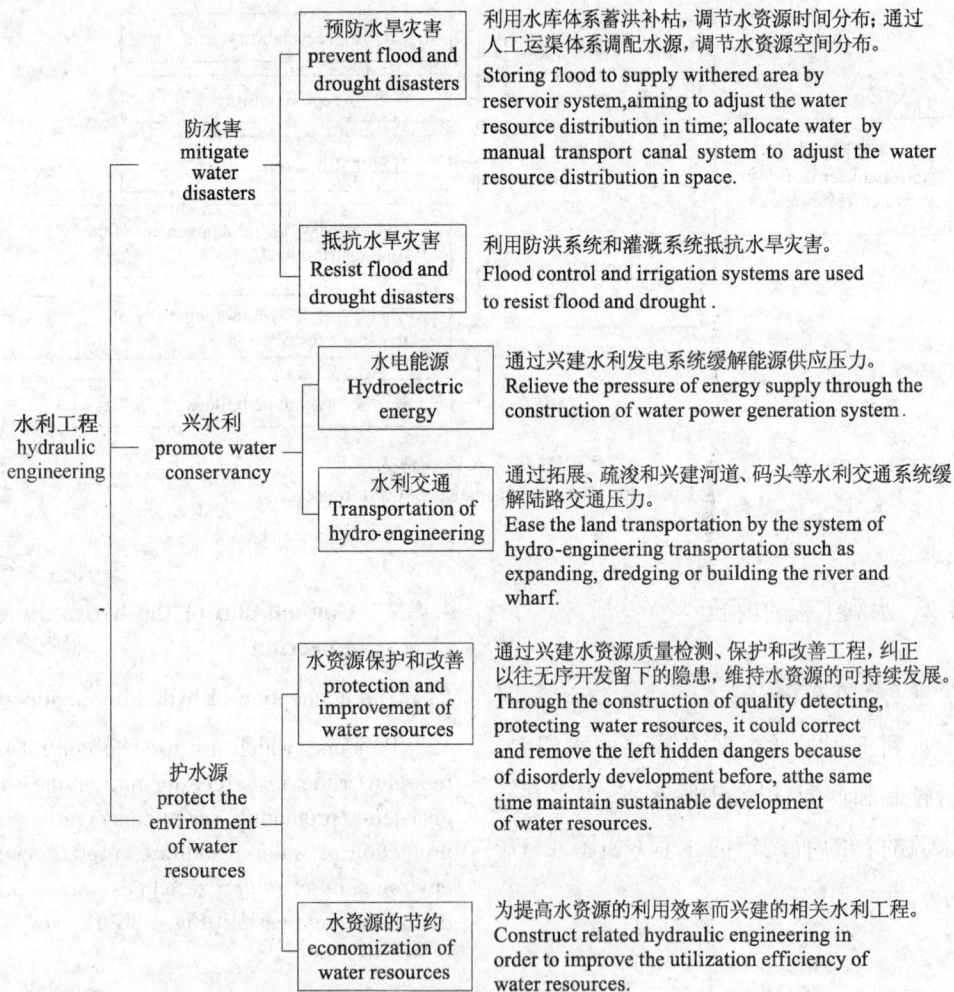

图 4.47 水利工程
Fig. 4.47 Hydraulic engineering

（2）水利工程的层次

水利工程按其规模自下而上可分为 水工建筑物、水利枢纽、体系化综合水利工程 3 个层次：① 水工建筑物是水利工程中采用的各种单体建筑物的统称。② 水利枢纽是在水系的某一区域内由若干个不同功能的水工建筑物协同工作所组成的有机综合体。③ 体系化综合化水利工程是若干个水利枢纽和水工建筑物以及天然水源有序结合组成的综合性、跨区域工程，如三峡工程、南水北调工程。

4.4.3 水工建筑物

（1）水工建筑物分类

水工建筑物按功能可以分成 6 类。

① 挡水建筑物，即用来拦截水流，抬高水位及调节蓄水量的建筑物。

② 泄水建筑物，即用于宣泄水库、渠道的多余洪水，排放泥沙和冰凌的建筑物。

③ 取水建筑物，即用以从水库或河流引取各种用水的水工建筑物。

④ 输水建筑物，即用以将水输送到用水地的水工建筑物。

⑤ 整治建筑物，即用以改善河流的水流条件、稳定河槽以及为防护河流、水库、湖泊中的波浪和水流对岸坡冲刷的建筑物。

⑥ 专门建筑物，即为灌溉、发电、过坝等需要兴建的建筑物。

（2）常见的水工建筑物

① 坝。大坝是重要的挡水建筑物，有

（2）The level of hydraulic engineering

According to its scale, hydraulic engineering is classfied into three levels. ① Hydraulic structure is the single building in hydraulic engineering. ② Hydro-junction is complex which is made up of several hydraulic structures with different functions. ③ Integration and systematism hydraulic engineering is the comprehensive and inter-regional engineering which is made up of several hydraulic structures and hydro-junctions, such as Three Gorges Dam and South-to-North water transfer project.

4.4.3 Hydraulic structures

（1）The classification of hydraulic structures

According to its function, hydraulic structure can be divided into six categories.

① Water retaining structure is used to intercept water, raise the water level and regulate the water storage.

② Water discharging structure is used to discharge extra flood in reservoirs and channels, and let out sediment and ice.

③ Water taking structure is a hydraulic structure that is used to take all kinds of water from reservoirs or rivers.

④ Water transporting structure is a hydraulic structure that is used to transport water to where water is needed.

⑤ Water regulating structure is used to improve river conditions, stable channel as well as protect the slope from erosion of rivers, reservoirs, waves and currents in the lakes.

⑥ Special structure is used to irrigate, generate power, or pass the dam and so on.

（2）Common hydraulic structures

① The dam is one of the most important water retaining structures. The dam also

时也兼有泄水功能,即通过坝顶溢流(允许坝顶溢流的大坝称为溢流坝)和坝身泄水孔泄水。大坝按坝体材料可分土石坝和混凝土坝。混凝土坝又可细分为重力坝(如图4.48所示)、拱坝(如图4.49所示)和支墩坝(如图4.50所示)。

sometimes combines drainage function by crest overflow and its draining hole. The dam that allows crest overflow are called overflow dam. The category of a dam includes earth-rock dam and concrete dam by material. The category of concrete dam includes gravity dam (Fig. 4.48), arch dam (Fig. 4.49) and buttress dam (Fig. 4.50).

图 4.48　重力坝
Fig. 4.48　Gravity dam

图 4.49　拱坝
Fig. 4.49　Arch dam

图 4.50　支墩坝
Fig. 4.50　Buttress dam

② 河岸式溢洪道。溢流坝设有溢流通道,但当坝体不适合溢流,或不能满足大量泄洪要求时,需在坝体外的河谷两岸适当位置单独设置溢洪道,称为<u>河岸式溢洪道</u>。河岸式溢洪道形式多样,图4.51所示为<u>井式溢洪道</u>。

② Overflow dam has its overflow channels. While the dam type is not suitable for overflow or cannot meet the requirements of large amount of flood discharging, spillway is set up separately at suitable location on both sides of the valley out of dam. That is called <u>chute spillway</u>. Chute spillway has diverse forms. Fig. 4.51 shows a <u>shaft spillway</u>.

图 4.51　井式溢洪道
Fig. 4.51　Shaft spillway

③ 堤。沿江河、沟渠、湖、海岸、水库等边缘修筑的挡水建筑物称为堤,图4.52

③ Water retaining structures at the edge of rivers, ditches, lakes, coasts and reservoirs is called <u>dike</u>(Fig. 4.52). In ad-

所示为河堤。此外,在引水枢纽中还有一种导流堤,其作用是将河流导入取水建筑物。

dition, there is a kind of diversion dike in water diversion hydro-junction, which can guide the water flow into water taking structures.

图 4.52　河堤
Fig. 4.52　Dike

④ 渡槽、倒虹吸管、水工涵洞和水渠。

渡槽是用以输送渠道水流跨越河渠、溪谷、洼地和道路的架空水槽,如图 4.53 所示。渡槽两端与渠道连接,是一种重要的输水建筑物。

④ Aqueduct, inverted siphon, hydraulic culvert, ditch.

Aqueduct (Fig. 4.53) is an overhead flume used to transport water of ditches across rivers, valleys, depressions and roads. Both sides of aqueduct are connected with ditches, and aqueduct is an important water transporting structure.

图 4.53　渡槽
Fig. 4.53　Aqueduct

倒虹吸管是用以输送渠道水流穿过河渠、溪谷、洼地、道路的压力管道(如图

Inverted siphon is a penstock used to transport water of ditches across rivers, valleys, depressions and roads (Fig. 4. 54).

4.54 所示）。倒虹吸管两端与渠道连接，也是一种重要的输水建筑物。

Both sides of inverted siphon are connected with ditches, and it is an important water transporting structure.

图 4.54　倒虹吸管
Fig. 4.54　Inverted siphon

水工涵洞是公路或铁路与沟渠相交的地方，使水从路下通过的水工建筑（如图 4.55 所示）。水工涵洞是一种常见的输水建筑物。

Hydraulic culvert is a hydraulic structure at the intersection of roads (or railways) and ditches (Fig. 4.55). It is a common transporting structure.

图 4.55　水工涵洞
Fig. 4.55　Hydraulic culvert

水渠是最常见的输水建筑物，是具有自由水面的人工水道（如图 4.56 所示）。

Ditch is the most common transporting structure, and it is the man-made ditches with free water surface(Fig. 4.56).

图 4.56 水渠
Fig. 4.56 Ditch

⑤ 其他。其他常见的水工建筑物还有水闸、水工隧洞等。水闸是一种利用闸门挡水和泄水的低水头水工建筑物。水工隧洞是穿山开挖建成的封闭式输水道。

4.4.4 水利枢纽分类

水利枢纽按功能可分为蓄水枢纽、引水枢纽和泵站枢纽,如图 4.57 所示。

⑤ Other common hydraulic structures include sluice, hydraulic tunnel, etc. Sluice is a low head hydraulic structure, retaining water and drainage water by gate. Hydraulic tunnel is a closed water pipeline building through the mountains.

4.4.4 The classification of hydro-junction

According to its function, the hydro-junction can be classified into water storage hydro-junction, water diversion hydro-junction and pumping station hydro-junction, as shown in Fig. 4.57.

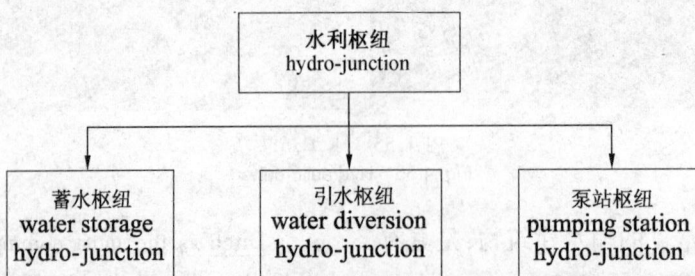

图 4.57 水利枢纽按功能分类
Fig. 4.57 The functional classification of hydro-junction

① 蓄水枢纽。蓄水枢纽的主要水工建筑物组成和功能如图 4.58 所示。除此之外,根据开发目标的不同,蓄水枢纽中

① Water storage hydro-junction
Parts of water storage hydro-junction and its functions are shown in Fig. 4. 58. Besides, special structures which meet the demands of navigation and water power ge-

还可能包含专门建筑物,以满足通航和水力发电的需求。为合理、有效的地利用水资源,蓄水枢纽中也可能包含 些整治建筑物。

neration according to the difference of development targets are included. In order to use water resources reasonably and effectively, regulating structures are included in water storage hydro-junction.

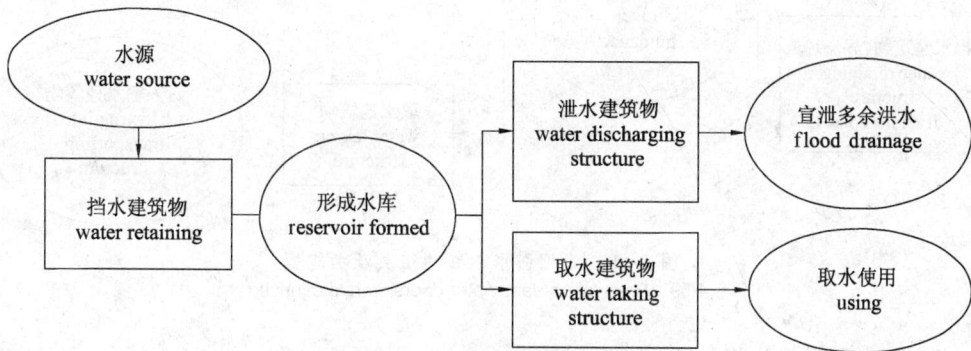

图 4.58　蓄水枢纽的组成及功能
Fig. 4.58　Parts of water storage hydro-junction

② 引水枢纽。引水枢纽包括有坝引水枢纽和无坝引水枢纽,处于水渠系统和自然水源交汇的渠首河段。无坝引水枢纽的主要水工建筑物组成和功能如图4.59 所示。除此之外,根据开发目标的不同,引水枢纽中还可能包含专门建筑物,以满足通航和水力发电的需求。为合理、有效地利用水资源,引水枢纽中也可能包含一些整治建筑物。

② Water diversion hydro-junction

Water diversion hydro-junction can be divided into no-dam and dam water diversion hydro-junction. It lies to the meeting of river and canal which is called canal head. Fig. 4.59 introduces the main parts and functions of a no-dam water diversion hydro-junction. Besides, the existence of special structures can meet the demands of navigation and water power generation depending on the different development targets. In order to use water resources reasonably and effectively, regulating structures are included in water diversion hydro-junction.

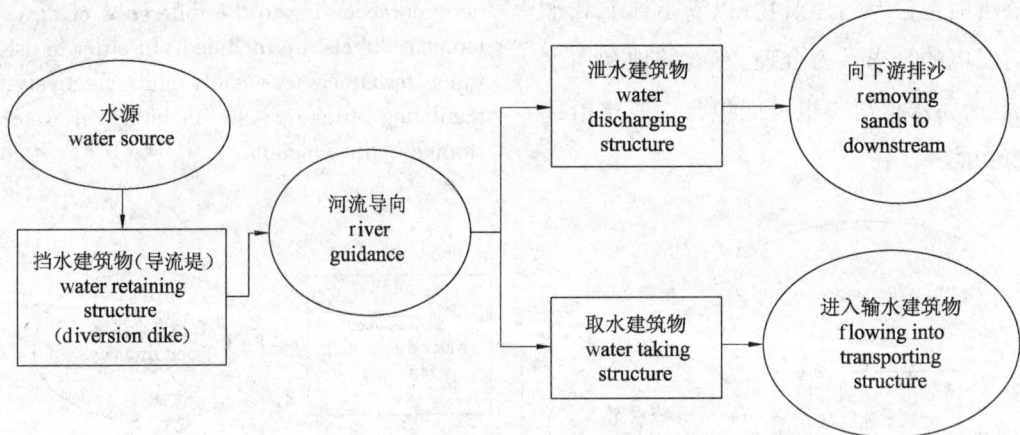

图 4.59 无坝引水枢纽的组成及功能
Fig. 4.59 Parts of no-dam water diversion hydro-junction

③ 泵站枢纽。为将低处水抽送到高处的水工建筑物综合体称为泵站枢纽，多以泵站及水闸为主体。

③ Pumping station hydro-junction. Pumping station hydro-junction is a complex structure used to pump water to higher places.

4.4.5 水利工程的设计标准

任何工程建筑都必须兼顾安全性和经济性要求，水利工程也不例外。安全性和经济性是一对矛盾体，为了正确处理好两者的关系，分等定级系统必须引入水利工程的设计中。

水利工程设计时，需将水利枢纽工程分等，并在此基础上对水工建筑物分级（如图 4.60 所示）。对不同等级的水工建筑物，按照不同的设计要求进行设计，以达到既安全又经济的目的。

4.4.5 Design standard of hydraulic engineering

Safety and economy should be considered at the same time in any engineering construction, and there is no exception in hydraulic engineering. Safety and economy are a pair of contradictory entity. In order to correctly handle the relationship between safety and economy, ranking and grading system must be initiated into the design of hydraulic engineering.

It is necessary to rank about hydro-junction project, and on this basis, we can rank on hydraulic structure, as shown in Fig. 4.60, in order to achieve the goal of safety and economy, different ranking hydraulic structure is designed according to different design requirements.

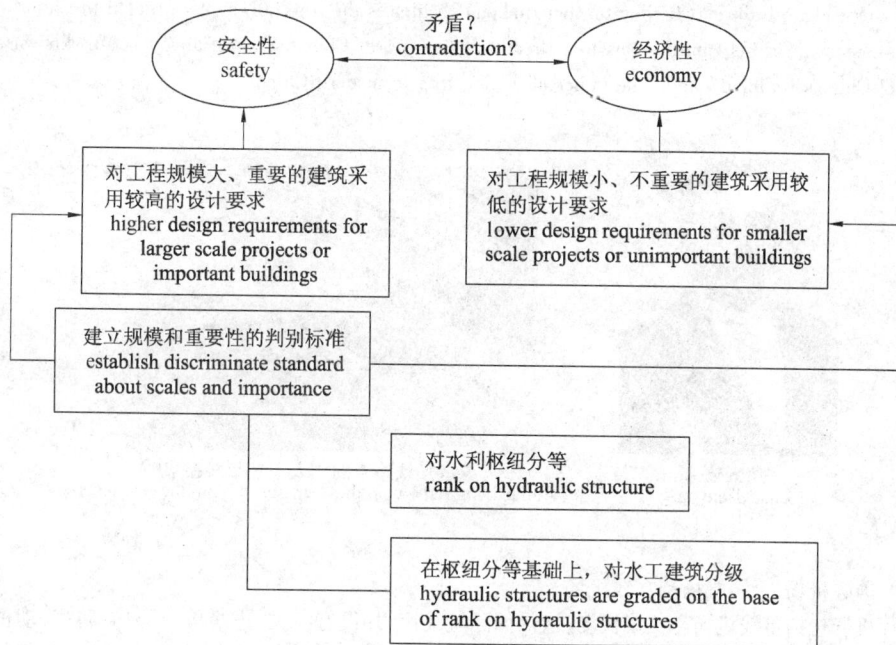

图 4.60　水利工程分等、级示意图

Fig. 4.60　Schematic diagram of ranking and grading of hydraulic engineering

注:本章图片均来源于网络。
Note: In this chapter, all pictures are from webs.

知识拓展
Learning More

相关链接　Related Links

(1) 中国高速公路网 http://www.china-highway.com/

(2) 中华铁道网 http://www.chnrailway.com/

(3) 中国机场网 http://www.chinaairports.cn/

(4) 中华人民共和国水利部 http://www.mwr.gov.cn/

小贴士　Tips

(1) 詹天佑和人字形铁路

詹天佑(1861—1919 年),字眷诚,号达朝,中国近代铁路工程专家。12 岁留学美国,1878 年考入耶鲁大学土木工程系,专习铁路工程。1905—1909 年主持修建我国自建的第一条铁路——京张铁路,创造"竖井施工法"和"人"字形线路,震惊中外,有"中国铁路之父"、"中国近代工程之父"之称。

Zhan Tianyou and Chevron railway

Zhan Tianyou(1861 – 1919), whose courtesy name is Juancheng, and styled Dachao, was a distinguished Chinese railroad engineer. Educated in Yale, he was the chief engineer responsible for construction of the Imperial Peking-Kalgan Railway (Beijing to Zhangjiakou), which is the first railway constructed in China without foreign assistance. For his contributions to railroad engineering in China, Zhan Tianyou is still known as the "Father of China's Railroad" and "the father of engineering in modern China".

詹天佑
Zhan Tianyou

京张铁路中的"人"字形线路段
Jingzhang railway in the "persons" glyph line segments

(2) 国际机场、干线机场和支线机场

国际机场包括干线机场和支线机场,指供国际航线用,并设有海关、边防检查、卫生防疫、动植物检疫和商品检验等联动机构的机场。干线机场是指省会、自治区首府及重要旅游、开发城市的机场。支线机场是指省、自治区内地面交通不便的地方所建的机场,其规模通常较小。

International airport, main airport, regional airport

International airport can accommodate international flights, typically equipped with customs, immigration facilities, quarantine of animals and plants, commodity inspection. Main airport is an airport located in the provincial capital, the capital of the autonomous region and the important development of tourism. Regional airport is an airport locating in the area where the transport is not well developed within the province and autonomous regions. Normally its size is smaller.

思考题 Review Questions

(1) 在山区修建高速公路,必须建造哪些结构设施?

When an expressway is being constructed in mountain area, what structural facilities should be built?

(2) 高速铁路与普通铁路有哪些区别,其结构有哪些特殊要求?

What are the differences between high-speed railway and ordinary railway, and what are the special requirements of its structure?

(3) 简述我国的城市轨道交通发展现状和对轨道结构的要求。

Please make a brief status of urban rail transit development in China and its requirements on track members.

(4) 你认为未来民用机场的结构设计应该向什么方面发展?

What is the development direction of the structural design in civil airport in your opinion?

(5) 三峡工程中采用了哪些常见的水工建筑物?

Which common hydraulic structures are used in Three Gorges Dam?

参考文献
References

［1］陈学军,江见鲸:《土木工程概论》,机械工业出版社,2006 年。

［2］H. C. M. Swamy. Elements of Civil Engineering. Laxmi Publications, 2008.

［3］叶志明:《土木工程概论》(第 3 版),高等教育出版社,2009 年。

［4］Raikar. R. V. Elements of Civil Engineering and Engineering Mechanics. Asoke K. Ghosh, PHI Leaming Private Limited, 2011.

［5］全国一级建造师执业资格考试用书编写委员会:《民航机场工程管理与实务》(第 3 版),中国建筑工业出版社,2011 年。

［6］中国交通年鉴社:《中国交通年鉴》,中国交通年鉴出版社,2012 年。

［7］佟立本:《高速铁路概论》(第 4 版),中国铁道出版社,2012 年。

［8］倪福全:《农业水利工程概论》,中国水利水电出版社, 2011 年。

［9］朱宪生,冀春楼:《水利概论》,黄河水利出版社, 2004 年。

交通与水利工程　Traffic Engineering and Hydraulic Engineering

第5章 市政工程

Chapter 5　Municipal Engineering

市政工程是指市政基础设施建设工程。市政基础设施是指在城市区、镇（乡）规划建设范围内设置的、基于政府责任和义务为居民提供有偿或无偿公共产品和服务的各种建筑物、构筑物、设备等的总称。市政工程是城市建设中的各种公共交通设施、给水、排水、燃气、城市防洪、环境卫生及照明等基础设施建设，是城市生存和发展必不可少的物质基础，是提高人民生活水平和对外开放的基本条件。

Municipal engineering refers to the municipal infrastructure project, which is the general term of buildings, structures and equipment that are set up within the scope of planning and construction in the urban area or town, in order to provide residents with paid or unpaid public services based on the government responsibility and obligation. Municipal engineering includes all kinds of public transport facilities, water supply and drainage, gas, flood control, environmental health, lighting and so on. It is not only the indispensable basis of city survival and development, but also the basic conditions of improving people's living standards and society developments.

5.1　城市道路和桥梁

公共交通是城市的大动脉，也是城市实现其效用的一个重要途径，它的发展直接影响着城市的经济发展和社会发展。城市公共交通设施主要是指道路和桥梁。

5.1.1　城市道路

城市道路是建在城市范围内，供车辆

5.1　Urban roads and bridges

Public transportation is not only the artery of a city, but also plays an important role to achieve its utility, which directly affects the city's economic development and social development. Urban public transport facilities mainly refer to road and bridge.

5.1.1　Urban road

Urban road is built within the scope of

及行人通行并具备一定技术条件和设施的道路,如图5.1所示。

the city for vehicles and pedestrians, which has a certain technical conditions and facilities, as shown in Fig. 5.1.

图5.1　城市道路
Fig. 5.1　Urban road

城市道路按其在城市道路系统中的地位、交通功能及对沿线建筑物的服务功能分为快速路、主干路、次干路和支路,见表5.1。

① 快速路是指为较高车速的远距离交通而设置的重要城市道路。

② 主干路是指在城市道路网中起骨架作用的道路。

③ 次干路是指城市中数量较多的一般交通道路,同时具有服务功能。

④ 支路是指城市道路网中干路以外联系次干路或者供区域内部使用的道路,用以解决局部地区交通和群众的使用要求。

According to its status in urban road system, transportation function and service function, the urban road can be divided into expressway, arterial road, secondary trunk road and branch road, as shown in Table 5.1.

① Expressway is an important city road for high speed transportation over a long distance.

② The arterial road plays a backbone role in urban road network.

③ The secondary trunk road with higher number in urban roads plays the service function.

④ In order to meet the requirements of local transportation, branch road is constructed to contact with secondary trunk road or used inside the area of urban road network in addition to arterial road.

表5.1　城市道路分类
Table 5.1　Classification of urban road

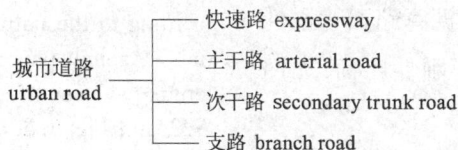

城市道路 urban road	快速路 expressway
	主干路 arterial road
	次干路 secondary trunk road
	支路 branch road

市政工程 Municipal Engineering

道路一般由路基、路面、排水系统、沿线设施组成,见表5.2。

① 路基是道路行车部分的基础,它承受路面传递下来的行车荷载,是由土、石按照一定的尺寸、结构要求所构成的带状土工结构物。路基要求稳定坚实。

② 路面是道路的行车部分,是用各种筑路材料分层铺筑在路基上的结构物,以供车辆在其上以一定车速,安全、舒适地行使。路面要求有足够的强度、较高的稳定性、一定的平整度、适当的抗滑能力,还要求行车时不产生过大的扬尘现象、减少对路面和车辆机件的损坏。

③ 排水系统主要是为了确保路基稳定,避免地面水及地下水等自然水的冲刷、侵蚀。它主要包括纵向排水和横向排水两个系统。

④ 沿线设施是道路沿线交通安全、管理、服务以及环保设施的总称,主要包括交通安全设施、交通管理设施、停车设施、绿化等。

Road is generally composed of roadbed, pavement, drainage system and other facilities, as shown in Table 5.2.

① Roadbed is the foundation of driving part of road. It bears the vehicle load from the pavement, and it is made up of soil and stone in accordance with the requirements of certain size and structure. Roadbed must be stable and solid.

② Pavement is the driving part of the road, paved on the roadbed, and is made up of all kinds of road materials, in order to have a safe and comfort driving. Pavement must meet the demands of strength, stability, flatness and appropriate anti-sliding ability. Besides, less dust is required to reduce the damage of pavement, and vehicle parts.

③ In order to ensure the roadbed steady and avoid the roadbed scoured or eroded by water, drainage system is set in the road system, mainly including both horizontal and vertical drainage system.

④ Other facilities are composed of traffic safety, management, service and environment protection facilities.

表5.2　城市道路组成
Table 5.2　Components of urban road

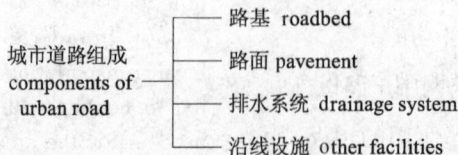

城市道路组成 components of urban road	路基 roadbed
	路面 pavement
	排水系统 drainage system
	沿线设施 other facilities

路面是道路最重要的组成部分,按路面材料的不同可分为沥青路面、水泥混凝土路面和其他路面,其中沥青路面为柔性路面,水泥混凝土路面为刚性路面,如图5.2、图5.3 所示。

Pavement is the most important part of the road, and it can be divided into asphalt pavement, concrete pavement and others according to the nature of the pavement materials. Asphalt pavement is flexible, and concrete pavement is rigid, as shown in Fig. 5.2 and Fig. 5.3.

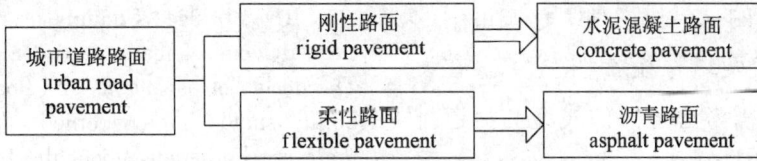

图 5.2　城市道路路面分类
Fig. 5.2　Classification of urban road pavement

(a) 沥青路面
Asphalt pavement

(b) 水泥混凝土路面
Concrete pavement

图 5.3　道路路面
Fig. 5.3　Urban road pavement

5.1.2　城市桥梁

城市桥梁指在城市范围内,修建在河道上的桥梁和道路与道路立交、道路跨越铁路的立交桥及人行天桥,包括永久性桥和半永久性桥。城市桥梁是城市道路的重要组成部分,如图 5.4 所示。每种桥型均由上部结构(支座以上的部分,包括桥跨结构和桥面构造)、下部结构(支座以下的部分,包括桥墩、桥台以及墩台基础)和支座三部分组成。

5.1.2　Municipal bridge

Municipal bridges refer to bridges, overpasses, pedestrian bridge above rivers, highways and railways, including permanent and semi-permanent ones, which is the essential part of the city's transport system (Fig. 5.4). Each kind of bridge consists of upper structure (above supports, including bridge structure and bridge deck), substructure (including piers, abutments and pier foundations), and supports.

(a) 跨河桥 River bridge　　　(b) 立交桥 Overpass　　　(c) 人行天桥 Pedestrian bridge

图 5.4　城市桥梁类型
Fig. 5.4　Classifications of city bridge

市政工程 Municipal Engineering

跨河桥的主要作用是跨越城市中的河流,将河流两端的道路连接起来。在设计跨河桥时需考虑道路的线型和河道排泄洪水的能力,对桥梁高度和下部的净空进行限制。

立交桥可以使平交路口的车流在不同高程上跨越,从空间上分开,各行其道,互不干扰,大大提高了车速和路口的通行能力。它可以充分利用城市空间,是大中型城市的一种重要交通方式。立交桥设计时需综合考虑路、桥之间的协调配合和桥下净空的要求。

人行天桥一般建造在车流量大、行人稠密的地段,或者交叉口、广场及铁路的上方。人行天桥只允许行人通过,用于避免车流和人流平面相交时的冲突,保障人们安全穿越,提高车速,减少交通事故。

River bridge is mainly used to connect two adjacent elements between two rivers. The design of pavements and flood discharge ability should be concerned. In addition, there are limitations about the bridge height and space under the bridge.

Overpass is a crossing of two highways or railroad at different levels where clearance to traffic on the lower level is obtained by elevating the higher level. Overpass benefits citizens and escapes traffic jams. The combination of highway and bridge as well as the limitations of underclearance need to be considered.

Pedestrian bridge is generally set up crowed areas, or intersections, squares and the railways. Pedestrian bridge is designed only for pedestrians to ensure citizens' safety, and benefit traffic.

5.2 城市给排水工程

5.2 Urban water supply and wastewater engineering

人类的生产、生活离不开水。可用水的及时供应与废水的及时排出是城镇得以正常运转的重要保障,而实现这一目的的一个重要条件就是设置给排水管道,其工程质量不仅影响城镇功能的充分发挥,而且对人居健康、道路交通、水环境保护和城市安全都有直接的影响。城市内的水循环如图 5.5 所示。

Our daily life and industrial production are inseparable from water. The installation of drainage pipelines is one of measurements to guarantee water supply and wastewater discharge timely. The quality of project has direct relationship with health of the residence, transportation, environment and city safety. Urban water cycle is shown in Fig. 5.5.

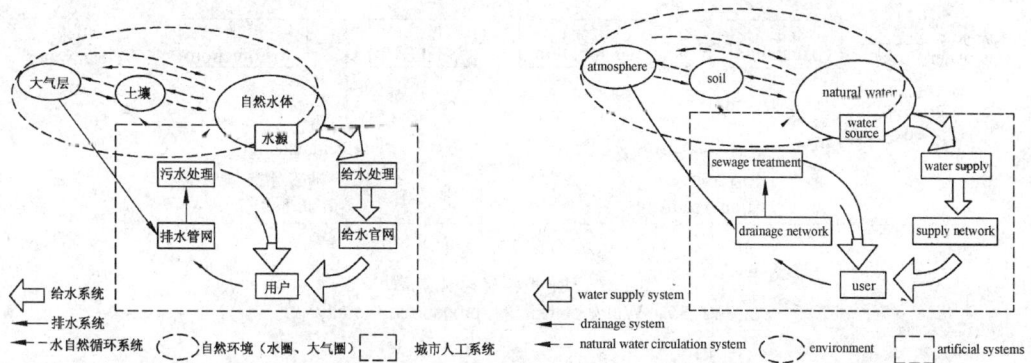

图 5.5　城市水循环示意图

Fig. 5.5　Schematic of water cycle in the city

5.2.1　城市给水工程

给水工程的基本任务是安全可靠、经济合理地供应城乡人民生活、工业生产、保安防火、交通运输、建筑工程、公共设施、军事建设等各项用水,满足用户对水量、水质和水压的要求。给水工程主要由给水水源、取水构筑物、原水管道、给水处理厂和给水管网组成,同时具有收集和输送原水、改善水质的作用,城市给水系统如图 5.6 所示,水质净化流程如图 5.7 所示。

5.2.1　Water supply engineering

The main function of water supply engineering is to ensure water supply for living, industrial production, fire security, transportation, public facilities, military construction etc. The supply must be safe, reliable, economical. Meantime, water supply engineering should meet the local requirements about water quality and relative properties. Generally, a whole water supply engineering consists of water source, water intake structures, water pipelines, treatment plants and water distribution network, and this engineering system can collect water, transport water, improve water quality. The sketch map is shown in Fig. 5.6 and Fig. 5.7.

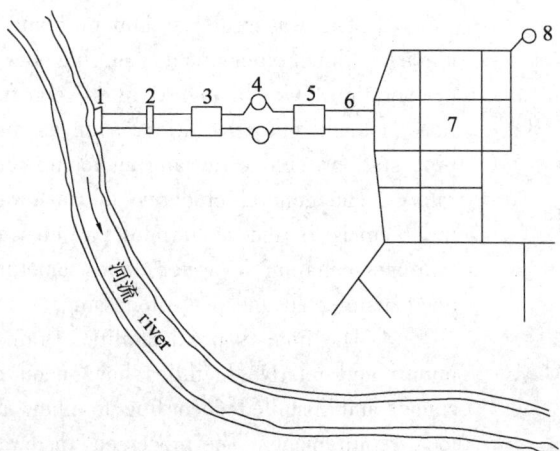

1—取水构筑物 water intake structure;

2—一级泵站 first pumping station;

3—水处理构筑物 treatment structure;

4—清水池 clean water tank;

5—二级泵站 second pumping station;

6—输水管 pipe;

7—管网 pipe network;

8—调节构筑物 regulating structure

图 5.6　城市给水系统示意图

Fig. 5.6　Schematic of water supply system in the city

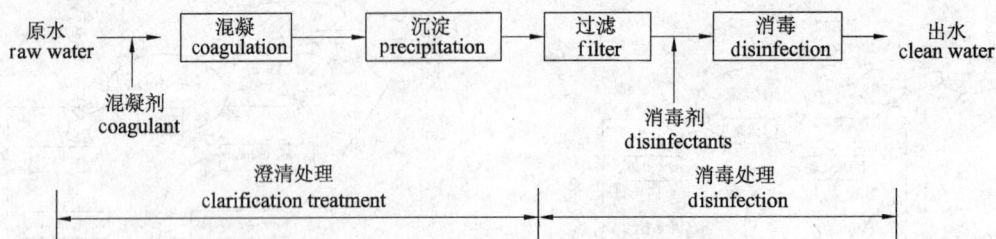

图 5.7　城市水质净化流程
Fig. 5.7　Water purification process in the city

5.2.2　城市排水工程

排水工程为排除人类生活污水和生产中的各种废水及多余地面水的工程,由排水管系(或沟道)、废水处理厂和最终处理设施组成,通常还包括抽升设施(如排水泵站)。

① 排水管系是收集和输送废水(污水)的管网,有合流管系和分流管系两种。合流管系只有一个排水系统,雨水和污水用同一管道排出。分流管系有两个排水系统,雨水系统收集雨水和冷却水等污染程度很低、不经过处理可直接排入水体的工业废水,其管道称雨水管道;污水系统收集生活污水及需要处理后才能排入水体的工业废水,其管道称污水管道。

② 废水处理厂包括沉淀池、沉沙池、曝气池、生物滤池、澄清池等设施及泵站、化验室、污泥脱水机房、修理工厂等建筑。废水处理的一般目标是去除悬浮物和改善耗氧性,有时还进行消毒和进一步处理。

③ 最终处理设施视不同的排水对象设有水泵或其他提水机械,将经过处理厂处理满足规定排放要求的废水排入水体

5.2.2　Urban drainage engineering

The drainage works are designed to eliminate wastewater and graywater, and puddles excess on the ground. Their main components are drainage system (or channels), wastewater treatment plant and disposal facilities, pumps (such as drainage pumps).

① The drainage system is used to collect and transport wastewater or sewage, which can be divided into combined systems and separate systems. Combined system is composed of a single drainage pipe to pour over rainwater and sewage together. While the separate system can be divided into two drainage systems: rainwater collections, which can be used to collect rainwater, chilled water, and wastewater with fewer chemical particles; sewage collections, which can be used to collect wastewater, graywater after treatment.

② The wastewater treatment plant includes sedimentation pool, settling basin, the aeration tank, biological filter, clarification pool and other facilities, such as pumping station, sludge dewatering room, laboratory. The general objective of wastewater treatment is to remove harmful particles and improve oxidation in water, and sometimes need disinfected and post-processing.

③ The final disposal facility. Drainage pumps and relative facilities have been designed and installed according to plans and local requirements. The processed wastewater is discharged into water or land when it meets

或土壤。

常用的排水管渠有钢筋混凝土管、陶土管、金属管、浆砌砖石管渠等。另外，为了排除污水，除管渠本身外，还需在管渠系统上设置某些附属构筑物，包括雨水口、连接暗井、检查井、跌水井、水封井、倒虹管、冲洗井、防潮门、出水口等。排水管渠在建成通水后，为保证其正常工作，必须经常进行养护和管理。排水管渠内常见污物淤塞管道、过重的外荷载、地基不均匀沉陷或污水的侵蚀作用等现象，使管渠损坏、开裂或被腐蚀。

5.3 城市燃(煤)气管道和热力管道安装工程

5.3.1 城市燃(煤)气工程

城市燃气输配系统按照用气量和所需压力将燃气输送和分配到城市各类用户。城市燃气输配系统一般由门站、燃气管网、储气设备、调压设施、管理设施、监控设施组成，如图5.8、图5.9所示。城市燃气输配系统的各种设施，应能满足各类用户的小时最大用气量，并能适应其波动情况。输配系统应该保证不间断地、可靠地给用户供气，运行管理安全，维修检测方便。此外，还应考虑在检修或发生故障时，可关断某些部分管段而不致影响全系统的工作。在一个输配系统中，宜采用标准化和系列化的站室、构筑物和设备，采用的系统方案应具有最大的经济效益，并

specifications and requirements.

Sewer pipes are usually made of reinforced concrete, clay, metal, bricks etc. Moreover, some auxiliaries have been installed for the purpose of eliminating the sewage, including rain inlets, connecting staples, inspection wells, drop wells, water wells, inverted siphon, flushing well, moisture-proof valves, and water outlets. In order to guarantee the performance, maintenance and management are needed after construction. The common failures of drainage systems are dirt blockage, external loads, uneven subsidence of foundation caused by overweight or water erosion.

5.3 Urban gas pipelines and heat pipelines installations

5.3.1 Urban gas supply engineering

Urban gas distribution system is designed for gas transportation and distribution in accordance with the required volume and pressure. Generally, the urban gas distribution system consists of valves, gas pipelines, storage tanks, pressure regulating facilities, management facilities, monitoring facilities, as shown in Fig. 5.8 and Fig. 5.9. All kinds of facilities of the urban gas distribution system should be able to ensure users' maximum hourly consumption, and can be adapted to changes of consumption. Gas distribution should be carried on continuously and efficiently. The management and maintenance of gas distribution system should be considered. In addition, turning off some part of the pipes cannot affect the whole performance of the system when maintenance, repair works needed. In a distribution system, serialized station room and equipment should be standardized. The design of system should

市政工程 Municipal Engineering

eliminate the costs, and can be construction and put into operation in stages.

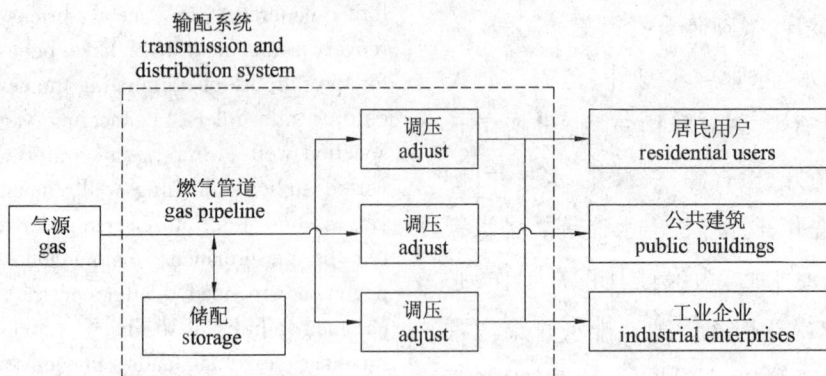

输配系统
transmission and
distribution system

气源
gas → 燃气管道
gas pipeline

储配
storage

调压
adjust → 居民用户
residential users

调压
adjust → 公共建筑
public buildings

调压
adjust → 工业企业
industrial enterprises

图 5.8　城市燃气供应示意图
Fig. 5.8　Schematic of gas supply system in the city

图 5.9　城市燃气管道
Fig. 5.9　Gas pipelines in the city

与其他管道相比,燃气管道有特别严格的要求,因为管道漏气可能导致火灾、爆炸、中毒等事故。燃气管道的压力越高,管道接头脱开、管道本身出现裂缝的可能性也越大。管道内燃气压力不同时,对管材、安装质量、检验标准及运行管理等要求亦不相同。我国城镇燃气管道按燃气设计压力(MPa)分为七级(见表 5.3)。

Compared with other pipelines, gas pipelines are particularly strict in case of fire, explosion, poisoning, etc. caused by gas leakage. The pressure of gas pipelines has a direct relationship with the performance of connections and pipes. The work of installation, construction, inspection and management should keep in pace with the standards and requirements according to types and pressure. In China, gas pipeline are divided into seven levels according to the pressure(Table 5.3).

土木工程导论

表 5.3 城镇燃气管道设计压力分级
Table 5.3 Ratings of designed pressure of gas pipeline in the city

名　称 Name		压　强 Pressure p/MPa
高压燃气管道 high-pressure gas pipeline	A	$2.5 < p \leqslant 4.0$
	B	$1.6 < p \leqslant 2.5$
次高压燃气管道 secondary-pressure gas pipeline	A	$0.8 < p \leqslant 1.6$
	B	$0.4 < p \leqslant 0.8$
中压燃气管道 medium pressure gas pipeline	A	$0.2 < p \leqslant 0.4$
	B	$0.01 < p \leqslant 0.2$
低压燃气管道 low-pressure gas pipeline		$p \leqslant 0.01$

燃气输配系统各种压力级别的燃气管道之间应通过调压装置连接。当有可能超过最大允许工作压力时,应设置防止管道超压的安全保护设备。

5.3.2 城市热力管道工程

热力管道是输送蒸汽或过热水等热能介质的管道,如图5.10所示。热力管道的特点是其输送的介质温度高、压力大、流速快,在运行时会给管道带来较大的膨胀力和冲击力。因此在管道安装中应解决好管道材质、管道伸缩补偿、管道支吊架、管道坡度及疏排水、放气装置等问题,以确保管道的安全运行。

Gas distribution system is needed to balance different pressure levels among pipes. Sometimes, safety protection device should be installed to prevent overpressure in the pipelines.

5.3.2 Urban heat pipeline installation engineering

Heat pipe is designed to transport steam or hot water, as shown in Fig. 5.10. The characteristics of heat pipe are high temperature, high pressure and high flow rate of the transported medium which accelerate expansion force and impact force of pipes. Therefore, materials, expansion compensation, supports, the slope and drainage, bleeding device of pipelines should be taken into account for the purpose of safe operation.

图 5.10　城市热力管道
Fig. 5.10　Heat pipe in the city

热力管道的敷设分为地上敷设和地下敷设,通常选用钢管,并尽量采用焊接连接;安装时,水平管道要具有一定的坡度。蒸汽管道的坡向最好与介质流向相同,这样管内蒸汽同凝结水流动方向相同,避免噪声。热水管道的坡向最好与介质流向相反,这样管内热水与气流方向相同,减少了热力流动的阻力,也有利于排气,防止噪声。热力管道的每段管道最高点或最低点分别安装排气和泄水装置。方形补偿器水平安装时,与管道坡度和坡向一致;垂直安装时,最高点应安装排气阀,最低点应安装排水阀,便于排水与放气。热力管道的排水、放气装置如图 5.11 所示。

The heat pipes are usually located on the ground or underground, and the pipes are made of steel. The method of connection is welding, and a certain slope is needed when horizontal pipelines are installed. Pipe slope of the steam should match the medium flow dimension, so that it can eliminate noise and make steam and water flow in the same direction. Pipe slope of the hot water should be on the opposite of the medium flowing, so that flow direction of the hot water and air are the same which could reduce thermal resistance, discharge exhaust and eliminate noise. The exhaust outlet device should be installed at the highest point and the lowest point of pipes respectively. Horizontal installation of square compensator should consist with pipe slope; exhaust valves should be installed at the highest point, and drain valves should be installed at bottom in the vertical installation. The distribution of pipes is shown in Fig. 5.11.

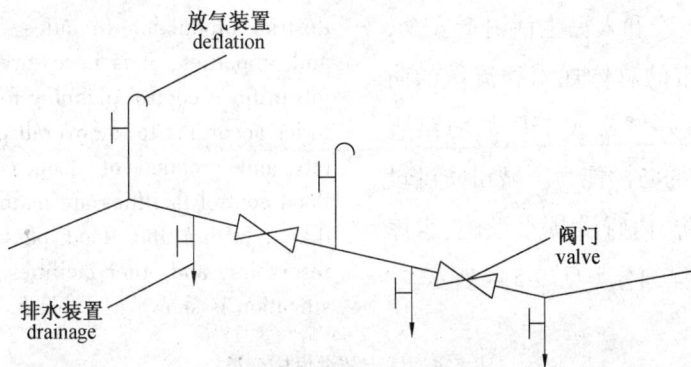

图 5.11　热力管道的排水、放气装置
Fig. 5.11　Drainage and deflation in heat pipe

热力管道的温度变形要充分利用转角管道进行自然补偿,如图 5.12 所示。自然补偿不能满足要求时,要加设补偿器补偿,常用的补偿器有方形补偿器、波形补偿器、套管式补偿器、球形补偿器等,使用时要按照敷设条件采用维修工作量小且价格较低的补偿器。

The sharp directions and corners of pipes can eliminate deformation caused by temperature, as shown in Fig. 5.12. Additional compensators are needed when natural compensation cannot meet requirements. Square compensator, wave compensator, sleeve compensator, and spherical compensator are commonly used according to the conditions and limitations.

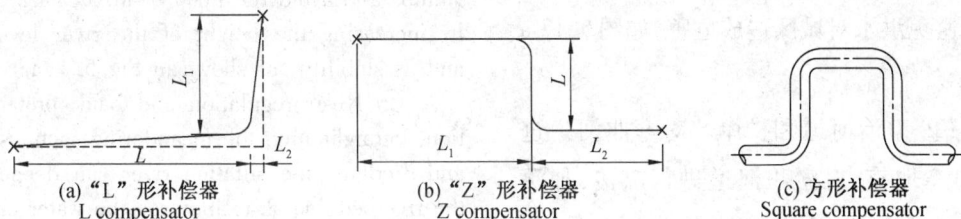

(a) "L" 形补偿器
L compensator

(b) "Z" 形补偿器
Z compensator

(c) 方形补偿器
Square compensator

图 5.12　热力管道补偿器
Fig. 5.12　Heat pipe compensator

5.4　城市防洪和防汛工程

5.4　Urban flood control projects

洪灾是一个十分复杂的灾害系统,它的诱发因素极为广泛,水系泛滥、风暴、地震、火山爆发、海啸等都可以引发洪水。在各种自然灾难中,洪水造成死亡的人口占因自然灾难死亡人口总数的 75% ,经济损失占到因自然灾难造成全部损失的40% 。为了尽量减少洪水造成的危害,保

The flood disaster is a very complex system, because of its extremely broad predisposing factors, such as flooding, storm, earthquake, volcanic eruptions and tsunami. In all natural disasters, death of the population due to floods accounted for 75% , and economic losses accounted for 40% of the total loss. In order to minimize the harm caused by the floods, and to protect the in-

护城市的工业生产和人民生活财产安全，有必要根据城市的总体规划和流域的防洪规划，认真做好城市或工厂的防洪规划，以提高城市的防洪能力。城市防洪设施主要包括堤防、内行洪排水设施、水库及其他设施，其详细分类见表5.4所示。

dustrial production in cities, people's lives and properties, it is necessary to do a good job in flood control planning for cities or factories according to the overall planning of the city and program of flood control. Urban flood control facilities are mainly include the dikes, within the flood passage drainage, reservoirs, and other facilities, and its classification is shown in Table 5.4.

表5.4 城市防洪设施分类
Table 5.4 Types of urban flood control facilities

城市防洪设施
urban flood
control facilities
- 堤防工程 dike project
- 整治河道护岸 river regulation and bank protection
- 设置防洪闸 floodgates
- 分(蓄)洪区、水库 flood diversion area, reservoir
- 生物工程 biological engineering
- 山洪拦蓄工程 flash floods impounding works
- 排涝措施 drainage measures
- 其他非工程措施 non-engineering measures

① 堤防工程。通过增加河流两岸大堤的高度和稳定性，提高河道安全泄洪量，避免洪水对城区造成危害，如图5.13 a所示。

② 整治河道和护岸。对弯曲河道进行裁弯取直，对淤积河道进行疏浚，加深河床以加大河道过水能力，降低水位，缩短河流里程。在河岸因水流冲刷容易造成河岸坍塌、影响河岸稳定和建筑物安全的地段采用护岸措施，如图5.13 b所示。

③ 防洪闸。河口城市和临江河城市，汛期外水水位高，往往形成江(河、湖、海)水倒灌，影响河流泄洪而造成洪涝灾害。在下河流出口处设防洪闸，是防止洪水、海水倒灌的一个重要措施，如图5.13 c所示。

④ 分(蓄)洪区和水库。在流经城市的河流上游修建水库拦洪蓄水，或将洪水

① Dike project. The purpose of dike project is to increase the safety discharge volume and avoid the flood of urban hazards by increasing the height of the river levee and its stability, as shown in Fig. 5.13a.

② River regulation and bank protection. Straightening the meandering sections, and dredging the siltation river can deepen the riverbed, so as to increase the water capacity of the river, lower the water level, shorten the river mileage. The riparian water erosion is to cause river bank collapse, and revetment measures taken in these lots will affect the river bank stability and building safety, as shown in Fig. 5.13b.

③ Floodgates. In the estuary and riverside city, the high water level of the flood results in the intrusion of the river (river, lake, sea) water, Which affects river flood discharge and causes the floods. Setting floodgate in the downstream can prevent floods, Which is an important measure, as shown in Fig. 5.13c.

④ Flood diversion and reservoirs are

引入低洼地,或用分洪道分洪,均可减小下游城市的洪水压力。

⑤ 生物工程措施。结合小流域治理,在流域上植树种草,增加流域下渗蓄水能力,从而减少进入河道中的径流和泥沙,起到蓄水防洪作用,如图 5.13 d 所示。

⑥ 山洪和泥石流的拦蓄、排导工程。在山坡上修建谷坊、塘堰、梯田,可以拦截泥沙、减少山洪危害,同时避免诱导泥石流发生。修建排洪沟、泥石流排导沟,可将山洪和泥石流引导至保护区范围以外。

⑦ 排涝措施。城市内涝可通过修建管渠排涝,一般采用自流排泄、高水排泄的方法,以上方法不能解决时修建泵站抽排,如图 5.13 e 所示。

⑧ 其他非工程措施。如洪水预报、洪水警报、蓄滞洪区管理、洪水保险、河道清障、灾后救济等,如图 5.13 f 所示。

build to control flood, or drawing the flood into low-lying land and using flood diversion in the upper reaches of the river can reduce the flood pressure.

⑤ Bio-engineering measures. Through small watershed management and planting trees and grass in the basin, you can increase the infiltration basin water storage capacity, thereby reducing runoff and sediment into the river, as shown in Fig. 5.13 d.

⑥ Storing and exhaust flash floods and mudslides engineering. Construction of check dams, ponds, terraces on the hillside can intercept sediment and reduce flash flood hazards, as well as avoid the occurrence of induced mudslides. The construction of flood discharge trench and the mudslides exhaust ditch can lead to flash floods and landslides outside the protected areas.

⑦ Drainage measures. General construction of pipe drainage can solve urban water logging. Using gravity excretion, high water excretion can achieve their goals. If the problems cannot be solved through above methods, we will build a pumping station as shown in Fig. 5.13 e.

⑧ Other non-engineering measures, such as flood forecasting, flood warning, flood storage management, flood insurance, river wrecker, post-disaster relief, are shown in Fig. 5.13 f.

(a) 城市防洪堤
Levee

(b) 河道整治
River training

(c) 城市防洪闸
Floodgate

(d) 生物工程防洪
Bio-engineering flood control

(e) 管渠排涝
Tube drainage drainage

(f) 洪水预报系统
Flood forecasting system

图 5.13　城市防洪设施
Fig. 5.13　Urban flood control facilities

同时,完善配套的城市防洪排涝设施是城市经济持续快速发展的重要保障。防洪工作需要与自然规律协调安排;要付出合理的投入,取得可能获得的最大收益;要进行科学规划,确定城市的防洪标准,修建城市防洪工程体系;保证城市在发生规划标准的常遇和较大洪水时,国家经济和社会活动不受影响;遇到超标准的大洪水和特大洪水时,有预定的分蓄行洪区和防洪措施,国家经济和社会不发生动荡,不影响国家长远计划的完成。

At the same time, the complete set of city flood control and drainage facilities is the important guarantee for the city sustained and rapid economic development. Flood prevention work should harmonize with the laws of nature. Reasonable investment should be made in the premise, so as to obtain the maximum benefit. The flood control standards and the urban flood control system should be made through scientific planning. Make sure that the economic and social aitivities will not be affected, when large flood happened. Building the flood area and flood control measures in case of the large floods and devastating floods, and the economy, social life as well as the completion of the national long-term plan, can not be affected.

城市园林和绿化工程

City garden and green engineering

城市园林和绿化工程是美化生活环境、提高人民身心健康水平的重要工程，是生态建设的重要组成部分，是有生命的基础设施，对保持经济社会发展和改善人民生活质量具有重要作用，已成为衡量城市文明和地区可持续发展能力的重要标志。随着社会经济的发展和人民生活水平的提高，居民对居住环境条件的要求越来越高，园林绿化工程有了更大的发展空间。

The city greening engineering cannot only beautify the living environment, improve people's physical and mental health, but also belongs to an important part of the ecological construction and life infrastructure. It plays an important role in maintaining economic and social development, and has become an important symbol to measure the sustainable development abilities of urban civilization. With the development of economy and the improvement of people's living standards, the living conditions have become higher increasingly, and landscaping work will have more development space.

5.5.1 城市园林工程

高质量、高水平的园林工程建设，既是改善城镇生态环境和建设投资环境的需要，又是人们高质量生存、生活、工作的环境基础。通过园林工程建设，构建完整的绿地系统和优美的园林艺术景观（如图 5.14 所示），是净化空气、防止污染、调节气候、改善生态、美化环境的需要。

5.5.1 Urban garden project

Garden constructions with high-quality and high-level cannot only improve the environment of urban ecology and the construction investment, but also belong to the basis of high-quality living and working environment. A complete green land system and beautiful landscape art are the requirements of cleaning the air, preventing pollution, regulating the climate, improving the ecology and beautifying the environment (Fig. 5.14).

图 5.14　城市园林
Fig. 5.14　Urban gardens

市政工程 Municipal Engineering

园林工程建设是集建筑科学、生物科学、社会科学于一体的综合性科学。现代园林工程建设学科已发展成为多学科边缘交叉的一门前沿科学体系。其设计主要包括如下几个部分(见表5.5):①园林地形工程设计,主要是根据园林性质和规划要求,因地制宜、因情制宜地塑造地形。②园路工程设计,即在园林中确定园路布局及园路结构设计的过程。③园林的给排水工程建设设计,主要是进行园林中的给水管网、排水系统和排水设施的设计。④园林植物造景工程设计,主要是进行园林中的植物种类、间距、形式的布置。⑤园林绿地喷灌工程设计,主要进行喷头的选型、管网的布置及灌水制度的制定等。⑥园林水景工程设计,主要是各种人工水体的营造设计,如湖泊、池塘、泉水等。⑦园林假山、置石工程设计,是综合运用力学、材料学、工程学及艺术学的知识再造自然山石的过程。⑧园林供电工程设计,主要是对园林输配电、照明及其他用电设备的设计和配备。⑨园林建筑、小品工程设计,主要是对园林中的景观建筑(如亭、廊、榭)和小型设施(如园椅、园凳、栏杆、小型雕塑)等进行的设计。

Garden construction is an integrated science, which combines building science, biological science, and social science. Modern landscape construction disciplines have developed into a frontier science system with the cross of many disciplines. Its design mainly includes the following parts (Table 5.5). ① Garden terrain design is to design the garden according to the nature of the landscape, the planning requirements, and the local conditions. ② Park road design is to determine the layout and the structures of the park road. ③ Garden water supply and drainage construction design is to design the main water supply network, drainage systems and drainage facilities in the garden. ④ Plant landscape design is to arrange garden plant species, spaces and forms. ⑤ Garden green irrigation is to select the nozzle, make the pipe network layout and make the irrigation system. ⑥ Garden waterfront design is to create a variety of artificial bodies, such as lakes, ponds, springs. ⑦ Design of garden rockery, stone home is to use mechanics, materials science, engineering and art learned recycling process of natural rocks. ⑧ Design of garden power supply focuses on power transmission and distribution, lighting and other designed and electrical equipment. ⑨ Landscape architecture and designed sketches focus on architecture (such as pavilions, corridors, and pavilion) and small facilities (like garden chairs, garden stools, railings, small sculptures).

表 5.5　城市园林工程的内容
Table 5.5　Contents in urban garden project

```
                              ┌── 地形工程  terrain engineering
                              ├── 园路工程  park road engineering
                              ├── 给排水工程  watersupply and drainage engineering
                              ├── 植物造景工程  plant landscape
园林工程                       ├── 绿地喷灌工程  green irrigation engineering
urban garden project ─────────┤
                              ├── 水景工程  waterfront engineering
                              ├── 假山置石工程  rockery,stone  engineering
                              ├── 供电工程  power engineering
                              └── 建筑、小品工程  landscape architecture
```

园林绿化工程建设的施工顺序,一般是先整理山水,改造地形,辟筑道路,铺装场地,营造建筑,构造工程设施,然后实施绿化,如图 5.15 所示。

The landscape construction is arranged in the sequence of arranging landscape, transforming terrain, making roads, paving sites, building, constructing facilities and greening (Fig. 5. 15).

图 5.15　园林绿化工程施工顺序
Fig. 5.15　Construction order of urban garden project

由于构成园林的要素极其复杂,既有地形、给排水、供电等工程方面的知识,又有植物、造景设计等生物方面的知识,还有各构成要素的布局、景观营造、色彩搭配等艺术方面的知识,所以园林工程的设计需要综合考虑上述各因素的影响进行详细规划。

The element of landscape is extremely complex, which contains topography, drainage, electricity and other engineering aspects, the plant landscaping design and other biological engineering, and some arts knowledge such as layout, landscape construction, and color matching. Therefore, a detailed consideration about the factors need to be made in landscape engineering design.

5.5.2　城市绿化工程

城市绿化是在城市中进行的,为提高城市居民生活质量,优化城市工作和生活环境,协调城市生态并创造优雅城市面貌,使其在一种健康和谐的基础上持续发展的种植植物的行为,是城市现代化建设的重要内容。城市绿化主要有人行道绿化带、分车绿化带、防护绿化带、基础绿化带、城市广场及公共建筑前的绿化、街头休息绿地、停车场绿化、立体交叉绿化、滨河路绿化、花园林荫路、建筑物绿化等,如图 5.16 所示。

5.5.2　Urban greening projects

Urban greening is to improve the living quality of urban residents, optimize the urban working and living environment, coordinate urban ecology, and create an elegant appearance. Urban greening project can make the city more healthy and harmonious on the basis of the sustainable development of the growing plants behavior, and it is an important part of urban modernization. The basic content of urban greening is sidewalk green, the drive green, green protection, basic green, city squares and public buildings green, street rest green, parking green, cloverleaf green, riverside road green, garden avenue building green, etc, as shown in Fig. 5.16.

(a) 道路绿化
Road green

(b) 建筑物绿化
Building green

(c) 停车场绿化
Parking green

(d) 广场绿化
Square green

图 5.16 城市绿化
Fig. 5.16 Urban greening

城市绿化工程需要定期维护,如浇水、清理垃圾、拔杂草、施肥、修剪、病虫害防治等。在当前资源紧张,环境污染较为严重的情况下,城市园林绿化应以最少的用地、最少的用水、最少的投资,选择对周围生态环境干扰最少的绿化模式,为城市居民提供最高效的生态保障系统。

Urban greening projects require regular maintenance, such as watering, cleaning up, pulling weeds, fertilization, pruning, and pest control. In the case of resource constraints and the more serious environmental pollution, the urban landscape should use the least amount of land, minimal water, investment, and disruption to the surrounding ecological environment, so as to provide the most efficient ecological protection systems for people.

5.6 市政工程发展前沿

5.6 Frontier development of municipal engineering

在城市的建设中,市政工程建设居于

Municipal engineering construction is a

主要地位,它代表着城市发展的主要趋势和形象,代表着城市居民的生活水平和精神面貌,因此根据社会的发展趋势对市政工程中所使用的材料、施工方法及管理措施进行更新是很有必要的。目前市政工程发展主要向以下几个方向推进。

5.6.1 新型材料在市政工程中的推广应用

随着社会的发展,传统材料在市政工程中可能会出现一些不利的影响,研发并推广新型材料是市政工程发展的首要问题,如高密度聚乙烯(HDPE)管在给排水工程中的应用(如图 5.17 所示),低频无极灯在城市道路照明中的应用等。

predominant aspect in the city's construction, and it represents the main trends of urban development, the standard living of urban residents and their mental outlook. It is necessary to update the materials used in municipal engineering, construction methods and management measures in accordance with the trend of social development. Currently, the development of municipal engineering contains following aspects.

5.6.1 The promotion and application of new materials in municipal engineering

With the development of society, the traditional materials in municipal engineering have some adverse effects. Researching and promoting the new materials is the primary issue of the municipal engineering development, such as the application of high-density polyethylene (HDPE) pipe in drainage (Fig. 5.17), and the induction lamp in city road lighting.

图 5.17　高密度聚乙烯管(HDPE)
Fig. 5.17　High-density polyethylene pipe

5.6.2 市政工程施工机械及方法的更新

市政工程在施工过程中会影响城市

5.6.2 Update of municipal engineering construction machinery and methods

Municipal engineering in the process of

的正常运行,如给水管道的维修会暂时停止供水、城市道路的改造会增加交通的拥堵和环境污染等。通过研发新的设备使市政工程的施工过程更绿色、更高效,且尽可能实现数字化是未来市政工程发展的另一个方向。

construction will affect the normal operation of the city. For example, the maintenance of the water supply pipeline will temporarily cut off the water supply, the transformation of urban roads will increase traffic congestion and environmental pollution. By the research and development of new equipment, municipal engineering construction process may become green, more efficient, and digitization as much as possible is another direction of the future development in municipal engineering.

5.6.3 市政设施管理的系统化及网络化

随着计算机和通信技术的发展,信息技术在市政工程管理过程中的作用越来越重要,通过开发市政设施管理的计算机集成应用系统并实现网络化,进而实现城市市政"无缝隙"、"精细化"管理,是未来市政设施管理的发展方向。

5.6.3 Systematic management of municipal facilities and network

With the development of computer and communication technology, the information technology plays an important role in the process of municipal engineering administration. Making the computer integrated application system through the development of municipal facilities management and network is the future development direction of the municipal facilities management and make the city municipal "seamless" and "fine".

注:本章图片均来源于网络。
Note: In this chapter, all pictures are from webs.

知识拓展
Learning More

相关链接 Related Links

如果想了解市政工程的详细知识、最新发展态势及相关政策,可访问中国市政工程网 http://www. zgsz. org. cn/。

If you want to obtain a detailed knowledge of municipal engineering, the latest development trend and related policies, please visit http://www. zgsz. org. cn/.

小贴士 Tips

市政工程的注册师制度:国内与市政工程有关的注册师有注册建造师、注册造价工程师、注册监理工程师、注册电气工程师、注册给排水工程师等,通过考试后可从事相应市政工程专业的设计、施工、管理等工作。

Registration system for municipal engineering: the registration division of municipal engineering include registered architect, registered cost engineer, registered supervision engineers, registered electrical engineer. They drainage engineers, registered by the appropriate examinations, and can be engaged in the design of mu-

土木工程导论

nicipal engineering, construction and management.

思考题　Review Questions

(1) 市政工程包括哪些内容?

What does municipal engineering include?

(2) 城市道路和桥梁可分为哪些类型?

What types can the urban roads and bridges be divided?

(3) 城市给排水系统的特点是什么?

What are the characteristics of the urban drainage system?

参考文献
References

［1］全国一级建造师执业资格考试用书编写委员会:《市政公用工程管理与实务》,中国建筑工业出版社,2010 年。

［2］毛惟德:《城镇排水工程》,中国建筑工业出版社,2009 年。

［3］蒋柱武,黄天寅:《给排水管道工程》,同济大学出版社,2011 年。

［4］张智:《城镇防洪与雨洪利用》,中国建筑工业出版社,2009 年。

［5］袁海龙:《园林工程设计》,化学工业出版社,2005 年。

［6］王俊安:《园林绿化工程估价》,机械工业出版社,2009 年。

［7］范慧方:《燃气供应》,华中科技大学出版社,2011 年。

［8］刑丽贞:《市政管道施工技术》,化学工业出版社,2004 年。

［9］姚时章,蒋中秋:《城市绿化设计》,重庆大学出版社,2000 年。

第6章 建筑环境与设备工程

Chapter 6 Building Environment and Equipment Engineering

长久以来,人们都渴望有温暖舒适的居住和工作环境。在远古时代,人类的祖先借山洞栖息,躲避风雨严寒。随着时代的进步,科学技术的发展,人们开始有能力建造房屋,为自己寻找更安全可靠的庇护之所。但是仅有一个处所仍然是不够的,人们还希望自己的家冬暖夏凉,能方便地使用水、电等生活设施,使其能在一个舒适的环境中度过愉快的时光,而这正是建筑环境与设备工程专业人员即公用设备工程师所从事的工作。

如今,公用设备工程师的任务已不是简单地控制室内各个环境参数的变化范围,而是应当站在人类可持续发展的高度,从保护环境、节约能源的角度出发,合理利用室外环境,创造出低能耗的各种建筑设备系统,满足人们生活和生产的需要。

For a long time, people are eager to live and work in warm and comfortable environment. In ancient times, our ancestors lived in caves to escape the cold weather. With the development of science and technology, people began to have the ability to build houses which are more secure and reliable for themselves. However, only a place to stay in is not enough. People want to have a comfortable and convenient house, as well as the basic facilities like water, electricity and other living facilities. These equipment and conditions provide a convenient living environment for people, and this is what the building environment and public facility engineers should do.

Nowadays, the tasks of public facility engineer's are not simply to control the variation range of each parameter in the indoor environment, they should stand on the height of sustainable development of mankind. From the view of environmental protection and energy conservation, they should use the outdoor environment rationally, as well as creat various construction equipment systems to meet the needs of living and production.

6.1 建筑环境

建筑的功能是在自然环境不能令人满意的条件下，创造一个微环境来满足居住者的安全以及生活生产的需要。因此从建筑出现开始，"建筑"与"环境"这两个概念就是不可分割的。建筑环境研究包括室内外的温度、湿度、气流、空气品质、采光与照明性能、噪声与室内音质等内容，为营造一个舒适、健康的室内外环境提供理论依据。建筑环境的保证一方面需要合理建筑物理设计，另一方面需要高效的建筑设备。建筑环境与设备工程的主要内容如图6.1所示。

6.1.1 建筑热湿环境

热舒适性是人体通过自身的热平衡条件和对环境的热感觉，经综合判断后得出的主观评价或判断。除了衣着、活动方式等个人因素外，影响人体热平衡进而影响热舒适性的环境因素主要是温度、湿度、气流速度和平均辐射温度，即室内热湿环境。

6.1 Building environment

The function of the building is to form a local environment for its resident's living and production demand, when the natural environment is not satisfying. The concept of "building" and "environment" cannot be separated. The study on building environment includes indoor and outdoor temperature, air flow, air quality, daylighting and lighting, noise and indoor acoustics, and building environment can provide the theoretical guidance for building a comfortable and healthy indoor and outdoor environment. The maintain of suitable building environment need reasonable physical design and effective building equipment. The main contents of building environment and equipment are shown in Fig. 6.1.

6.1.1 Building thermal and humidity environment

Thermal comfort is the subjective evaluation obtained by thermal balance condition, thermal feeling, and personal judgments. Besides dressing, activity and other personal factors, the main environment factors which affect the thermal comfort are temperature, humidity, air flow speed and average radiant temperature, namely, indoor thermal and humidity environment.

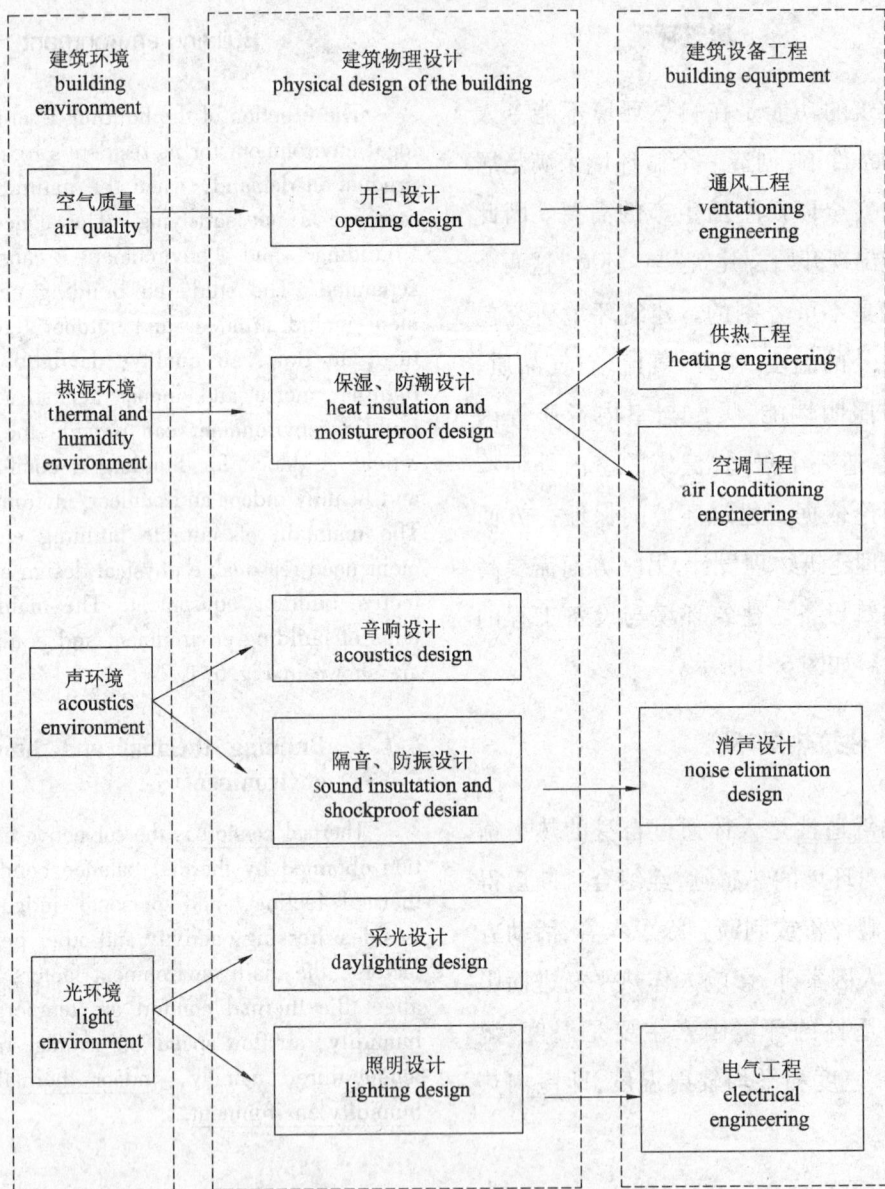

图 6.1 建筑环境与设备工程主要内容

Fig. 6.1 Content of the building environment and equipment

建筑室内热湿环境受室外气象参数与建筑围护结构的影响。通过非透明外围护结构的热传递方式为热传导,通过透明围护结构的热传递方式包括热传导与日射得热两种。通过围护结构的湿传递与室内外水蒸气分压力有关,稳定情况

Indoor thermal and humidity environment are affected by weather conditions and surrounding structures. The heat transfer mode through the envelop enclosure of non-transparent surrounding structure is heat conduction, and the heat transfer mode through the envelop enclosure of transparent surrounding includes heat conduction and solar

土木工程导论

下,单位时间内通过单位面积围护结构的水蒸气量与两侧空气中的水蒸气压力差成正比。除围护结构外,建筑室内热湿环境受室内设备、照明以及人体等室内热湿源的影响。建筑室内热湿环境的形成如图6.2所示。

heat gain. The moisture transfer through the surrounding structure is affected by the indoor and outdoor water vapor pressure, and at the steady state, the quantity of the water vapor transferred through unit area of the surrounding structure during unit time is proportional to the difference of the water vapor pressure in the air of the two sides. Forming of the indoor thermal and humidity environment is shown in Fig. 6.2.

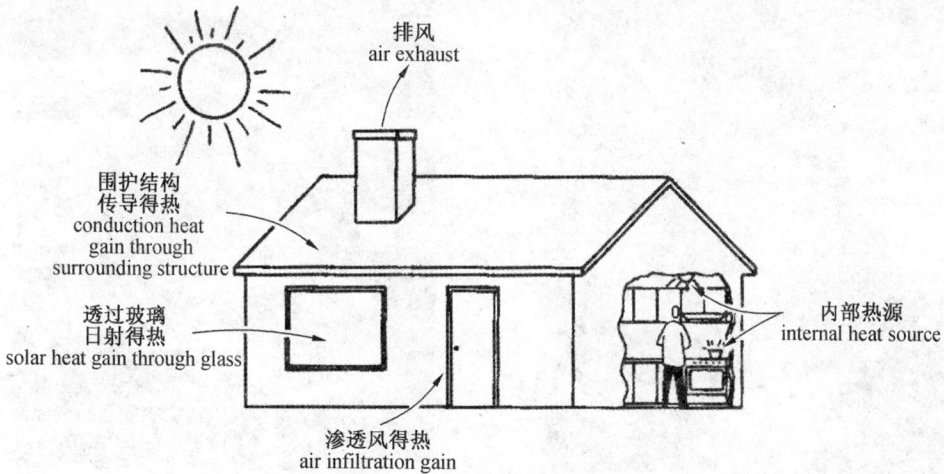

图 6.2　建筑室内热湿环境形成示意图
Fig. 6.2　Diagram of the indoor thermal and humidity environment

为维持舒适的室内热湿环境以及降低暖通空调系统能耗,许多国家均对围护结构的热性能指标以及暖通空调系统室内温湿度设计参数做出了规定。

In order to keep suitable indoor thermal and humidity environment, and decrease the energy consumption of the heating ventilating and air condition (HVAC) system, the thermal performance of the surrounding structure and the design parameters of the HVAC system for temperature and humidity are normalized in many countries.

6.1.2　室内空气品质

随着信息化的发展,越来越多的人长期在建筑内生活、学习和工作。现代化建筑的功能越来越丰富,加之建筑非常封闭,造成室内环境恶化且建筑能耗增加。典型的现代化建筑及其室内办公环境如图6.3、图6.4所示。长期在现代建筑中

6.1.2　Indoor air quality

With the development of information technology, more and more people are living, studying and working in buildings. Modern buildings have more and more functions, but they are very closed, which will deteriorate the indoor environment and increase the energy consumption of these buildings. Typical modern buildings and in-

生活和工作的人群,表现出越来越严重的病态反应,包括眼睛发红、流鼻涕、嗓子疼、困倦、头痛、恶心、头晕、皮肤瘙痒等,这种症状称为病态建筑综合征。

图 6.3　现代化的封闭式建筑
Fig. 6.3　Modern closed buildings

　　室内空气品质问题已引起许多国家、地区和组织的重视,各国先后制定了相关的标准。引起空气品质的问题主要来自暖通空调系统和室内污染物作用两个方面。暖通空调系统方面包括通风与气流组织不好,新风量不足等。室内空气污染源主要包括来自建筑装饰材料、复合木建材及其制品所散发的有机挥发性化合物,灰尘、纤维尘和烟尘等物理污染,以及细菌、真菌和病毒引起的生物污染。

　　室内污染的控制可通过以下 3 种方式实现:① 源头治理;② 通风稀释和合理

door office environment are shown in Fig. 6.3 and Fig. 6.4. People living and working in these modern buildings have more and more serious morbidity, such as red eyes, runny noise, sore throat, sleepy, headache, nausea, dizziness, and skin pruritus. These dislocations are called sick building syndrome.

图 6.4　现代建筑中的办公环境
Fig. 6.4　Working environment in modern buildings

　　Importance has been attached to indoor air quality in many countries, regions and organizations, and its relative standard has been enacted in many countries. The reasons for poor indoor air quality are mainly HVAC system and indoor pollutant. The reasons of the HVAC system include the unreasonable organizing of ventilation and air flow, insufficient free air supply etc. The indoor pollutants mainly come from the volatile organic compound emitted from the construction and decorating material, compositing board and their produce, the physical pollution from the dust, fiber, soot, and the biologic pollution from the bacilli, epiphyte and virus.

　　Three methods can be used to control of the indoor pollution: ① source controlling; ② ventilation diluting and reasonable organi-

组织气流;③ 空气净化。

zing air flow; ③ air purification.

6.1.3 建筑声环境

建筑声环境主要研究室内声音控制问题,包括三方面内容,即室内音质设计、建筑隔声和噪声控制。

音质设计问题一般只限于各类厅堂,如影剧院、音乐厅、体育馆、报告厅等。音质设计的主要内容包括:确定房间容积、体型设计、噪声控制、扩声系统设计等。中国国家大剧院音乐厅的体型设计如图 6.5 所示。

6.1.3 Building acoustic environment

The main content of building acoustic environment is the indoor acoustic controlling, including indoor tone design, sound insulation design and noise control.

Indoor tone design is needed only in hall buildings, such as theater, odium, gymnasium, reporting hall. The indoor tone design mainly includes cubage confirm, shape design, noise control design, sound-reinforcement system, etc. The shape design of the odium of Nation Centre for the Performing Arts in China is shown in Fig. 6.5.

图 6.5 中国国家大剧院音乐厅
Fig. 6.5 The odium of Nation Centre for the Performing Arts in China

随着城市化程度的日益加剧,隔声和噪声控制成为各类建筑面临的一个普遍性问题。噪声主要来源于建筑外部环境,建筑内部其他房间,室内设备噪声,空调通风噪声等。不同噪声需采取不同的控制措施,主要的措施有隔声、吸声与消声。

With the urbanization development, sound insulation and noise controlling have become a common problem in all buildings. The sources of noise include the outer environment of the building, other rooms, indoor equipment, air flow, etc. Different noises need different methods to control. Main noise control methods include insulation, absorption and elimination.

6.1.4 建筑光环境

建筑光环境设计包括建筑采光设计

6.1.4 Building light environment

Building light environment include day-

和建筑照明设计两部分内容。建筑采光设计就是设法通过采光口使光线进入室内。建筑照明设计通过人工光源的应用，改善建筑的功能效益和环境质量，提高人们的视觉功效。随着建筑照明技术的进步和社会经济的发展，公共照明、景观照明和夜景照明正在逐步发展和完善。图6.6、图6.7为采光设计实例。图6.8为人民大会堂顶部LED照明设计。

lighting design and light design. Building daylighting design is about lighting opening design. Building light design is to improve the function and environment of the building, enhance people's vision through artificial light source. With the advancement of lighting technology and the development of economy, public lighting and landscape lighting and landscape lighting have become developed and perfect. Fig. 6.6 and Fig. 6.7 show the example of daylighting design. Fig. 6.8 shows the LED lighting design of the Great Hall of the People.

图 6.6　教室自然采光
Fig. 6.6　Daylighting of a classroom

图 6.7　商场天井采光
Fig. 6.7　Daylighting of a shop through skylight

图 6.8　人民大会堂顶部 LED 照明
Fig. 6.8　Top LED lighting in the Great Hall of the People

6.2　供暖工程

供暖系统的目的是满足人们日常生活和社会生产所需要的大量热能，它是利

6.2　Heating engineering

The purpose of heating system is to meet the requirements of large heat in people's dai-

用热媒(如热水、水蒸气或其他介质)和热力管道将热能从热源输送至各个热用户的工程技术。

6.2.1 供暖系统分类与组成

供暖系统主要由<u>热源</u>、<u>热媒输配</u>和<u>散热设备</u>3个部分组成。根据这3个部分的相互位置关系,供暖系统又可分为局部供暖系统和集中供暖系统。热源、热媒输配、散热设备设置在一起的供暖系统为局部供暖系统;热源、散热设备分别设置,用热媒输送管道相连,由热源向各个部分供给热量的供暖系统为集中供暖系统。目前,集中供热已成为现代化城镇的重要基础设施之一,是城镇公共事业的重要组成部分。另外,按采用热媒方式不同也可将供暖系统分为<u>热水供暖系统</u>、<u>蒸汽供暖系统</u>等。供暖系统的分类及组成如图6.9所示。

ly life and social production. It is an engineering technology using the heat medium (such as hot water, steam and other medium) to transfer the heat from the heat source to each user through the heating pipe line.

6.2.1 Classification and composition of heating system

Heating system is mainly composed of three parts: the heat source, the thermal coal distribution and the cooling device. The heating system can be divided into local heating and central heating system according to the mutual position relations hips among the three parts. The heating system has heat source, heat medium conveying and cooling device on the structure composed the local heating system. The heat source and the cooling device should be set respectively, then connected with the heat medium conveying pipes, so as to transmit heat to each part from the heat source. This system is the central heating system. At present, the central heating has become one of the important infrastructures of modern towns and an important part of urban public utilities. In addition, the heating system can also be divided into hot-water heating system and steam heating system according to the different heat medium using way. The classification and composition of heating system is shown in Fig.6.9.

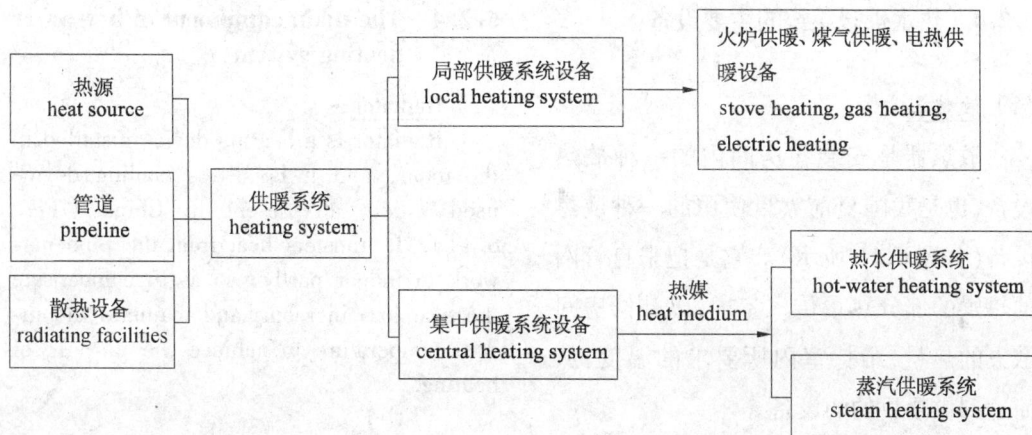

图 6.9 供暖系统的分类及组成
Fig. 6.9 Classification and composition of heating system

6.2.2 热水供暖系统

以热水为热媒的供暖系统，称为热水供暖系统。

6.2.3 热水供暖系统的分类

按照热水供暖循环动力的不同，可分为<u>自然循环系统</u>和<u>机械动力系统</u>（如图6.10所示）。

6.2.2 Hot-water heating system

The heating system using hot water as heat medium is known as the hot-water heating system.

6.2.3 Classification of hot-water heating system

According to the different power supply for hot water cycle, the hot water cycle can be divided into <u>natural circulation system</u> and <u>mechanical system</u> (Fig. 6.10).

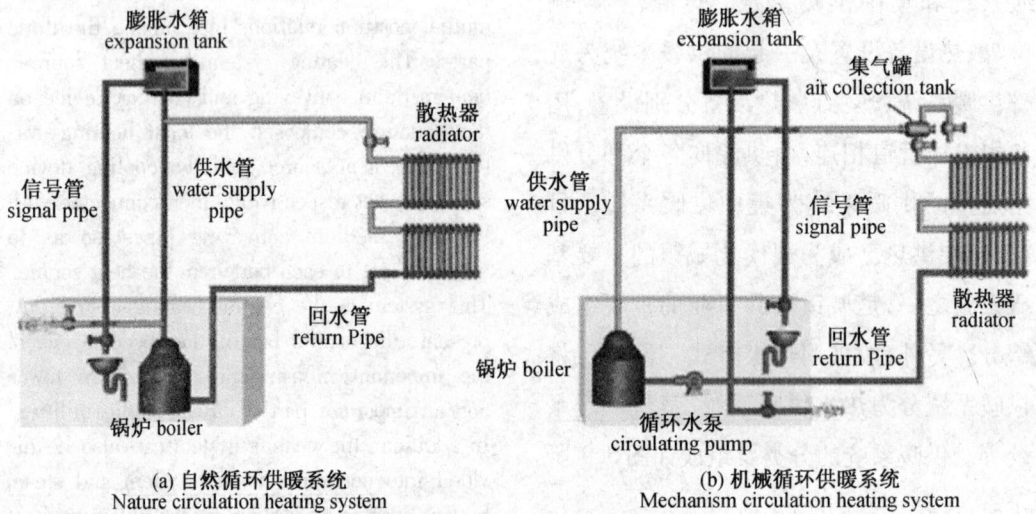

(a) 自然循环供暖系统
Nature circulation heating system

(b) 机械循环供暖系统
Mechanism circulation heating system

图 6.10 热水供暖系统
Fig. 6.10 Hot-water heating system

6.2.4 热水供暖系统的主要设备

（1）<u>散热器</u>

散热器是安装在房间内的一种放热设备，也是我国目前大量使用的一种散热设备（如图6.11所示）。它是把来自管网的热媒的部分热量传入室内，以补偿房间散失的热量，维持室内所要求的温度，从而达到供暖目的的设备。

6.2.4 The main equipment of hot-water heating system

（1）<u>Radiator</u>
Radiator is a heating device installed in the room, and it is also a cooling device used widely at present in China (Fig. 6.11). It transfers heat from the pipe network to indoor partly, so as to compensate the heat loss in rooms and maintain the indoor temperature to achieve the purpose of heating.

图 6.11　散热器
Fig. 6.11　Radiator

(2) 膨胀水箱

膨胀水箱一般安装在系统的最高点，用来容纳系统加热后膨胀的体积水量，并控制系统的充水高度，保证系统压力稳定（如图 6.12 所示）。

(3) 排气设备

排气设备是及时排除供暖系统中空气的重要设备，在不同的系统中可以使用不同的排气设备，如集气罐、自动排气阀（如图 6.13 所示）、手动放气阀等。

(2) Expansion tank

Expansion tank is installed on the highest point of the system, which is used to accommodate the inflated volume of water after the system heated has been and control the water level (Fig. 6.12).

(3) Exhaust equipment

Exhaust equipment is an important equipment to remove the air timely in the heating system. Different exhaust equipment can be used in different systems, such as tank, automatic exhaust steam valve (Fig. 6.13), the manual air bleed valve.

图 6.12　膨胀水箱
Fig. 6.12　Expansion tank

图 6.13　自动排气阀
Fig. 6.13　Automatic exhaust steam valve

(4) 散热器控制阀

散热器控制阀安装在散热器入口管上,是根据室温和给定温度之差自动调节热媒流量的大小来控制散热器散热量的设备(如图6.14所示)。

(4) Radiator control valve

Radiator control valve is the equipment installed on the inlet pipe of a radiator. According to the difference between the room temperature and the given temperature, this device can regulate the rate of the heat medium flow automatically and control the radiator heat release automatically (Fig. 6.14).

图 6.14　温控阀
Fig. 6.14　Temperature control valve

6.2.5　蒸汽供暖系统

以蒸汽为热媒的供暖系统,称为蒸汽供暖系统,其应用极为普遍。图6.15为蒸汽供暖系统示意图,蒸汽从热源(锅炉)沿蒸汽管路进入散热设备(散热器),蒸汽凝结放出热量后,凝水通过疏水器靠重力流至凝结水箱,再靠凝水泵使其返回热源重新加热。

6.2.5　Steam heating system

Steam heating system is the heating system using steam as the heating medium, and it has a very broad application. Fig. 6.15 is a schematic diagram for steam heating system. Steam from the heat source (boiler) flows along the steam pipeline into the heat radiation device (radiator). After releasing condensation heat, steam transforms into condensate water and flows to the condensed water tank through the steam trap of gravity, and then relys on the condensate pump back to the heat source.

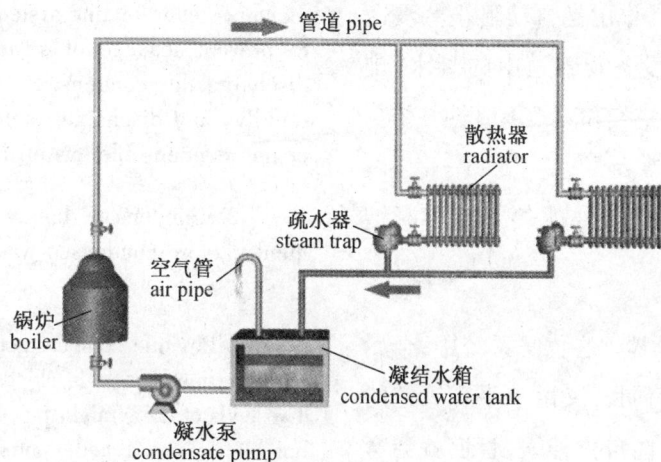

图 6.15　蒸汽供暖系统
Fig. 6.15　Steam heating system

(1) 蒸汽供暖系统的分类

按照供汽压力的大小,蒸汽供暖可分为两类:供汽的表压力 > 70 kPa 时,称为高压蒸汽供暖;供汽的表压力 ≤ 70 kPa 时,称为低压蒸汽供暖。

(2) 蒸汽供暖系统主要设备

蒸汽供暖系统的主要设备如图 6.16 所示。

(1) Classification of steam heating system

According to the steam pressure, steam heating can be divided into two categories, namely supplied steam gauge pressure > 70 kPa, and supplied steam gauge pressure ≤ 70 kPa. The former is also named as high pressure steam heating, the latter is called low pressure steam heating.

(2) The main equipment of steam heating system

The main equipment of steam heating systems are shown in Fig. 6.16.

(a) 疏水器
Steam trap

(b) 凝结水箱
Condensed water tank

(c) 管道补偿器
Pipeline compensator

图 6.16　蒸汽供暖系统的主要设备
Fig. 6.16　Main equipment of steam heating system

① 疏水器。疏水器是蒸汽供暖系统

① Steam trap. It is an important equip-

建筑环境与设备工程　Building Environment and Equipment Engineering

中的重要设备,其作用是自动阻止蒸汽逸漏,迅速排出热设备及管道中的凝水,排除系统中积留的空气和其他不凝性气体。

② 凝结水箱。凝结水箱是贮存凝结水的设备。

③ 管道补偿器。在供暖系统中,金属管道会因受热而伸长,又由于平直管道的两端都被固定不能自由伸缩,管道就会弯曲变形,严重时发生破裂,因此需要在管道上设管道补偿器。

6.3 通风工程

通风就是把室内被污染的空气直接或净化后排至室外,把新鲜空气补充进来,从而保持室内的空气环境符合卫生标准或满足生产工艺的需要,并在一定程度上改善室内的热湿参数。

6.3.1 通风系统的分类

通风系统按照工作动力可分自然通风和机械通风两种。

自然通风依靠室内外空气的温度差(实际是密度差)造成的热压,或者是室外风造成的风压,促使房间内外的空气进行交换,从而改善室内的空气环境。自然通风不需要另外设置动力设备,是一种经济、有效的通风方法。其缺点是无法处理进入室内的外空气,也难于对排出的污浊

ment in steam heating system. Its function is to prevent leakage of steam automatically, discharge the condensed water in the pipe rapidly, and discharge retention air and other non-condensable gas of the system.

② Condensate tank. It is the equipment to store condensed water.

③ Pipeline compensator. In heating system, the metal pipe is elongated due to heat. Two ends of the straight pipe are fixed and can not be expanded and contracted freely. The pipeline will be deformed and ruptured seriously. Therefore, pipeline compensator must be arranged on the pipe.

6.3 Ventilation engineering

Ventilation is to exhaust the indoor polluted air directly or after cleanse, and to supply the fresh air, so as to keep the indoor environment within the healthy standard or production requirement, and improve the indoor thermal and humidity parameters at a certain extent.

6.3.1 Classification of ventilation system

Ventilation system can be classified into natural ventilation and mechanical ventilation according to the working power.

In natural ventilation, air exchanges between indoor and outdoor depending on thermal pressure from the temperature difference. Actually, it is the density difference between indoor and outdoor, or wind pressure forming from the outdoor winding moving. These exchanging improve indoor air environment. Natural ventilation does not need any power equipment, and it is an economical and effective ventilation method.

空气进行净化处理;其次,自然通风受室外气象条件影响,通风效果不稳定。图6.17为某厂房的自然通风设计。

One disadvantage of natural ventilation is that it cannot handle the air into indoor, and it can't cleanse the exhausted air. The other disadvantage is that natural ventilation is affected by outdoor weather, and the ventilation effect is unstable. Fig. 6.17 shows the natural ventilation design of some plant.

图 6.17　某厂房的自然通风设计
Fig. 6.17　Natural ventilation design of some plant

机械通风依靠风机作用使空气流动,使房间进行通风换气。由于风机的风量和风压可根据需要确定,这种通风方法能保证所需要的通风量,控制房间内的气流方向和速度,并可对进风和排风进行必要的处理,使房间空气达到要求。因此,机械通风方法得到了广泛应用。

In mechanical ventilation, air exchanging depends on fans. Because the air quantity and air pressure of fans can be decided according to demand, mechanical ventilation can meet the requirements of the air quantity and control the air flow direction and velocity in rooms. It can also handle the air intake and outtake, make the indoor environment parameters meet the requirements. Therefore, mechanical ventilation is applied widely.

6.3.2　通风系统的主要设备

（1）风机

风机是依靠输入的机械能提高气体的压力并排送气体的机械。按照工作原理不同,风机可分为离心式、轴流式和贯流式 3 种。离心式风机和轴流式风机如图 6.18 和图 6.19 所示。

6.3.2　Main equipment of ventilation system

（1）Fans

Fans are used to increase air pressure and transport the air depending on inputted mechanical energy. According to working principle, fans can be classified into three types: centrifugal, axial flow and cross flow. Centrifugal fan and axial flow fan are shown in Fig. 6.18 and Fig. 6.19 respectively.

图 6.18　离心式风机
Fig. 6.18　Centrifugal fan

图 6.19　轴流式风机
Fig. 6.19　Axial flow fan

（2）风管

风管是输送空气的管道。按其横截面形状，风管可分为圆形风管、矩形风管等，如图 6.20 和图 6.21 所示。

（2）Air duct

Air duct is used to transport air. According to its cross section shape, air duct can be classified into rounded one, and rectangle one, as shown in Fig. 6.20 and Fig. 6.21.

图 6.20　圆形风管
Fig. 6.20　Rounded air duct

图 6.21　矩形风管
Fig. 6.21　Rectangle air duct

（3）送排风口

送排风口是用于室内送排风的装置。送排风口的位置和形式将影响到室内气流组织。典型的送排风口形式如图 6.22 所示.

（3）Supply and exhaust outlet

Supply and exhaust outlet is used to supply and exhaust air. The position and form of the supply and exhaust outlet affect the organization of the indoor air. Typical forms of supply and exhaust outlet are shown in Fig. 6.22.

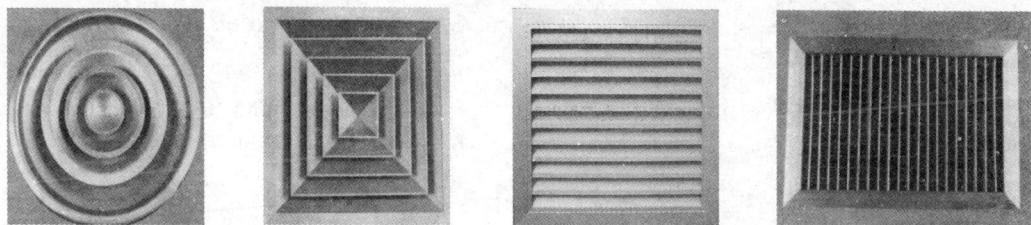

图 6.22　送排风口形式
Fig. 6.22　Forms of supply and exhaust outlet

（4）排风罩

排风罩是用于收集被污染气体的装置,按其结构形式可分为侧吸式和顶吸式。顶吸式排风罩如图 6.23 所示。

(4) Exhaust hood

Exhaust hood is used to collect the polluted air. According to its structure, exhaust hood can be classified into side hood and top hood. Fig.6.23 shows the top exhaust hood.

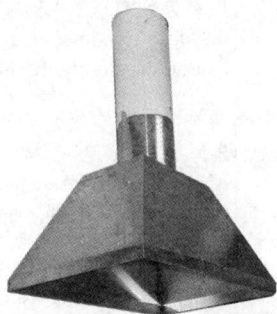

图 6.23　顶吸式排风罩
Fig. 6.23　Top exhaust hood

6.4　空调工程

空调工程是把特定空间内部的空气环境控制在一定的状态下,使其满足人体舒适或生产工艺要求的工程技术。它所控制的内容包括空气的温度、湿度、流速、压力、洁净度、噪声等。以生产或科学实验服务为目标的空调系统称为"工艺性空调",而以人体舒适及健康为目标的空调系统称为"舒适性空调"。

6.4　Air-conditioning engineering

Air-condition engineering is the technology used to control the air environment at a special state, so as to satisfy the demand of the people and production technology. Air-conditioning engineering includes temperature, humidity, velocity, pressure, cleanliness, noise etc. The air-conditioning system used for production and science experiment is called "industrial air-conditioning". The air-conditioning system used for people's comfort and healthy is called "comfortable air-conditioning".

6.4.1 空调系统分类

空调系统的分类见表6.1。

6.4.1 Classification of air-conditioning system

The classification of air-conditioning system is shown in Table 6.1.

表6.1 空调系统分类
Table 6.1 Classification of air-conditioning system

分类方法 Classification method	空调系统 Air-conditioning system	系统特征 Characteristics of the system
按空气处理设备的设置情况分类 classification according to the arrangement of air handling units	集中式系统 centralized system	空气处理设备集中在机房内,空气经处理后,由风管送入各房间。 All air handling unit are equipped in machine room, and air is handled and transported to each room through air dust.
	半集中式系统 half-centralized system	除了有集中的空气处理设备外,在各个空调房间内还分别设有处理空气的"末端装置"。 Besides centralized air handling unit, there are end units in each room to handle the air further.
	全分散式系统 dispersed system	每个房间的空气处理分别由各自的整体式(或分体式)空调器承担。 Air handle are taken on by the incorporate (or separated) air conditioners in each room.
按负担室内空调负荷所用的介质分类 classification according to the media taking on the air conditioning load	全空气系统 air system	全部由处理过的空气负担室内空调负荷 All air-conditioning load is taken on by handled air
	空气 – 水系统 air and water system	由处理过的空气和水共同负担室内空调负荷。 Air-conditioning load is taken on by handled air and water.
	全水系统 water system	全部由水负担室内空调负荷。 All air-conditioning load is taken on by water.
	制冷剂系统 refrigerant system	制冷系统的蒸发器直接放在室内承担空调负荷。 All air-conditioning load is taken on by the evaporator of the refrigerator directly.

6.4.2 空调系统主要设备

(1) 冷热源

冷热源是空调系统冷量和热量的来源。典型的冷源包括电动压缩式冷水机组、吸收式冷水机组;典型的热源包括热泵、锅炉、城市热网。图6.24为空调系统中广泛应用的电动压缩式冷水机组。

6.4.2 Main equipment of air-conditioning system

(1) Heat and cold source

Heat and cold source is the source of the air-conditioning system. Typical cold sources include electric compressing chiller and absorbing chiller. Typical heat sources include a heat pump, a boiler and a heat supply network. Fig. 6.24 shows the widely used electric compressing chiller.

图 6.24 电动压缩式冷水机组
Fig. 6.24 Electric compressing chiller

（2）水系统相关设备

空调中的水系统包括冷冻水系统,冷却水系统和冷凝水系统。冷冻水系统承担室内的空调负荷,冷却水系统承担冷水机组的冷却,冷凝水系统承担冷冻除湿形成的冷凝水的输送。其中冷冻水系统和冷却水系统一般需要水泵驱动。图 6.25 所示为冷冻水系统及其水泵。图 6.26 所示为冷却水系统中用的冷却塔。

（2）Water system and its related equipment

Water system in the air-conditioning system include chilled water system, cooling water system and condensate water system. Chilled water system is used to take on the air load. Cooling water is used to cool the chiller, and condensate water system is used to transport the condensate water formed by the cold dehumidification. Chilled water system and cooling water system need water pumps. Fig. 6.25 shows the chilled water system and its water pump. Fig. 6.26 shows the cooling tower used in cooling water system.

图 6.25 冷冻水系统及水泵
Fig. 6.25 Chilled water system and its water pump

图 6.26 冷却塔
Fig. 6.26 Cooling tower

(3) 风管及送排风口

空调中的风管以及送排风口与通风工程中的设备类似。根据风管内空气的用途，风管可分为送风管道、回风管道、排风管道以及新风管道等。图6.27所示为空调送风管道及送风口。

(3) Air duct, supply and exhaust outlet

Air duct and supply and exhaust outlet in air-conditioning system is similar to those in the ventilation engineering. According to the air function in the air duct, the duct can be classified into supply air duct, return air duct, exhaust air duct, and fresh air duct. Fig. 6.27 shows the air duct and supply outlet in the air-conditioning system.

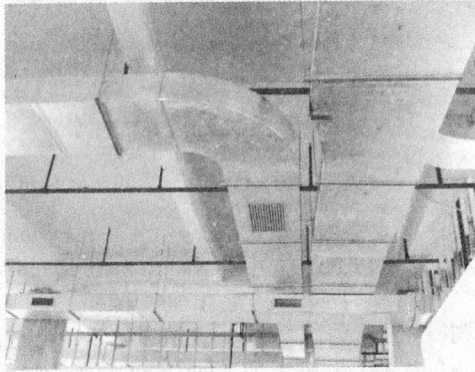

图 6.27 风管及送风口
Fig. 6.27 Air duct and supply outlet

(4) 空气处理设备

空气处理设备用于对空调系统中的空气进行热湿处理以及过滤净化。其中应用最广泛的热湿处理设备为喷水室和表面式换热器。将空气过滤、热湿处理以及风机组合在一个箱体中实现空气综合处理的设备称为组合式空气处理机组，如图6.28所示。将冷却盘管和风机组合在一起对房间空气直接处理的设备称为风机盘管机组，如图6.29、图6.30所示。

(4) Air handling equipment

Air handling equipment is used to heat, cool, humidify, dehumidify and purify the air in the air-conditioning system. The most widely used thermal and humidity handling equipment are spraying chamber and surface heat exchanger. The unit that contains air purify, thermal and humidity and fan is called combined air handling unit, as shown in Fig. 6.28. The unit that is equipped with a cooling coil and a fan in one box to directly handling the air in the room is called fans coil unit, as shown in Fig. 6.29 and Fig. 6.30.

图 6.28　空调箱
Fig. 6.28　Air box

图 6.29　卧式风机盘管
Fig. 6.29　Horizontal fan-coil

图 6.30　卡式风机盘管
Fig. 6.30　Cassette fan-coil

（5）单元式空调机

带有制冷压缩机、冷凝器、直接膨胀式空气冷却器、空气过滤器、风机和自控系统等整套装置的空气处理机组，称为单元式空调机组。典型的单元式空调机如图 6.31、图 6.32 和图 6.33 所示。

(5) Air conditioning unit

The air handling unit with a compressor, a condenser, a directly expanded air cooler, an air filter, a fan and an automatic control system is called air-conditioning unit. Typical air-conditioning unit is the air condition as shown in Fig. 6.31, Fig. 6.32 and Fig. 6.33.

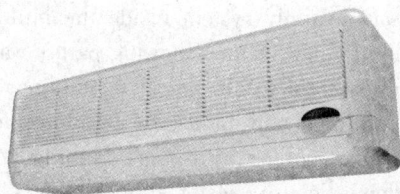

图 6.31　卧式室内机
Fig. 6.31　Horizontal indoor unit

图 6.32　室外机
Fig. 6.32　Outdoor unit

图 6.33　立式室内机
Fig. 6.33　Vertical indoor unit

6.5　建筑给水排水工程

6.5.1　建筑内部给水系统

建筑内部给水系统是将城镇给水管网或自备水源给水管网的水引入室内,选用适用、经济、合理的最佳供水方式,经配水管送至室内各种卫生器具、生产装置和消防设备,并满足用水点对水量、水压和水质要求的供应系统。

（1）建筑内部给水系统的分类

建筑内部的给水系统按用途可分为民用建筑、工业建筑内的饮用、烹调、盥洗、洗涤、沐浴等的生活用水系统,供生产设备冷却、原料产品洗涤、产品制造过程中所需生产用水的生产给水系统,供消防设备灭火用水的消防给水系统。

6.5　Building water supply and drainage engineering

6.5.1　Construction of internal water supply system in the building

Construction of internal water supply system is to bring the water from the urban water supply network or self-provided water network to the room, using the suitable, economical and reasonable way to supply water and deliver water to a variety of indoor sanitary ware, water nozzle, producing device and firefighting equipment through water distribution pipe. This water supply system must meet the requirements of users for water dosage, water pressure and water quality.

（1）Classification of internal water supply system in the building

According to different functions, the water supply system inside the building can be divided in the following parts: water supply system within the civil buildings, public buildings and industrial buildings for drinking, cooking, washing, bathing and other domestic use, water supply system for cooling production equipment, washing raw material products, supplying water in the process of product manufacturing, water supply system for fire-fighting.

（2）建筑内部给水系统的组成

建筑内部给水系统一般由引入管、给水管道、给水附件、给水设备、配水设施和计量仪表等组成，如图6.34所示。

（2）Composition of water supply system in the building

Internal water supply system is composed of an introducing pipe, water supply pipes, water supply accessories, water supply equipment, water distribution facilities and measurement instruments and other components, as shown in Fig. 6.34.

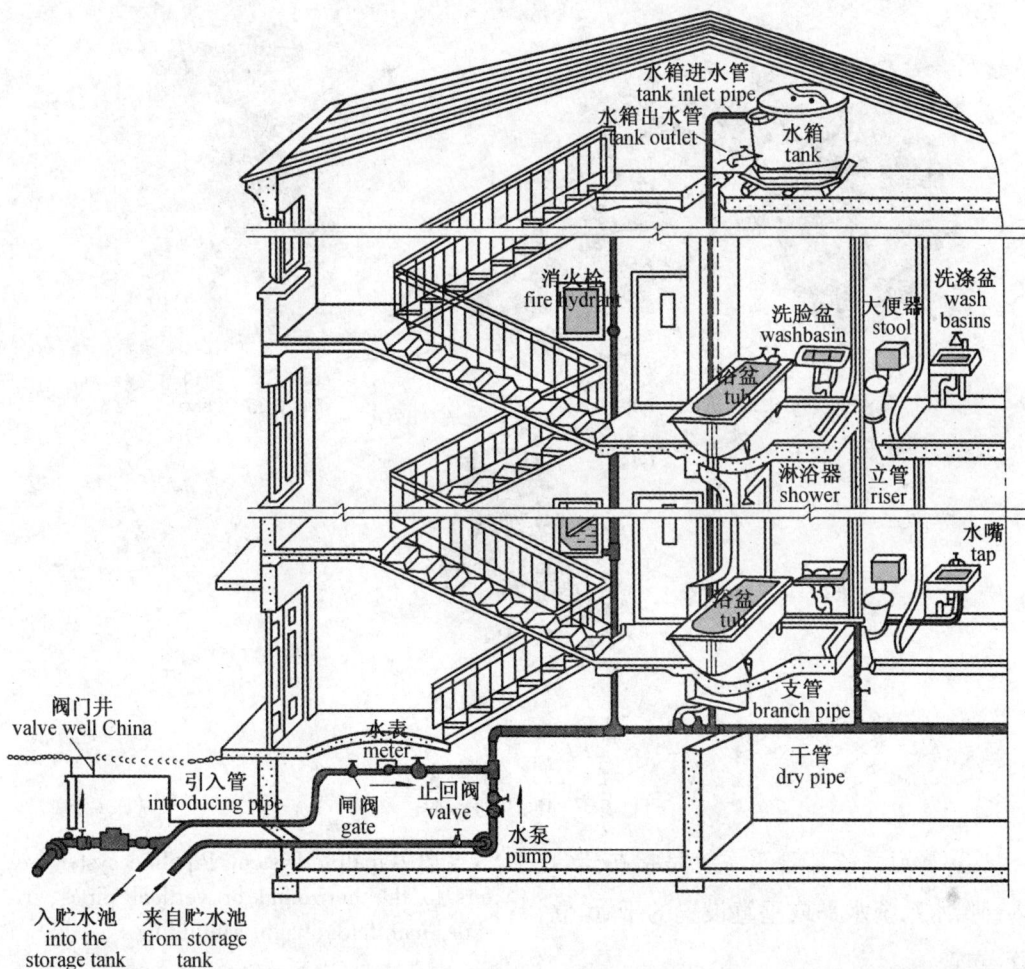

图6.34 建筑内部给水系统
Fig. 6.34   Construction of internal water supply system

① 引入管。对一幢单独建筑物而言，引入管是室外给水管网与室内管网之间

① Introducing pipe. As for a separate building, the introducing pipe is the connec-

建筑环境与设备工程 Building Environment and Equipment Engineering

的联络管段,也称进户管。对于一个工厂、一个建筑群体、一个学校区,引入管系指总进水管。

② 水表节点。水表节点是指引入管上装设的水表(如图 6.35 所示)及其前后设置的闸门(如图 6.36 所示)、泄水装置(如图 6.37 所示)等的总称。

tion part between the outdoor water supply pipe network and the indoor pipe network, and it is known as a tube into the household. As for a factory, a construction group and a school district and the introducing pipe is represented for the total inlet pipe.

② Water meter node. Water meter node contains a water-meter (Fig. 6. 35), a valve (Fig. 6. 36) and a discharge device (Fig. 6. 37) set before or after the water-meter.

图 6.35　水表
Fig. 6.35　Water meter

图 6.36　阀门
Fig. 6.36　Valve

图 6.37　泄水阀
Fig. 6.37　Discharge valve

③ 管道系统。管道系统是指在建筑内部给水系统水平或垂直设置的干管、立管、支管。

④ 给水附件。给水附件指管路上的闸阀等各式阀门及各式配水龙头(如图 6.38 所示)。

③ Pipeline system. Pipeline system refers to the horizontal or vertical pipes, risers, manifolds inside a building.

④ Water attachments. Water attachments refer to the valve in the pipe system, all kinds of other valves and taps (Fig. 6.38).

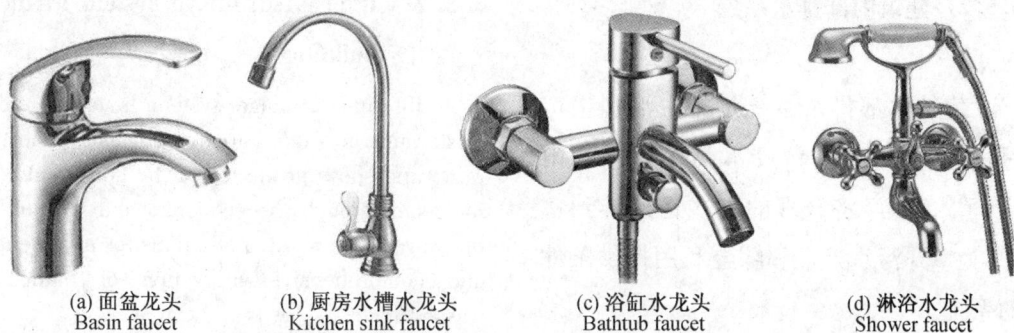

(a) 面盆龙头
Basin faucet

(b) 厨房水槽水龙头
Kitchen sink faucet

(c) 浴缸水龙头
Bathtub faucet

(d) 淋浴水龙头
Shower faucet

图 6.38　各式配水龙头
Fig. 6.38　Faucets

⑤ 增压和贮水设备。在室外给水管网压力不足或建筑内部对安全供水、水压稳定有要求时,需设置各种附属设备,如水箱、水泵、气压设备、水池等增压和贮水设备。

⑥ 室内消防。按照建筑物的防火要求及规定需要设置消防给水时,一般应设消火栓消防设备(如图 6.39、图 6.40 所示)。有特殊要求时,另专门装设自动喷水灭火或水幕灭火设备等。

⑤ Booster and storage equipment. In the condition of low pressure of outdoor water supply network or the special requirements for water safety and steady water pressure, a variety of ancillary equipment such as tanks, pumps, pressure equipment, pools, other pressurization and storage devices are needed.

⑥ Indoor fire fighting. In accordance with the provisions of the building and fire safety requirements, we need to set hydrant fire fighting equipment (Fig. 6. 39 and Fig. 6. 40). If there are special requirements, other specialized installation of automatic sprinklers or water curtain fire-fighting equipment should be installed.

图 6.39　消火箱
Fig. 6.39　Eliminate fire box

图 6.40　消火栓
Fig. 6.40　Fire hydrant

6.5.2 建筑内部排水系统

建筑内部排水系统是将生活和生产过程中所产生的污、废水及房屋顶的雨水、雪水,用经济合理的方式迅速排到室外,为室外污水处理和综合利用提供条件的系统。

(1) 排水系统的分类

按照所排的污、废水性质,建筑内部排水系统可分为排除居住、公共、工业建筑生活间污、废水的生活排水系统,排除工艺生产过程中产生的污、废水的工业废水排水系统,排除多跨工业厂房、大屋面建筑、高层建筑屋面上雨、雪水的屋面雨水排除系统。

(2) 排水系统的组成

建筑内部排水系统一般由卫生器具和生产设备的受水器、排水管道、清通设施、通气管道以及污、废水的提升设备和局部处理构筑物等组成,如图 6.41 所示。

6.5.2 Internal drainage system in the building

Internal drainage system is to let the wastewater generated in the process of life, rain, and snow on the top of the house out of the room through the economic and reasonable way, and provide conditions for treatment and comprehensive utilization of outdoor wastewater.

(1) Classification of the drainage system

According to the properties of the sewage, the internal drainage system can be divided into drainage system to exclude wastewater of residential, public and industrial buildings , drainage system to exclude wastewater generated in the process of industrial, drainage system to exclude wastewater of multi-span industrial plant, large roof buildings, high-rise building roof.

(2) Composition of the drainage system

Building internal drainage system generally consists of water heater, drain pipes, clear communication facilities, and ventilation pipes in sanitary ware and production equipment, as well as the upgrade sewage and waste treatment facilities, local structures and other components, as is shown in Fig. 6.41.

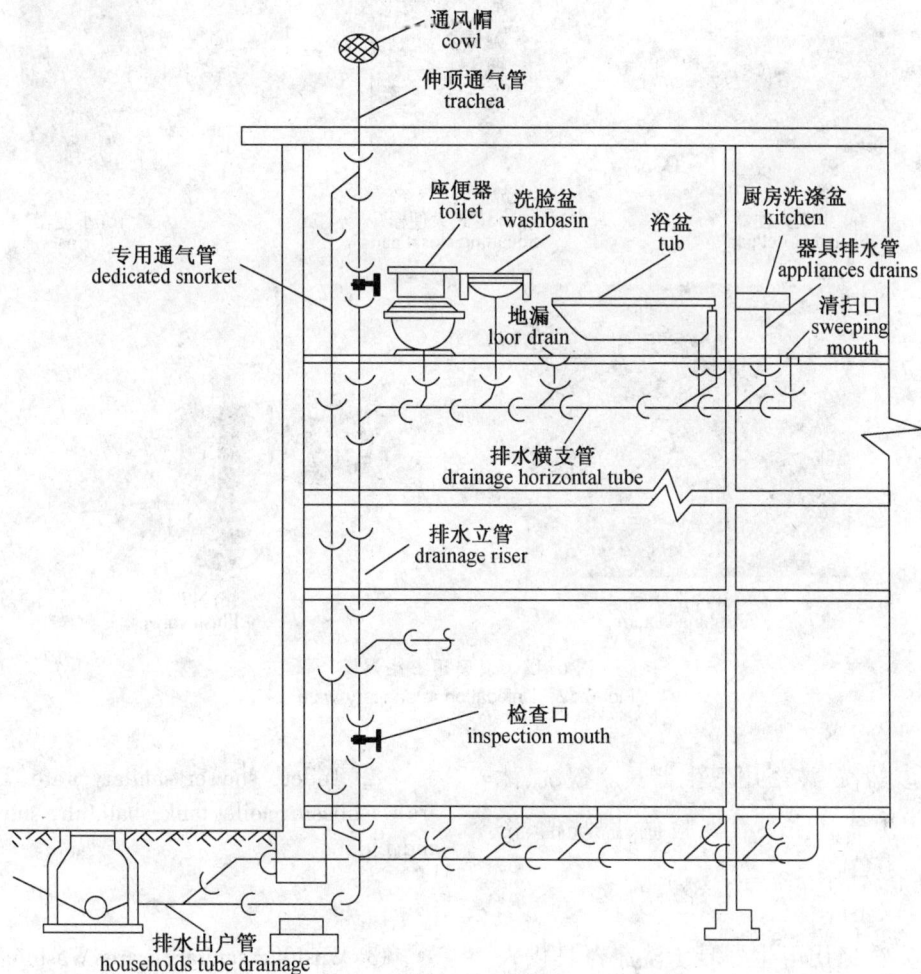

图 6.41　建筑内部排水系统
Fig. 6.41　Internal drainage system in the building

1）卫生器具和生产设备受水器

　　① 便溺用卫生器具。便溺用卫生器具设置在卫生间和公共厕所，用来收集生活污水。便溺用卫生器具（如图 6.42 所示）主要包括大便器、小便器和冲洗设备。

1）Water heater in sanitary ware and production equipment

　　① Urinating and defecating sanitary ware. Urinating and defecating sanitary ware is set in the washing room, and is used to collect sewage. Urinating sanitary ware (Fig. 6.42) includes stools, urinals and washing facilities.

(a) 坐式大便器
Seated closet pan

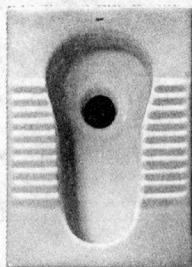

(b) 蹲式大便器
Squatting closet pan

(c) 小便器
Urinal

(d) 冲洗水箱
Flushing cistern

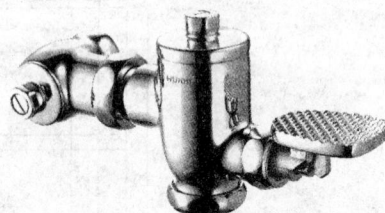

(e) 冲洗阀
Flush valve

图 6.42　便溺用卫生器具
Fig. 6.42　Defecation in sanitary ware

② 盥洗、沐浴用卫生器具。盥洗、沐浴用卫生器具主要有洗脸盆、盥洗槽、浴盆、淋浴器、净身盆等。

③ 洗涤用卫生器具。洗涤器具供人们洗涤器物之用,主要有污水盆、洗涤盆、化验盆等。

④ 地漏及存水弯(如图6.43、图6.44所示)。

2) 排水管道

建筑内部排水管道包括器具排水管道、排水横支管、立管、埋地干管和排出管。

② Toilet, shower sanitary ware. There are washbasin, toilet tank, bathtub, shower, and bidet.

③ Washing sanitary ware. Washing appliances are used for people to wash utensils, as sewage basins, sinks, pots and other laboratory tests.

④ Floor drain and trap are shown in Fig. 6.43 and Fig. 6.44.

2) Drains

Internal drainage pipelines include appliances, horizontal drainage pipes, risers, buried mains and discharge pipe.

3）清通设施

为疏通建筑内部排水管道,保障排水畅通,需设置 3 种清通设施,即清扫口、检查口、检查口井。

4）通气管道

为防止因气压波动造成水封破坏,而将排水管内臭气和有害气体排到大气中,需在建筑内部排水系统中设置通气管道,与大气相通(如图 6.45 所示)

3）Clearing facilities

In order to clear the construction of internal drainage channels and guarantee the smooth drainage, we need to set three clear communication facilities: clean mouth, inspection openings, and check wells.

4）Ventilation pipe

In order to prevent the seal damage due to air pressure fluctuations, and prevent draining harmful gases inside from venting to the atmosphere, we should use a ventilation pipe to connect the internal drainage system with atmosphere(Fig. 6. 45)

图 6.43　地漏
Fig. 6.43　Floor drain

图 6.44　存水弯
Fig. 6.44　Trap

图 6.45　通气管
Fig. 6.45　Vent

5）提升设备

各种建筑地下室中的污、废水不能自流排至室外检查井,需设集水池和水泵等局部提升设备,将污水排到室外排水管道中去。

6）污水局部处理构筑物

建筑物内部污水未经处理不允许直接排入市政排水管网或水体时,需设污水局部处理构筑物。

5）Lifting equipment

Sewage, wastewater in the basement of various buildings can not be discharged to outside inspection artesian wells, so it is needed to set up sump pumps and other partial lifting equipment to discharge sewage drains into the outdoor drainage system.

6）Local sewage treatment structures

Internal untreated wastewater is not allowed to directly discharge into the municipal sewer or water system. In this case, the local sewage treatment structure needs to be set up.

6.5.3 建筑内部热水供应系统

建筑内部热水供应系统主要供给生产、生活洗涤及盥洗用热水,应能保证用户限时可以得到符合设计要求的水量、水温和水质。

(1) 热水供应系统的分类

按照热水供水范围的大小,建筑内部热水供应系统分为区域热水供应系统、集中热水供应系统和局部热水供应系统。

(2) 热水供应系统的组成

热水供应系统的组成因建筑类型和规模、热源情况、用水要求、加热和储存设备的情况、建筑对美观和安静的要求等不同情况而异。

建筑内部热水供应系统通常由加热设备(如锅炉、太阳能热水器、直燃机、各种热交换器等),热媒管网(蒸汽管或过热水管,凝结水管等),热水储存水箱,热水输配水管网与循环管网,其他设备和附件组成。图6.46为一典型的局部热水供应系统。图6.47为一新型的太阳能集中热水供应系统。

6.5.3 Hot water supply system in the building

The hot water supply system inside the building mainly supplies production, life washing and toilet water. It should ensure that users can get the amount of designed water with the right temperature and quality in time.

(1) Classification of the water supply system

In accordance with the range of hot water supply, hot water supply systems inside the building can be divided into regional water supply system, centralized hot water supply system and local water supply systems.

(2) Composition of the water supply system

The composition of the hot water supply system is different due to building type, size of heat situation, water requirements, heating and storage of the device, and the demands of construction.

Hot water supply system inside the building is generally consisted of the following parts: heating equipment (such as boilers, solar water heaters, direct gas turbine, variety of heat exchangers), the heat medium pipe (steam or superheated water, condensation water, etc.), hot water storage tanks, water distribution networks and water circulation pipe network, other devices and accessories. Fig. 6.46 shows a typical local water supply system. Fig. 6.47 shows a new type of concentrating solar hot water supply system.

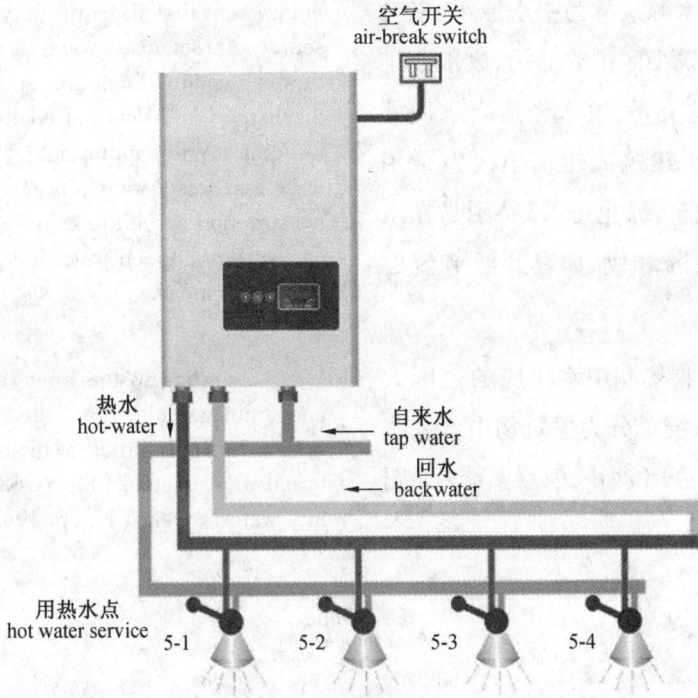

图 6.46　局部热水供应系统
Fig. 6.46　Local hot water supply system

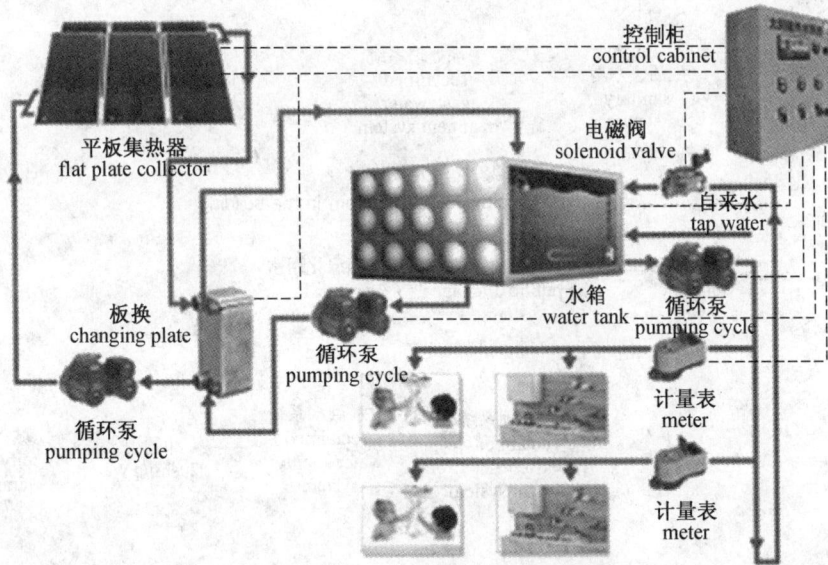

图 6.47　太阳能集中热水供应系统
Fig. 6.47　central heating system

6.5.4　建筑中水工程及其他水系统

建筑中水工程技术最早应用于日本

6.5.4　Reclaimed water system and other water systems in the building

Construction of water engineering tech-

东京。中水的水源又称为中水原水,来自于建筑物或建筑小区排放的污、废水或排放的冷却水,这类污、废水经适当水质处理后,能应用于建筑或建筑小区内杂用(如冲厕所、洗车、绿化用水)特别是在水资源缺乏的地区,中水具有开源节流的作用。

根据排水收集和中水供应的范围大小,建筑中水系统又分为建筑物中水系统(如图6.48所示)和小区中水系统(如图6.49所示)。

nology was first used in Tokyo, Japan. The source of reclaimed water is also known as Suwon, coming from the sewage and waste discharge or emission of cooling water in the building. With appropriate treatment of sewage and waste water, it can be used in the construction or building (e. g. flush toilets, car washing, green water) especially in water-scarce areas.

According to the range of water supply and drainage collection, the reclaimed water system in the building is divided into the internal water system (Fig. 6.48) and community water system (Fig. 6.49).

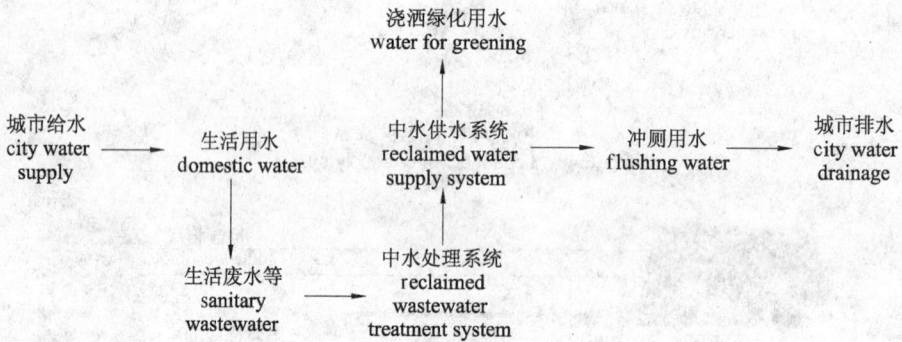

图 6.48　建筑物中水系统
Fig. 6.48　Reclaimed water system in the building

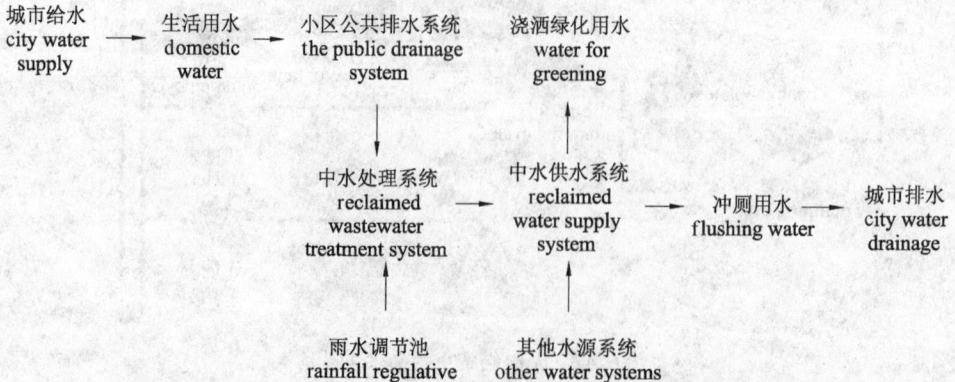

图 6.49　建筑小区中水系统
Fig. 6.49　Reclaimed water system in residential district

其他的水系统有雨水系统、特殊建筑

Other forms of water systems include

给水排水系统(如游泳池、洗衣房用水)、直饮水供应系统、喷池等景观建筑水系统等,这些水系统对完善建筑功能、改善建筑环境具有重要的作用。如在宾馆、公寓、医院等公共建筑中常设有洗衣房,用于洗涤床上用品、各类工作服等,以增加建筑的竞争力;水景不仅可以美化环境、装饰厅堂,还可以起到增加空气湿度、增加负氧离子浓度、净化空气、降低气温等改善小区气候的作用,也可以兼作消防、冷却喷水的水源。

storm water system, special building water supply and drainage system (such as swimming pools, laundry water), drinking water supply system, landscape architecture system such as spray pool water system. These water systems play an important role in improving the function and environment of the building. Hotels, apartments, hospitals and other public buildings usually have laundry room for washing quilt, all kinds of clothes, etc. which can improve the competitiveness of architecture. Water landscape can beautify the environment, decorate hall, increase the air humidity and the concentration of negative oxygen ions, purify the air, reduce the temperature and change the climatic of the area. It also can be used as the source of firewater.

6.6 未来展望

健康、能源、环境已成为备受人类关注的三大主题,建筑环境与设备工程这3个方面有着密切的关系。土木工程行业将越来越关注建筑的"可持续发展"技术,关注节能环保,控制建筑设计施工以及使用过程中对自然环境造成的影响,降低室内外建筑环境控制中建筑设备的能耗。例如,国外广泛使用的被动式太阳能采暖及降温装置,为采暖、通风、空调技术提供了新型的冷源和热源;使用程序控制装置调节建筑的通风空调系统,可以使建筑物的通风量随气象参数自动调节,保证室内卫生舒适条件;使用自动温度调节器,可以保证室内采暖及空调的设计温度,并节约能源。节能建筑与绿色建筑将成为土木工程行业的重点发展方向。

6.6 Future prospects

Health, energy, and environment have become three topics recognized by people all over the world, and building environment and equipment engineering relates these three topics closely. Civil engineering industry will pay more and more attention to the sustainable development technology. Besides, this industry will concentrate on the energy saving and environment protecting technology, so as to control the effect on nature and reduce the energy consumption of building equipment during controlling the indoor and outdoor building environment. For example, passive solar heating and cooling equipment has been widely used overseas. It provides a new type of cold and heat source for heating, ventilation and air-conditioning technology. By using the program control device regulating building ventilation and air conditioning system, we can adjust the ventilation quantity automatically along with the meteorological parameters and ensure health and comfortable indoor conditions. By using automatic temperature regulator, we can guarantee the indoor heating and the given air-conditioning temperature and save energy. Energy saving buildings and green buildings will be the important development directions of this industry.

知识拓展
Learning More

相关链接　Related Links

（1）美国供热、制冷、空调工程师学会（ASHRAE）http://www.ashrae.org/

（2）国际室内空气品质和气候学会（ISIAQ）http://www.isiaq.org/

（3）中国建筑学会（ASC）http://www.chinaasc.org/

（4）中国环境科学学会（CSES）http://www.chinacses.org/cn/index/html/

（5）中国建筑业协会（CCIA）http://www.zgjzy.org/

常用应用软件介绍　Brief introduction of common application softwares

目前与本行业相关的专业软件数目众多，大致可以分为三类：

第一类是工程设计类，包括计算机绘图和计算软件，如 AutoCAD，天正建筑设计软件，鸿业空调设计系列软件；

第二类是能耗和环境模拟软件，如 FLUENT 模拟软件、DOE 能耗模拟软件、EnergyPlus、eQuest 快速能耗模拟软件、Ecotect 生态建筑分析大师、DeST；

第三类是建筑设备诊断软件和工具等，如麦克维尔公司开发的制冷机组诊断软件等。

Currently, large numbers of professional softwares related to this industry can be divided into three categories.

The first category is engineering design categories, including computer graphics and computational software, such as AutoCAD, Tengen architecture design software, Hongye air-conditioning design series software.

The second category is the energy consumption and environmental simulation software, such as FLUENT simulation software, DOE energy simulation software, EnergyPlus, eQuest quick energy simulation software, Ecotect analysis of ecological building master, and DeST.

The third category is construction equipment diagnostic software and tools, such as chiller diagnostic software developed by McQuay.

专业执业资格考试　Professional qualification examination

目前,与建筑环境与设备工程专业相关的注册工程师种类主要有注册公用设备工程师、注册监理工程师和注册建造师等。

At present, the main types of registered engineers related to the building environment and equipment engineering are registered public facility engineer, registered supervision engineers and registered architect, etc.

小贴士　Tips

(1) 事故通风

在拟定工业厂房的通风方案时,对于可能突然产生大量有害气体的车间,除应根据卫生和生产要求设置一般的通风系统外,还要另设一个专用的全面机械排风系统,以便在发生上述情况时能够迅速降低有害气体的含量。这样的机械排风系统叫做"事故排风系统"。

Emergency ventilation

When designing ventilation system for plant, a special mechanical ventilation system is needed which is likely to emit large amount of polluted air suddenly, besides the common ventilation system for health and production requirements. The content of the polluted air can be decreased quickly. The mechanical ventilation is called "emergency ventilation".

(2) 蓄冷空调系统

蓄冷空调系统在建筑物不需冷量或需冷量少时(如夜间),利用制冷设备将蓄冷介质中的热量移除,进行蓄冷,并将此冷量用在空调用冷的高峰期。蓄冷空调转移了制冷设备的运行时间,可以利用夜间的廉价电降低运行成本,同时减少白天的峰值电负荷,达到电力"削峰填谷"的目的。目前许多国家已将空调蓄冷技术作为重点的建筑节能技术进行推广。

The cold storage air-conditioning system

The cold storage air-conditioning system removes the heat in cold storage media and accumulate cold using refrigerator when the building does not need cold or needs a little cold (for example at night). Therefore, the building can use these stored cold in the rush hour. The cold storage air-conditioning system shifts the running time of refrigerator, and reduces operating cost, because it uses the cheap electricity at night, and decreases the peak electricity load at day at the same time. Cold storage air-conditioning system has been extended in many countries as an important energy saving technology.

(3) 地源热泵系统

地源热泵是一种利用大地能量,包括土壤、地下水、地表水等天然能源作为冬季热源和夏季冷源,然后再由热泵机组向建筑物供冷供热的系统,是一种利用可再生能源的既可供暖又可制冷的新型中央空调系统。

但由于技术限制,抽取地下水水源热泵很难实现全部回灌,监督实施也比较困难,而且容易造成地下水污染。目前,国外大面积推广使用的是埋管式地源热泵技术,是充分利用浅层地热的最佳技术途径。

目前埋管式地源热泵在欧美国家已得到普遍应用,已被充分证明是成熟可行的技术。在我国,一些省市的建筑节能政策中也明确提出了要推广使用地源热泵。

The ground-source heat pump system

The ground-source heat pump is a new central air-conditioning renewable energy system using the earth energy, including soil, groundwater, surface water and other natural energy as a heat source in winter and cold source in summer.

建筑环境与设备工程　Building Environment and Equipment Engineering

It is not easy to recharge the entire pipe with hot groundwater for the groundwater source heat pump because of the technical limitations, and the implementation is also difficult. It is likely to cause groundwater contamination. Currently, most areas in foreign countries promote the use of buried ground-source heat pump technology, and it is the best technical way making full use of shallow geothermal.

Currently, buried ground-source heat pumps have been widely used in the United States and Europe, and have been proved to be the most sufficiently mature and viable technology. In China, some building energy polices in some provinces clearly required to promote the use of ground-source heat pumps.

思考题　Review Questions

(1) 建筑、能源与环境的关系是什么？

What is the relationship among the building, the energy and the environment?

(2) 建筑环境与设备工程有哪些新技术应用？

What new technology applications have been used in the field of building environment and equipment engineering?

参考文献
References

[1] 霍达:《土木工程概论》,科学出版社,2007 年。

[2] 刘俊玲,庄丽:《土木工程概论》,机械工程出版社,2009 年。

[3] 卢军:《建筑环境与设备工程概论》,重庆大学出版社,2003 年。

[4] 王增长:《建筑给水排水工程》,中国建筑工业出版社,2010 年。

[5] 张国强,李志生:《建筑环境与设备工程专业导论》,重庆大学出版社,2007 年。

[6] 朱颖心:《建筑环境学》(第 3 版),中国建筑工业出版社,2010 年。

[7] 柳孝图:《建筑物理环境与设计》,中国建筑工业出版社,2008 年。

[8] 赵荣义:《空气调节》(第 4 版),中国建筑工业出版社,2011 年。

[9] 孙一坚,沈恒根:《工业通风》,中国建筑工业出版社,2010 年。

[10] ASHRAE Handbook. American Society of Heating, Refrigerating and Air-Conditioning Engineers. Inc, Atlanta, GA. 2010.

[11] CIBSE Guide B- Heating, Ventilating, Air conditioning and Refrigeration. Chartered Institution of Building Services Engineers,2005.

[12] 中华人民共和国建设部:《采暖通风与空气调节设计规范》(GB 50019—2003),中国标准出版社,2004 年。

[13] 中华人民共和国公安部:《建筑设计防火规范》(GB 50016—2006),中国计划出版社,2006 年。

[14] 中国建筑科学研究院:《夏热冬冷地区居住建筑节能设计标准》(JGJ134—2001),中国建筑工业出版社,2001 年。

[15] 中国建筑科学研究院:《夏热冬暖地区居住建筑节能设计标准》(JGJ75—2003),中国建筑工

业出版社,2003 年。

[16] 中华人民共和国建设部:《公共建筑节能设计标准》(GB 50189—2005),中国建筑工业出版社,2005 年。

[17] 中华人民共和国建设部:《建筑节能工程施工质量验收规范》(GB 50411—2007),中国建筑工业出版社,2007 年。

[18] 上海城乡建设和交通委员会:《建筑给排水及采暖工程施工质量验收规范》(GB 50242—2002),中国建筑工业出版社,2002 年。

[19] 上海城乡建设和交通委员会:《建筑给水排水设计规范》(GB 50015—2009),中国建筑工业出版社,2009 年。

[20] ASHRAE Standard 55-2004, Thermal Environment Conditions for Human Occupancy, Atlanta: ASHRAE, 2004.

建筑环境与设备工程

Building Environment and Equipment Engineering

第7章 土木工程防灾与减灾

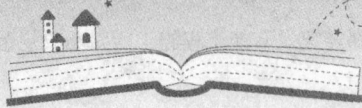

Chapter 7 Civil Engineering Disaster Prevention and Mitigation

土木工程防灾减灾学,是以防止和减轻灾情为目的,综合运用地震学、气象学、爆炸学、水力学、地质学、工程材料学、经济学和社会科学等多种科学理论和技术,为社会安定与经济可持续发展提供可靠保障的一门交叉学科。其研究内容包括:① 土木工程(城市)防灾规划;② 土木工程结构抗灾理论及应用;③ 土木工程结构防灾、抗灾技术及应用;④ 土木工程减灾技术;⑤ 土木工程结构在灾后的检测与加固;⑥ 高新技术在土木工程防灾减灾中的应用。

Civil engineering disaster prevention and mitigation is to prevent and mitigate the disaster. The integrated uses of seismology, meteorology, explosion, hydraulics, geology, engineering materials science, economics, social sciences and other scientific theories and technologies provide a reliable guarantee for the interdisciplinary study of the social stability and economic sustainable development. The research includes: ① civil engineering disaster prevention planning; ② civil engineering disaster-resistance theory and application; ③ civil engineering disaster prevention, disaster-resistance technology and application; ④ mitigation techniques of civil engineering; ⑤ The detection and reinforcement of post-disaster of civil engineering structure; ⑥ application of the high and new technology in disaster prevention and mitigation in civil engineering.

7.1 防灾减灾概论

7.1.1 灾害的含义与类型

灾害是指那些由于自然的、人为的或人与自然综合的原因,对人类的生存和社

7.1 Introduction to disaster prevention and mitigation

7.1.1 Definition and types of disaster

Disaster is the various phenomenon which can cause damage to human beings and social development due to natural rea-

会发展造成损害的各种现象。它一般具有危害性、突发性、永久性、频繁性、广泛性与区域性。土木工程灾害,是指由于人们的不当活动——选址、设计、施工、使用和维护导致所建造的土木工程不能抵御突发的荷载,而致使土木工程失效和破坏,乃至倒塌而造成的损失。

全世界每年都发生很多的灾害,严重的灾害会造成建筑物、构筑物的毁坏,交通通信、供水供电等工程中断,并引发次生灾害,导致大量人员伤亡,引起社会动荡,造成严重的经济损失,甚至使一个区域、一个城市在顷刻之间消失。土木工程灾害主要分为自然灾害和人为灾害。

自然灾害是自然界中物质的变化、运动造成的损害,包括地震灾害、风灾害、洪水灾害、滑坡灾害、泥石流灾害等。例如,强烈的地震,可使一座上百万人口的城市在顷刻之间化为废墟。2008 年 5 月 12 日,四川汶川县发生里氏 8.0 级地震,导致大批房屋倒塌和破坏(如图 7.1 所示),造成巨大人员伤亡和经济损失。

人为灾害是由于人的过错或某些丧失理性的失控行为给人类自身造成的损害,包括火灾、爆炸、地陷(人为地大量开采地下水造成)以及不适当的工程设施对环境造成的隐患,或者工程质量低劣造成的工程事故等。例如,2001 年 9 月 11 日,美国纽约世界贸易中心大厦在飞机撞击

sons, man-made reasons or comprehensive reasons. Disaster is destructive, sudden, permanent, frequent, wide and regional. The civil engineering disaster is caused by people's improper activities, such as location, design, construction, use and maintenance of civil engineering lead to the building failure against sudden load, which would make big losses caused by the civil engineering failure, even collapse.

Disasters often occur every year all over the world, and the serious one will destroy the whole of building and its structure, the transportation and communication, the water supply and the electric supply. Serious disasters may trigger a secondary disaster with heavy casualties, social unrest, serious economic losses, and even the disappearance of a region or a city in an instant. Civil engineering disaster is mainly divided into natural disaster and man-made disaster.

Natural disasters are caused by physical changes and movements in nature. These disasters include earthquake, windstorm, flood, landslide, debris flow and so on. For example, a strong earthquake can make a city with millions of people into ruins in an instant. A magnitude 8 earthquake occurred in Wenchuan County of Sichuan on May 12, 2008, resulting in a large number of houses collapsed and damaged (Fig. 7.1), huge casualties and economic losses.

Man-made disasters are always caused by fault or irrational behaviors of human being, which include fire, explosion, subsidence, inadequate facilities on the environment caused by the hidden danger, and poor construction quality accidents. For example, the United States World Trade Center Building in New York was in fire after plane crash on September 11, 2001. The two world landmark skyscrapers collapsed in a short

后起火,在很短的时间内造成两栋世界标志性摩天大楼的整体倒塌,给美国造成了巨大的灾害(如图7.2所示)。

图 7.1 汶川地震灾害
Fig. 7.1 Wenchuan earthquake disaster

图 7.2 纽约世贸中心大厦火灾
Fig. 7.2 The New York World Trade Center Building in fire

7.1.2 土木工程防灾减灾

我国是世界上自然灾害最为严重的国家之一,灾害种类多、分布地域广、发生频率高、造成损失重。近年来,在全球气候变化和我国经济社会快速发展的情况下,我国自然灾害损失不断增加,重大自然灾害乃至巨灾时有发生,灾害风险进一步加剧。在这种背景下,防灾减灾具有十分重要的意义。防灾减灾系统工程是一个由多种防灾减灾措施组成的有机整体,主要由以下几个环节组成。

① 灾害监测。灾害监测指监视测量与灾害有关的各种自然因素变化数据的工作,获取的监测资料用来认识灾害的发生规律并进行预防、预报。如监视地下岩石的运动和应力变化可以预测地震、滑坡等灾害。

7.1.2 Civil engineering disaster prevention and mitigation

China is one of the countries which suffers the natural disaster seriously. There are many types of disasters with wide distribution and high frequency. Under the situation of global climate change and rapid development of China's economy and society, the loss caused by natural disaster is increasing. Major natural disasters and catastrophes occur at any time, and further disasters aggravate. In this context, it has an important significance for disaster prevention and reduction. Disaster prevention and mitigation system is composed of a variety of disaster prevention and mitigation measures, which can be divided into the following parts.

① Disaster monitoring. The acquisition of monitoring data related to natural factors is always used to recognize the regulation of disaster occurrence, give prevention measures and predictions. For example, the monitoring of underground rock movement and the stress changes can predict earthquakes, landslides and other disasters.

② 灾害预报。灾害预报是指根据灾害的周期性、重复性、灾害间的相关性、致灾因素的演变和作用、灾害发展趋势、灾源的形成、灾害载体的运移规律，以及灾害前兆信息和经验类比，对灾害未来发生的可能性做出估计或判断。

③ 防灾。防灾是在灾害发生前采取的避让性预防措施，这是最经济、最安全又十分有效的减灾措施。防灾的主要措施有规划性防灾、工程性防灾、技术性防灾、转移性防灾和非工程性防灾等。

④ 抗灾。抗灾是指人类面对自然灾害的挑战做出的反应，如抗洪、抗震、抗风、抗滑坡和泥石流等，它主要包括工程结构的抗灾和工程结构灾后的检测和加固等。工程抗灾是防灾总体工作中的关键环节和重中之重。

⑤ 救灾。救灾是指灾害已经发生后迅速采取的减灾措施。救灾实际上是一场动员全社会、甚至国际社会力量对抗自然灾害的战斗，从指挥运筹到队伍组织，从抢救到医疗，从生活到公共安全，从物资供应到维护生命线工程，构成了一个严密的系统。

⑥ 灾后重建。灾后重建是指遭受毁灭性的自然灾害，如地震、洪水、飓风等之后，在特殊情况下的建设。

② Disaster prediction. The possibilities of the disaster are estimated, according to its periodicity, repeatability, relationship among disasters, its evolution and influence, its development trend, formation, movement, and precursory information and experience analogy.

③ Disaster prevention. Disaster prevention measures are taken before disasters, which is the most economical, safe and effective measure. The main measures include planning of disaster prevention, engineering prevention, technology prevention, transfer of disaster prevention and other non-engineering prevention.

④ Disaster resistance. Disaster resistance is the human response in front of natural disasters. It includes engineering control, such as earthquake resistance, flood resistance, landslide and debris flow resistance, wind resistance, as well as the detection of disasters and structural reinforcement. Engineering disaster resistance is a key sector in disaster prevention.

⑤ Disaster relief. Disaster relief is the mitigation measure which is taken quickly after the disaster occurred. Disaster relief is actually a movement of the whole society and even the international community power against natural disasters. From command management to team organization, from the rescue to medical treatment, from life to public security, from the material supply to maintain the lifeline engineering, it formed a tight system.

⑥ Post-disaster reconstruction. Post-disaster reconstruction is the construction in special conditions after the devastating natural disasters, such as earthquake, flood, hurricane, etc.

土木工程防灾减灾是综合防灾减灾的重要组成部分,是防灾减灾中最有效的对策和措施。图 7.3 所示的不仅是土木工程与抗灾、减灾的关系,更主要的是表明:

① 几乎所有自然灾害甚至人为灾害(如战争、核泄漏)都与土木工程有关;② 土木工程几乎对所有灾害都具有极强的积极主动性和不可替代性。例如,在建设时尽可能地提高工程的地震烈度可以大大地减少地震灾害时的损失。再如,筑堤、建坝可蓄洪,事先锚固可防止滑坡、泥石流;建造安全壳可防止核泄漏。几乎没有一个行业能像土木工程这样对抗灾、减灾具有如此巨大的积极主动性和不可替代性。

Civil engineering disaster prevention and mitigation is the most effective strategy and measure for disaster prevention and reduction. Fig. 7.3 shows the relationship among civil engineering, disaster resistance and disaster reduction, and other information are shown as follow: ① Almost all natural disasters and man-made disasters (such as war, nuclear leakage) are related to civil engineering. ② Civil engineering is active and irreplaceable to almost all disasters. For example, increasing earthquake intensity in the construction, so as to greatly reduce the loss caused by earthquake. We can build canal irrigation to resist drought, build embankment dams to resist flood, and make anchorage to resist landslide and debris.

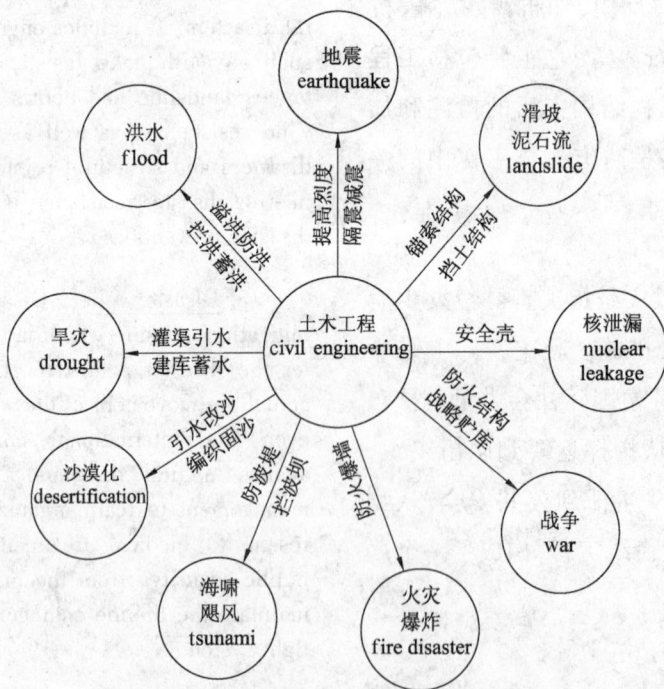

图7.3 土木工程与防灾减灾的关系示意图
Fig. 7.3 The diagram of the civil engineering and disaster prevention and mitigation

7.2 工程灾害与防灾减灾

7.2.1 地震灾害及抗震

地震是由于地壳破坏引发的地面运动,这种地面运动具有突发性和不可预测性,可能对土木工程结构造成严重破坏。全世界每年发生地震五百万次左右,其中1%为有感地震。我国平均每年发生30次5级以上地震,6次6级以上强震,1次7级以上大震,是世界上地震活动水平最高、地震灾害最重的国家之一。表7.1列出了21世纪以来的灾难性地震灾害。

7.2 Engineering disaster and disaster prevention and mitigation

7.2.1 Earthquake disaster and anti-seismic design

Earthquake is the result of the ground movement caused by the destruction of crust. The ground movement is sudden, unpredictable and can cause serious damage to civil engineering structure. Earthquakes occur around the world about five million times per year, including 1% felt earthquakes. On average, 30 earthquakes above Richter magnitude 5 are monitored every year in China, in which 6 earthquakes are above Richter magnitude 6, 1 earthquake is above Richter magnitude 7. China is one of the highest seismic active country in the world. Table 7.1 lists the earthquake disasters since the 21st Century.

表 7.1 21 世纪以来的灾难性地震

Table 7.1 The catastrophic earthquakes happened since the 21st century

发生时间和地点 Time and place	震级 Magnitude	造成损失 Losses
2013 年四川雅安市 Ya'an, China in 2013	7.0 7.0	死亡 196 人,伤 11 470 人,200 万人受灾 196 people dead, 11 470 people injured, and two million people affected
2011 年日本本州岛海域 Honshu island, Japan in 2011	9.0 9.0	死亡 15 844 人,失踪 3 450 人 15 844 people dead, 3 450 people missing.
2010 年海地太子港西部 west of port-au-prince, Haiti in 2010	7.3 7.3	死亡 27 万人,370 万人受灾 270 000 people dead, 3 700 000 people affected
2010 年智利康赛普西翁市 City of conception, Chile in 2010	8.8 8.8	死亡 802 人,近 200 万人受灾,经济损失达 200 亿美元 802 people dead, nearly two millon people affected, and with economic losses of $20 billion
2008 年中国四川省汶川县 Wenchuan county, China in 2008	8.0 8.0	死亡 7 万人,失踪 2 万人,伤 37 万人,损失 1300 亿美元 70 000 dead, 20 000 people missing, 370 000 people injured
2005 年巴基斯坦克什米尔地区 Pakistani Kashmir in 2005	7.6 7.6	死亡 7.3 万人,数百万人无家可归 73 000 people dead and millions of people become homeless
2004 年印尼苏门答腊岛海域 The sea of Sumatra, Indonesia in 2004	7.9 7.9	死亡或失踪 20 多万人。 More than two hundred thousand people dead or missing

地震造成的灾害可分为直接灾害和次生灾害。直接灾害主要表现为地面裂缝、错动、塌陷、喷砂冒水、山崩、滑坡等地表破坏；房屋倒塌、桥梁断落、水坝开裂等工程结构破坏；供水、供电、交通等生命线工程系统破坏。地震的次生灾害是指由地震间接引发的灾害，如地震诱发的火灾、水灾、有毒物质污染、海啸、瘟疫等。图7.4为2011年日本本州岛附近海域9.0级地震造成的灾害。

Earthquake disasters can be divided into direct disasters and secondary disasters. Direct disasters mainly display as follow. ① The surface damage, such as the ground crack, dislocation, collapse, landslide, sand blasting. ② Engineering structure damage, such as houses collapse, bridges and dams cracking off. ③ Lifeline engineering system damage, such as water supply, power supply, transportation supply disruptions. Secondary disasters are indirectly produced by earthquake, such as fire, flood, toxic pollution, tsunami, pestilence. Fig. 7. 4 shows the disasters caused by the Honshu island earthquake in Japan in 2011.

(a) 建筑物毁坏
Buildings destroyed

(b) 地面裂缝
The ground fissures

(c) 桥梁毁坏
Bridges destroyed

(d) 地震引发海啸
Tsunami

(e) 炼油厂爆炸燃烧
Blast combustion

(f) 核泄漏
The nuclear leakage

图7.4 2011年日本本州岛附近海域9.0级地震造成的灾害
Fig. 7.4 The disasters caused by the Honshu island earthquake in Japan in 2011

地震灾害不仅造成了众多结构物的倒塌、生命线工程的破坏、财产的重大损失，而且还夺去了众多的生命，对人们产生了重大的心理影响，引发了众多的社会

Earthquake disasters not only cause a lot of structure collapse, lifeline engineering damage, the loss of property, but also exert a significant psychological impact on people, and produce a number of social problems.

土木工程导论

问题。因此,抗震防灾工作具有十分重大的意义。目前,减轻地震灾害的对策从宏观上可分为三方面:地震预测预报、地震转移分散和工程抗震。

① 地震预测预报主要是根据地震地质、地质活动性、地震前兆异常和环境因素等多种情况,通过多种科学手段进行预测研究,对可能发生的地震进行预报。目前,地震预报还存在着许多难以解决的问题。

② 地震转移分散,是把可能在人口密集的大城市发生的大地震,通过能量转移,诱发至荒无人烟的山区或远离大陆的深海,或通过能量释放把一次破坏性的大地震转化为无数次非破坏性的小震。这种方法目前尚在探索研究初期。

③ 工程抗震是通过工程技术提高城市综合抗御地震的能力和各类建筑的耐震能力,当突发性地震发生时,把地震灾害减少至最轻的程度。工程抗震包括地震危险性分析和地震区划、工程结构抗震、工程结构减震控制等。

在工程抗震方面,通过重新修订各地区的抗震设防烈度,明确提出抗震设防目标,提高了工程抗震设计和检验的标准。近十年来,结构振动控制的研究和应用成为工程抗震领域的热点。传统结构抗震设计方法是依靠增加结构自身的强度和

Therefore, the earthquake disaster prevention is of a great significance. At present, countermeasures to mitigate earthquake disasters can be divided into: earthquake prediction, seismic shift dispersion, and anti-seismic engineering.

① Earthquake predictions are mainly based on seismic geology, geological activity, earthquake precursory anomaly, environmental factors and other conditions. Through a variety of scientific means, the potential earthquakes can be predicted. At present, there are still many problems in earthquake predictions.

② The seismic dispersion is to transfer the earthquake which may occur in a dense city, through the energy transfer, to the mountains or the continental sea, or transfer from a large destructive earthquake to numerous non-destructive earthquakes through releasing its energy. Relative method is still in research stage.

③ Anti-seismic engineering is to improve the city's comprehensive seismic capability and the seismic capacity of buildings through engineering and technology, and reduce the damage to the lightest degree when a sudden earthquake occurs. It includes earthquake risk analysis and seismic zoning, earthquake resistance of engineering structure, engineering structural vibration control.

Anti-seismic engineering improves the seismic design and inspection standard by revising the seismic fortification intensity in each area. Research and application of structural vibration control in nearly ten years has become a hot topic in the earthquake engineering field. Traditional anti-seismic design methods are mainly by increasing strength, deformation capacity of

变形能力来抗震,而减震控制方法则是采用隔震、消能、施加外力、调整结构动力特性等方法来消减结构地震反应,具有安全可靠、方便有效、经济节省和适用范围广等优点,是土木工程防灾减灾积极有效的方法和技术。

隔震及消能减震是目前土木工程中技术较成熟且应用较广的方法。在建筑物基础与上部结构之间设置隔震装置形成隔震层,把房屋结构与基础隔离开来,利用隔震装置来隔离和耗散地震能量以避免或减小地震能量向上部结构传输,以减小建筑物的地震反应,实现地震时建筑物只发生轻微运动或变形的目的,从而使建筑物在地震作用下不损坏或倒塌的抗震方法称为房屋基础隔震。传统抗震结构房屋与隔震房屋在地震中的情况对比,如图7.5所示。隔震系统一般由隔震器、阻尼器组成,它具有竖向刚度大、水平刚度小,能提供较大阻尼的特点。

structure to resist earthquake. Seismic control method is mainly by seismic isolation, energy dissipation, applied force, dynamic characteristics of structural adjustment, which has the advantages of safe and reliable, convenient and efficient, economical and applicable to a wide range.

At present, seismic isolation and energy dissipation are a more mature technology, and have a wide application in civil engineering. Isolation devices are arranged between the building foundation and upper structure (or system) to form the isolation layer. The structure and base are isolated, and the isolation device is used to isolate and dissipate the seismic energy in order to avoid or reduce the earthquake energy transmitted to the upper structure and reduce the earthquake response of buildings. At the same time, seismic structures only have slight movement or deformation. The buildings were not damaged or collapsed in the earthquake. This anti-seismic method is called the housing base isolation, as shown in Fig. 7.5. Isolation system is generally composed of isolator, damper, which has a great vertical stiffness and small lateral stiffness and can provide a larger damping.

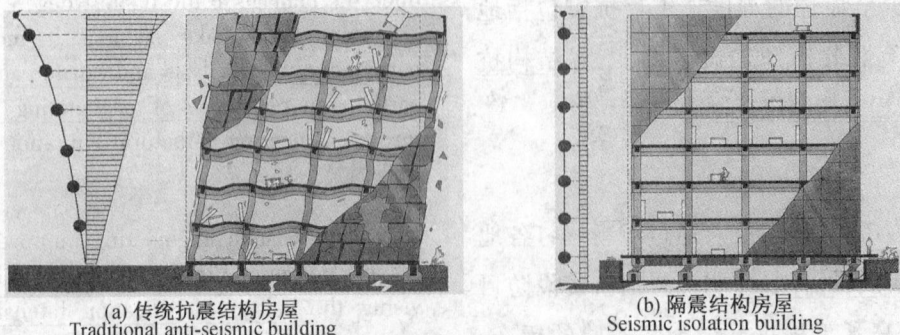

(a) 传统抗震结构房屋
Traditional anti-seismic building

(b) 隔震结构房屋
Seismic isolation building

图7.5 传统抗震房屋与隔震房屋在地震中的情况对比
Fig. 7.5 The contrast between the traditional houses and isolation houses in the earthquake

结构消能减震是通过采用一定的消能装置,消耗输入主体结构的地震能量,

We can reduce the vibration and damage of the structure by installing a certain energy dissipation device in structure. Energy dissipa-

从而减轻结构的振动和破坏。消能装置不改变主体承载结构体系,可同时减少结构水平和竖向的地震作用,在新建和建筑抗震加固中均可采用。消能装置包括各种消能支撑、消能剪力墙、摩擦阻尼器、软钢阻尼器、黏弹性阻尼器、黏滞流体阻尼器和组合式消能减震体系等。图7.6为采用橡胶支座和黏滞阻尼器组合消能减震体系的同济大学钢框架土木大楼。

tion device does not change the main bearing structure system, but can reduce the horizontal and vertical seismic behaviors of the structure. Therefore, it can be used in the reinforcement construction and building. Energy dissipation device seismic reinforcement and new construction include various energy dissipation supports, energy dissipation shear wall, friction damper, viscous elastic damper, viscous fluid damper and a combined energy dissipation system. Fig. 7. 6 shows the steel frame building in Tongji University, which using energy dissipation system that combined rubber bearings and viscous dampers.

| (a) 消能减震体系位置 | (b) 消能减震体系施工安装 |
| The location of energy dissipation system | The location of energy dissipation system |

图 7.6 橡胶支座和黏滞阻尼器组合消能减震体系
Fig. 7.6 Energy dissipation system combined by rubber bearing and viscous damper

7.2.2 风灾及抗风

风是大气层中空气形成的压力作用运动。由于地球表面不同地区的大气层所吸收的太阳能量不同,造成了各地空气温度的差异,从而产生气压差。气压差驱使空气从气压高的地方向气压低的地方流动,这就形成了风。风速就是风的前进速度。相邻两地间的气压差越大,空气流动越快,风速越大,风的力量自然也就越大。因此,通常都以风力来表示风的大小。根据风速大小,国际上将风力划分为

7.2.2 Wind damage and wind resistance

Wind is produced by interaction and movement of the air pressure in the atmosphere. Since the atmosphere in different regions of the surface absorb different solar energy, the air temperature and the air pressure are also different. The pressure difference makes the air flow from the high pressure region to the low pressure region, and the air flow forms the wind. Wind speed is the forward speed of the wind. The greater the pressure difference between two adjacent places is, the faster the air flows, so the higher the wind speed is, the greater the wind power is. Therefore, the wind is usually measured by wind power. The wind is divided into 18 le-

18 个等级(见表 7.2)。

vels according to the wind speed (Table 7.2).

表 7.2　风力等级表
Table 7.2　Wind scale table

等级 Scale	名　称 Name	距地 10 m 高处相当风速 Equivalent wind speed at the height of 10 meters		陆地地面现象 Phenomenon of the land surface	海面浪高/m Wave height
		风速/(m/s) Wind Speed	风速/(km/h) Wind Speed		
0	静风 calm	0.0 ~ 0.2	<1	静烟直上 smoke straight up	0.0
1	软风 light air	0.3 ~ 1.5	1 ~ 5	烟示风向 smoke shows wind direction	0.1
2	轻风 light breeze	1.6 ~ 3.3	6 ~ 11	感觉有风 feel the wind	0.2
3	微风 gentle breeze	3.4 ~ 5.4	12 ~ 19	旌旗展开 flags expand	0.6
4	和风 moderate breeze	5.5 ~ 7.9	20 ~ 28	吹起尘土 blowing dust	1.0
5	清风 fresh breeze	8.0 ~ 10.7	29 ~ 38	小树摇摆 trees sway	2.0
6	强风 strong breeze	10.8 ~ 13.8	39 ~ 49	电线有声 wire audio	3.0
7	疾风 near gale	13.9 ~ 17.1	50 ~ 61	步行困难 difficulty in walking	4.0
8	大风 gale	17.2 ~ 20.7	62 ~ 74	折毁树枝 branches destroyed	5.5
9	烈风 strong gale	20.8 ~ 24.4	75 ~ 88	小损房屋 small loss of the house	7.0
10	狂风 storm	24.5 ~ 28.4	89 ~ 102	拔起树木 uprooted trees	9.0
11	暴风 violent storm	28.5 ~ 32.6	103 ~ 117	损毁重大 significant damage	11.5
12	飓风 hurricane	32.7 ~ 36.9	118 ~ 133	摧毁极大 great destroy	14.0
13 ~ 17		≥37.0	≥134		

风灾是全球最常见和最严重的自然灾害之一,年复一年地给人类社会带来巨大的生命和财产损失。风灾具有发生频率高、次生灾害大(如暴雨、巨浪、风暴潮、洪水、泥石流等),持续时间长等特点。一

Wind disaster is one of the most common and most serious natural disasters in human society with huge losses of life and property year after year. Wind disaster has the property of high frequency, huge secondary disasters (such as rainstorm, billow, storm surge, flood, debris flow) and long dura-

般 6 级以下的风不会引起大的危害，6 级及 6 级以上较强的风有时会造成房屋、桥梁、车辆、船舶、树木、农作物以及通信系统、电力设施破坏及人员伤亡，由此造成的灾害称为风灾。

常见的导致灾害的风型主要有暴风、台风、龙卷风等。2009 年台风"莫拉克"造成我国 500 多人死亡、近 200 人失踪、46 人受伤（如图 7.7 所示）。1999 年 5 月 3 日，强劲龙卷风袭击美国俄克拉何马州和邻近的堪萨斯州，共造成 49 人丧生，摧毁了 2 600 间房屋，导致 8 000 多建筑物受损，经济损失达 12 亿美元（如图 7.8 所示）。

tion. Generally speaking, winds under level six do not cause great damage, while others at the six level or above may cause damage of houses, bridges, vehicles, ships, trees, crops and facilities of communication systems and powers, and it may cause casualty as well. The disasters caused by the above reasons is called wind disaster.

Common wind disasters are storms, typhoons, tornadoes. The typhoon Morakot in 2009 caused more than 500 people dead, nearly 200 people missing and 46 people injured in China (Fig. 7.7). On May 3rd, 1999, a strong tornado hit Oklahoma and its neighboring state Kansas, leading to 49 casualties, 2 600 houses destroyed, more than 8 000 buildings damaged, and the economic losses was $ 1.2 billion (Fig. 7.8).

图 7.7 台风"莫拉克"灾害
Fig. 7.7 Losess caused by the typhoon Morakot

图 7.8 龙卷风袭击美国俄克拉何马州
Fig. 7.8 Tornado hitting Oklahoma in America

从自然风所包含的成分看，风对构筑物的作用包括平均风作用和脉动风作用，从结构的响应来看，包括静态响应和风致振动响应。平均风既可引起结构的静态响应，又可引起结构的横风向振动响应。脉动风引起的响应则包括结构的准静态响应、顺风向和横风向的随机振动响应。当这些响应的综合结果超过结构的承受能力时，结构将发生破坏。风对构筑物的

The effect of wind on the structure can be divided into average wind effect and fluctuating wind effect according to the component of natural wind, and it can be divided into static response and wind-induced vibration response according to the response of the structure. The average wind can result in not only the static response, but also across-wind vibration response of the structure. The fluctuating wind response includes quasi-static response of the structure, random vibration response of wind direction and across-wind

破坏主要有以下几个方面。

① 对房屋建筑结构的破坏,主要表现在对多高层结构的破坏,对简易房屋,尤其是轻屋盖房屋的破坏,对外墙饰面、门窗玻璃及玻璃幕墙的破坏。2005 年 8 月 29 日,飓风卡特里娜吹毁了新奥尔良凯悦酒店等许多建筑的窗户、幕墙和外墙饰面,并砸毁了大量停在楼下的汽车以及其他物品(如图 7.9 所示)。

② 对高耸结构的破坏。风对高耸结构的破坏主要涉及桅杆和烟囱、电视塔等塔式结构。1988 年,美国一座高 610 m 的电视桅杆受阵风倒塌,造成 3 人死亡。

③ 对大跨结构的破坏。体育场馆、会展中心、汽车收费站等大跨结构也经常遭受风灾。2004 年河南省体育中心围护结构在 8 ~ 9 级的瞬时风袭击下严重受损(如图 7.10 所示)。

direction. When the combined result of these responses exceed the bearing capacity of the structure, the structure will be destroyed, which are shown in the following forms:

① Destruction of the housing construction. This destruction mainly displays in the high-rise structure, simple houses, especially in light roof houses, external wall finishes, doors, windows, and glass curtain wall. On August 29th, 2005, hurricane Katrina ruined the windows of Hyatt Regency in New Orleans and many other buildings, curtain wall and external wall finishes, and smashed a large number of cars parked downstairs and other items (Fig. 7.9).

② Destruction of the towering structure. The towering structure mainly include the mast, chimney tower, TV tower, and many other tower structures. In 1988, a TV mast with the height of 610 m collapsed due to gust, and three people were killed.

③ Destruction of long-span structures. Stadiums, convention centers, automotive toll stations and other long-span structures often suffer wind disasters. In 2004, envelope structure of Henan Sports Center building was severely damaged under the instantaneous winds of eight to nine levels (Fig. 7.10).

图 7.9　飓风致窗户损坏
Fig. 7.9　Windows damaged by hurricane

图 7.10　体育馆遭风灾破坏
Fig. 7.10　One gymnasium destroyed by typhoon

④ 对桥梁结构的破坏。风对桥梁的破坏作用也是非常巨大的。1940 年,美国华盛顿州塔科玛海峡建造的科马悬索桥,主跨 853 m,建好后不到 4 个月,就在一场风速不到 20 m/s 的风灾下,因产生上下和来回扭曲振动而倒塌了(如图 7.11 所示)。

⑤ 对电厂冷却塔的破坏。冷却塔也容易遭受风灾。1965 年 11 月 1 日的一场平均风速为 18 ~ 20 m/s 的大风把英国渡桥热电厂 8 个冷却塔中的 3 个都摧毁了(如图 7.12 所示)。

④ Destruction of the bridge structure. The wind damaging effects on bridge are very huge. In 1940, Coma suspension bridge built on Tacoma Narrows in the state of Washington U. S., whose main span is 853 m, collapsed because of the twisting back and forth at a hurricane with the wind speed of less than 20 m/s(Fig. 7.11).

⑤ Destruction of the power plant cooling towers. The cooling tower is also vulnerable to storms. On November 1st, 1965, a strong wind with average speed of 18 ~ 20 m/s destroyed three of the eight cooling towers in UK Du Bridge Thermal Power Plant (Fig. 7.12).

图 7.11　美国塔科玛海峡桥风灾毁坏
Fig. 7.11　Tacoma Narrows Bridge (USA) Destroyed by Wind

图 7.12　冷却塔风灾毁坏
Fig. 7.12　Cooling tower Destroyed by wind

⑥ 对输电系统等的破坏。供电线路的电杆埋得浅,在大风中容易被刮倒,造成停电事故,严重影响生产和生活(如图 7.13 所示)。

⑦ 对港口设施的破坏。2002 年 8 月 31 日,强台风"鹿莎"使韩国釜山港遭受重创。

⑧ 对海洋工程结构的破坏。2005 年秋季的"卡特里娜"和"丽塔"两个飓风毁

⑥ Destruction of the power transmission system. If the burial depth of the supply line pole is shallow, the pole is easy to be blown down by strong winds, causing power outages which seriously affect the production and life (Fig. 7.13).

⑦ Destruction of the port facilities. On August 31st, 2002, violent typhoon Rusa swept throughout South Korea. A large number of cranes and other equipment were damaged, and the port of Rusa suffered a big loss.

⑧ Destruction of marine engineering structures. In the autumn of 2005, the hurricane Katrina and Rita destroyed 113 oil rigs

坏了墨西哥湾地区113座石油钻井平台以及457条油气管道(如图7.14所示)。

and 457 oil and gas pipelines in Gulf of Mexico (Fig. 7.14).

图7.13 高压输电塔风灾折断
Fig. 7.13 High-voltage transmission tower destroyed by cyclone

图7.14 石油钻井平台风灾毁坏
Fig. 7.14 Oil rig destroyed by typhoon

风工程就是研究风对结构作用和结构对风的响应以及减小结构风致响应的一门多领域交叉学科。风工程所涉及的范围很广,包括大气科学、空气动力学、结构力学、实验力学等。图7.15是由加拿大学者达文波特提出的一个风荷载链。为了求得结构对风荷载的响应,该荷载链上的每一环都是不可或缺的。

Wind engineering is a multi-field interdiscipline. It mainly focuses on the study of wind effects on the structure, the responses of the structure under the wind, the measures to reduce structural wind-induced response and the wind damage accident. It covers a wide range, including atmospheric science, aerodynamics, structural mechanics, experimental mechanics, etc. Fig. 7.15 is a wind load chain presented by Canadian scholar Davenport. In order to obtain the response of the structure under the wind load, each ring on the load chain is indispensable.

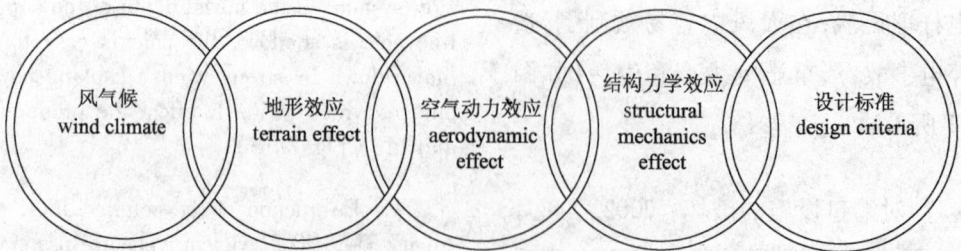

图7.15 达文波特风灾害链
Fig. 7.15 Wind disaster chain presented by Davenport

第一环"风气候"要确定不同地理区域气候意义上的平均风的一般特性。

The first ring "wind climate" is to determine the general characteristics of the average wind in the sense of different geographical regions climate.

第二环"地形效应"要确定受到地表不同地形影响的低层大气的局部风特性。对于低层大气中的风,由于空气运动受地表摩擦阻滞的影响,不仅其平均风速随高度的降低而降低,而且还表现出较强的紊流特性和随机性,因此,对它的描述或处理显得更加困难。通常只能比较粗略地对不同地区的地形进行分类,并用一个统计意义上被称为"粗糙长度"的参数来表征。

第三环"空气动力效应"要确定的是由风产生的作用在结构的上荷载,包括静力荷载和动力荷载。作用在一个结构上的风荷载不仅与风的特性有关,而且在很大程度上受到结构本身的几何外形和相邻建筑物的影响,而这种影响一般可以通过风洞试验来确定。风荷载一般随时间和结构物表面的空间位置而变化。

第四环"结构力学效应"要确定由风荷载引起的结构响应,包括静力响应和动力响应。确定结构的静力响应相对较简单,而确定结构的动力响应要复杂得多。

第五环"设计标准"要解决的问题是如何把前四个环节中的研究成果总结成尽可能简洁、准确的标准条文,应用到实际结构的抗风设计上。

风工程研究方法有现场测试、风洞试

The second ring "terrain effect" is to determine the local wind characteristics of the lower atmosphere under the influence of different surface topography. Since air movement is affected by the surface friction block, the wind in the lower atmosphere not only decrease the average wind speed along with the reduction of height, but also show the strong turbulent characteristic and randomness. Therefore, it seems to be more difficult to describe or deal with it. Usually, we can only classify the terrain in different parts roughly, and use the parameter in a statistical sense known as "roughness length" to characterize it.

The third ring "aerodynamic effect" is to determine the load on the structure produced by the wind, including the static load and the dynamic load. The wind load acting on a structure is not only related to the characteristics of the wind, but also largely influenced by the geometric shape of the structure itself and the neighboring buildings, and this kind of influence can generally be determined through the wind tunnel tests. The wind load generally varies with the time and the surface spatial position of the structure.

The fourth ring "structural mechanics effect" is to determine the structure response caused by the wind load, including the static response and the dynamic response. The determination of the static response of structure is relatively simple, but the determination of the structural dynamic response is much more complex.

The fifth ring "design criteria" aims to solve the problem on how to summarize the above four research results as standard provisions as concise and accurate as possible and apply them to the actual structure of wind-resistance design.

Wind engineering research has three

验和理论计算 3 种。理论计算包括解析计算和数值计算。现场测试方法不适用于对风工程现象的规律性和机理的研究，也无法在工程建设实施前解决相关的实际问题。现场测试方法常被作为一种有效的验证理论计算和风洞试验结果的手段。风洞试验不仅保留了直观性的优点，可以节约人力、物力和时间，而且在风洞试验中可以在很大程度上人为地控制、调节和重复一些试验条件，因此它是一种很好的研究风工程现象变参数影响和机理的手段（如图 7.16 和图 7.17 所示）。

methods, including field test, wind tunnel test and theoretical calculation. Theoretical calculation includes analytic calculation and numerical calculation. The field test method is not applicable to the research of regularity and mechanism of wind engineering phenomena, and it cannot solve the practical problems before the implementation of engineering construction. The field test method is often used as an effective means to verify the theoretical calculations and wind tunnel test results. Wind tunnel test not only retains the advantage of intuitive, but also saves manpower, material and time relatively. Moreover, we can artificially control, adjust and repeat some test conditions to a large extent in wind tunnel test. It is a very good means to research variable parameters effect and the mechanism of the phenomenon in wind engineering (Fig. 7.16 and Fig. 7.17).

图 7.16 大跨悬索桥风洞试验
Fig. 7.16 Long-span suspension bridge wind tunnel test

图 7.17 上海中心大厦风洞试验
Fig. 7.17 Shanghai tower wind tunnel test

目前，防止风灾害的主要措施有以下几种：

① 重点研究各地区的风荷载特性。例如，研究地区风压分布、地面粗糙度划分、高层建筑风效应、大跨建筑和桥梁结构风效应等，为制定和修正荷载及相关规范提供依据。

② 充分考虑风灾因素，加强工程结构的抗风设计。

At present, the measures to prevent wind disaster include:

① Focus on the wind load characteristics of different regions.

② Take full account of the wind disaster factors and strengthen the wind-resistance design of engineering structures.

③ Construct windbreak, sand fixation and wind revetment vegetation to reduce wind damage to the city and coast.

④ Establish the mechanisms of forecast and early warning in the regions which often

③ 建造防风固沙林和防风护岸植被，以减少风力对城市和海岸的破坏。

④ 在经常遭受风灾害的地区，建立预报、预警机制。

⑤ 城市应编制风灾害影响区划，建立合理有效的应对策略。

传统的结构抗风对策是首先保证强度，然后验算位移，若位移过大则应通过增强结构自身刚度和抗侧力能力来抵抗风荷载作用，这是一种被动的、不经济的方法。近30多年来发展起来的结构振动控制技术开辟了结构抗风设计的新途径。结构振动控制技术就是在结构上附设控制构件和控制装置，在结构振动时通过被动或主动地施加控制力减小或抑制结构的动力反应，从而减少动力位移，以满足结构的安全性、适用性和舒适度的要求。图7.18所示为台北101大楼采用调谐质量阻尼器（TMD）进行风振舒适度控制。图7.19所示为桥梁黏滞阻尼器风振控制。

suffer the wind disaster.

⑤ Compile regionalization affected by wind and establish reasonable and effective coping strategies.

The traditional structural countermeasure against wind is to ensure the strength and then check the displacement. However, when displacement is too large, we will strengthen the structure stiffness and lateral resistance to resist the effect of wind loads. This is a passive and uneconomical way. The structural vibration control technology developed in nearly 30 years has offered a new approach to structural wind-resistance design. The structural vibration control technology is used to attach the control members and control devices to the structure, by applying passively or actively control to reduce or inhibit the dynamic response of structures, when the structure vibrates thereby reducing the dynamic displacement in order to meet the requirements of structural safety, applicability and comfort. Fig. 7.18 is Taipei 101 building which adopts tuned mass damper (TMD) for wind-induced comfort control. Fig. 7.19 is the wind vibration control of bridge with viscous damper.

图7.18　台北101大楼调频质量阻尼器风振舒适度控制

Fig. 7.18　TMD wind vibration comfort control in Taipei 101 building

图7.19　桥梁斜拉索黏滞阻尼器风振控制

Fig. 7.19　Wind vibration control of bridge with viscous damper

7.2.3 冰雪灾害及抗灾

在冬季,各类土木工程还会受到雪荷载的作用,大的雪荷载及由此引起的冰雨还会引起建筑结构、道路、桥梁及输电线塔等结构产生破坏。2008 年 1 月期间,我国南方大部分地区相继出现了持续的大范围灾害性冰雪天气,此次雨雪冰冻天气过程影响范围大、持续时间长、涉及面广、危害程度大,给当地的交通、电力、通信和人民生活带来严重影响(如图 7.20 和图 7.21 所示)。

7.2.3 Disaster of snow and resisting disaster

All kinds of civil engineering are affected by the snow load in winter, and the sleet caused by the large snow load may destroy the building structures, roads, bridges and transmission line towers. In January 2008, most regions in southern China continuously suffered a wide range of severe snow and ice. The results of the freezing rain and snow weather had a wide influence, long duration, big harm degree and serious impact on transportation, electricity, communication and people's life (Fig. 7.20 and Fig. 7.21).

图 7.20 雪荷载引起的厂房倒塌
Fig. 7.20 The collapse of factory building caused by snow load

图 7.21 雪荷载及冻雨引起的输电线塔破坏
Fig. 7.21 The damage of transmission tower caused by snow load

《建筑结构荷载规范》中所采用的雪荷载,是根据历史记录,经统计分析,结合结构设计原则与方法确定的。一方面,随着极端灾害的发生,需要对所采用的标准及具体数值进行不断地修正和调整。另一方面,随着人们对工程安全性、适用性与耐久性等功能要求的不断提高,工程抗灾能力的设防标准也要不断提高。

According to the historical data and the statistical analysis, snow load used in *Building Structures Load Specification* is determined by combining with structural design principles and methods. On the one hand, with the occurrence of extreme disasters, the adopted standards and the specific values need to be continuously correct and adjust. On the other hand, people's needs for the functional requirements of the engineering safety, suitability and durability are continuously improved, and the fortification standards of engineering resilience must be continuously improved.

7.2.4 火灾及防火

火灾是指时间和空间上失去控制的火,在其蔓延发展过程中将给人类的生命财产造成损失的一种灾害性的燃烧现象。火灾是各种灾害中发生最频繁、影响面最广的灾种之一。火灾可以分为自然火灾和建筑火灾两大类。随着城市化的发展,建筑火灾及其危害越来越严重。这些火灾不仅带来了重大的人员伤亡和财产损失,也严重影响了建筑结构的安全。例如,上海静安区胶州路公寓大楼特大火灾(如图 7.22 所示)以及中央电视台新址配楼火灾(如图 7.23 所示)。

7.2.4 Fire disaster and fire prevention

Fire disaster is a disastrous combustion phenomenon. In its spreading process, the fire is out of control with the loss of life and property. Fire disaster is the most frequent disaster and has the greatest impact. Fire disaster can be divided into two major categories: natural fire and building fire. With the development of urbanization, building fire is more and more serious. These fires not only brought heavy casualties and property losses, but also seriously affected the structural safety. For example, the catastrophic fire in Jiaozhou Road, Jing'an District, Shanghai Condominium (Fig. 7.22), the new CCTV side building fire disasters (Fig. 7.23).

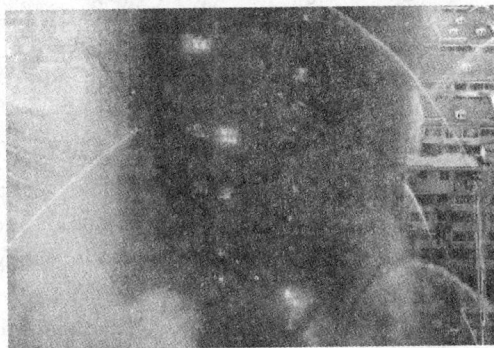

图 7.22 上海静安区胶州路公寓大楼特大火灾
Fig. 7.22 Catastrophic fire in Shanghai

图 7.23 央视新址配楼大火
Fig. 7.23 The new CCTV side building fire disaster

火灾是一个燃烧过程,要经过发生、蔓延和充分燃烧几个阶段。火灾的严重性取决于持续时间和温度,而这两者又受建筑类型、燃烧荷载等诸多因素的影响。控制和改善影响燃烧的各种因素是建筑防火设计首先要考虑的问题。

对于建筑结构构件,在受火时,随着

Fire disaster is a combustion process which has to go through the stages of occurence, spreading and full combustion. The seriousness of the fire depends on the duration and temperature, both of which are affected by building types, combustion load, ect. Controlling and improving various factors that affect the burning are the first thing to consider when design the building fire safety.

With the lengthening of temperature and duration in the fire, the mechanical proper-

温度的升高和持续时间的增加,构件的力学性能下降到不足以承受设计规定的荷载,此时该构件将部分或全部失去正常工作的能力。为避免火灾对建筑结构安全性的影响,防止结构在火灾中发生破坏或坍塌,必须对结构进行抗火设计;火灾发生后,应当及时、有效地进行扑救,并在灾后对建筑结构进行损伤鉴定和加固修复,降低火灾的危害。

① 建筑防火,包括建筑火灾基础科学、建筑总体布局、建筑内部防火隔断、防火装修以及消防扑救、安全疏散路线、自动防排烟系统的设计和研究。在建筑防火方面,我国已制定《建筑设计防火规范》(GB 50016—2012)。

② 建筑结构抗火性能,主要包括结构材料的抗火性能,结构在火灾高温下的强度、刚度、变形、承载能力,建筑结构耐火时间以及结构抗火构造等内容。目前,欧洲各国颁布了钢结构耐火设计技术规范,法国还制定了混凝土结构的耐火强度计算方法,日本建立了建筑结构耐火设计数据库,而我国至今还没有建筑结构抗火设计规范。

③ 火灾后建筑结构的损伤鉴定和加固修复。火灾后建筑物的损伤诊断主要包括现场调查及火灾温度判断;火灾后建筑材料及结构性能的检测;受损分析和剩余承载力的计算;结构受损综合评定。火灾后的结构加固方法,目前常用的有喷射

ties of building structural member will not be able to hold the design loads, and the member will partially or totally loss the ability to work at that time. In order to avoid the fire influence to the building structure and prevent the damage or collapse of the structure in the fire, we need to make fire resistance design of structures. When the fire disaster happening, we should fight the blaze promptly and effectively, also we need to make a damage identification, the reinforcement and restoration of the structure so as to reduce the fire hazard.

① Building fire protection includes the basic science of building fire, the overall building layout, the cut-off of building interior fire, fire decoration and firefighting, safe evacuation routes, automatic anti-exhaust system design and research. In building fire protection, we have *Code for Fire Protection Design of Buildings* (GB 50016–2012).

② Fire-resistance performance of building structures includes fire-resistance performance of structural materials, the strength, stiffness, deformation and bearing capacity of the structure under the high temperature in the fire, fire resistance time of building structure, etc. European countries have enacted steel fire design and technical specifications, and France has also developed a fire-resistant strength calculation method on the concrete structure. Japan has established a database about fire resistance design of building structures.

③ Damage identifying, reinforcement and restoration of building structure after fire disaster. Damage identifying of building structures includes field investigation and determination of fire temperature, detection of building materials and structural properties, the damage analysis and residual bearing capacity calculation, and the integrated assess-

混凝土法、粘钢加固法、碳纤维增强复合材料加固法等。

7.2.5 地质灾害及防治

自然的变异和人为的作用都可能导致地质环境或地质体发生变化,当这种变化达到一定程度,其产生的后果便给人类和社会造成危害,称为地质灾害。由于我国处于特殊的地质构造部位,且2/3地区属于山地,地质灾害分布广、类型多、频率高、强度大,每年都造成众大人员伤亡和严重经济损失,已成为影响我国城乡建设和人民生存环境的重大问题。地质灾害包括滑坡、泥石流、崩塌、地面沉降等。

滑坡是指斜坡上大量土体或岩体由于种种原因在重力作用下,沿一定的滑动面整体向下滑动的现象(如图7.24所示)。滑坡的防治途径主要有以下几种:

① 终止或减轻诱发滑坡的外部环境条件,如截流排水、卸荷减载和坡面防护。

② 改善边坡内部力学特征和物质结构,如土质改良。

③ 设置抗滑工程直接阻止滑坡的发展,如抗滑桩、挡土墙和预应力锚固等。

泥石流是一种工程动力地质现象,它是一种水与泥沙、石块混合在一起流动的

ment of structural damage. The structural reinforcement method includes sprayed concrete, bonded steel reinforcement, carbon fiber reinforced composites reinforced.

7.2.5 Geological disaster and prevention

Natural variations and human activities may lead to changes of the geological environment or geological body. When such changes reach a certain level, their consequences will cause harm to humans and society, which called geological disasters. Since China is in a special geological structure parts, and 2/3 of the area belongs to the mountain, geological disasters in China are widely distributed, multi-type, high frequency and intensity, and caused many casualties and serious economic losses, which has become the major issues affecting the urban and rural construction and people's living environment. Geological disaster includes landslide, debris flow, collapse and ground subsidence.

Landslide is the phenomenon that the overall soil or rock mass on the slopes slide downward under gravity along the sliding surface due to various reasons (Fig. 7.24). The main ways of landslide prevention and treatment are as follow:

① Terminate and reduce the external environmental conditions which caused the landslide, such as closure and drainage, unloading and slope protection.

② Improve the internal mechanical characteristics and the structure of the matter in a slope, such as soil improvement.

③ Set anti-sliding engineering which prevent the development of the landslide directly, such as pile, retaining wall and prestressed anchor.

Debris flow is an engineering geological phenomenon, and it also belongs to a special torrent mixed with water, mud and rock. It's

特殊洪流,具有突然爆发、流速快、流量大、物质容量大和破坏力强的特点。泥石流爆发时,大量泥沙石块沿山沟奔腾而下,在很短的时间内,冲进乡村和城镇,冲毁铁路、公路和航道等交通设施,淹没农田,造成灾害(如图 7.25 所示)。泥石流的防治措施主要有:

①跨越。在泥石流地段的铁路、公路线路,可采用桥梁、隧道等方式跨越泥石流。

②排导。修筑排导沟、急流槽、导流堤等工程,将泥石流顺利排走。

③拦挡。在泥石流的沟中修筑一系列的低矮小坝。

④拦截。修筑拦淤库和储淤场,减弱下泄物质总量和洪峰流量。

⑤水土保护,包括生物措施和工程措施两类防护方法。

feature include sudden outbreak, high flow velocities, and strong destructive. When debris flow happens, large amount of sands and stones swoop down along the ravine, rush into the villages and towns, destroy rails, roads and waterway transportation facilities, and flood farmland in a very short time (Fig. 7.25). The prevention and control measures of debris flow are as follows:

① Cross. Railway and highway lines can step across debris flow by bridge and tunnel.

② Exhaust. We can drain away debris flow smoothly by building exhaust ditch, rapids slot and diversion dike.

③ Resist. We can build a series of low dams.

④ Intercept. We can build silt library and silt storage field, which can reduce the total discharged substances and peak flow.

⑤ Soil and water conservation, including biological and engineering measures.

图 7.24　山体滑坡对建筑物造成破坏
Fig. 7.24　Damages caused by landslide

图 7.25　2010 年甘肃舟曲特大泥石流灾害
Fig. 7.25　Extremely heavy debris flow disaster in Zhouqu,Gansu

崩塌是指较陡斜坡上的岩土体在重力作用下突然脱离母体崩落、滚动、堆积在坡脚的地质现象(如图 7.26 所示)。崩塌是多山地区及黄土高原常见的自然灾害之一。崩塌可以单独发生,也可能与滑坡一起发生,甚至可能和泥石流同时发生。防治崩塌的常用措施有:

Collapse is a geological phenomenon that rock and soil on the steeper slopes separated from the body, collapsed, scrolled and piled up at the foot of the slope (Fig. 7.26). Collapse is one of the common natural disasters in the mountainous area and loess plateau. Collapse can occur separately, and it may also occur related with landslide, and even occur with debris flow. Commonly

① 清除坡面危岩。

② 加固坡面。在易风化剥落的边坡地段，修建护墙、挡墙。

③ 拦截、遮挡。通过设置拦石网、拦石墙、落石槽等方法进行拦截。

④ 修筑排水构筑物，防止水流大量渗入掩体而恶化斜坡的稳定性。

地面沉降是指由于自然动力因素，如地壳的下降运动、地震、火山活动等，或受地下开采、地下施工或灌溉等人工活动的影响，造成地下空洞或使地下松散土压缩固结，导致地面标高下降的现象（如图7.27所示）。地面沉降可能引起地面建筑物、市政管道和交通设施等城市基础设施的损坏。例如，上海市由于过度开采地下水及高楼密度过大，自20世纪90年代起，地面平均每年下降1cm。

measures about prevention and treatment of collapse are as follow:

① Clearing unstable rocks on slope.

② Reinforcing slope.

③ Intercepting rocks through setting block stone network and retaining wall.

④ Building draining structures.

Ground subsidence is a phenomenon of ground elevation drop caused by downward movement of the earth's crust, earthquakes, volcanic activity, or by underground mining, underground construction or irrigation and other human activities (Fig. 7.27). Ground subsidence can damage the ground buildings, municipal pipelines, transport facilities and other urban infrastructures. Shanghai, for example, due to over-exploitation of groundwater and excessive high-rise building, the average annual decline is 1cm since the 1990s.

图 7.26　重庆彭水山体崩塌毁坏道路
Fig.7.26　The road destroyed by collapse of the mountain

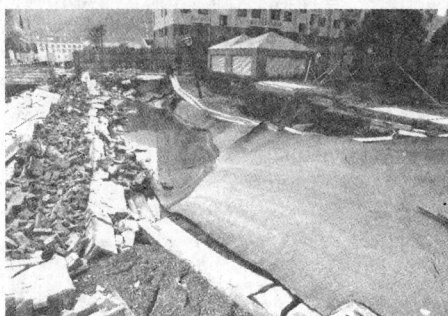

图 7.27　浙江温岭保障房小区地面沉降
Fig.7.27　Ground subsidence in a residential district

当前对地面沉降的控制和治理措施分为表面治理和根本治理两大类。表面治理是对已有地面沉降的地区，通过地面整治和改善环境的手段，减弱地面沉降造成的损失。根本性治理是从研究消除引起地面沉降的根本因素入手，谋求缓和，

Currently, the control measures of ground subsidence can be divided into surface control and fundamental control. Surface control can reduce the losses by ground remediation and environmental improvement. Fundamental control is used to ease the subsidence until control or end it by researching how to eliminate the fundamental factors.

直至控制或终止地面沉降的措施。其主要方法有：人工补给地下水；限制地下水开采，调整开采层次，以地面水源代替地下水源；限制或停止开采固体矿物。

7.2.6　工程事故灾害

　　工程事故灾害是指由于勘察、设计、施工和使用过程中存在重大失误造成工程倒塌（或失效）而引起的人为灾害。1999 年 1 月 4 日，重庆市綦江县旧彩虹桥发生整体垮塌（该桥建成不足 3 年），造成40 人死、14 人伤的惨剧和 600 多万元的重大经济损失（如图 7.28 所示）。事故发生后，专家对彩虹桥工程进行质量鉴定指出，该工程存在多处致命的质量问题：① 拱架钢管焊接存在严重缺陷，焊接质量不合格；② 钢管混凝土抗压强度不足，所用混凝土平均强度低于设计强度的 1/3；③ 桥梁构造设计不合理，致使连接拱架与桥梁、桥面的钢绞线拉索、锚具、锚片严重锈蚀；④ 压力灌浆不密实。

The main methods include three aspects：artificial recharge of groundwater，restrict groundwater extraction，and adjust the level of exploitation of ground water instead of groundwater sources，restrict or stop the mining of solid minerals.

7.2.6　Engineering accident disaster

　　Engineering accident disaster is the man-made disaster caused by the collapse of the project or failure due to the significant errors in the process of survey，design，construction and use. On January 4th，1999，the overall collapse of the old Rainbow Bridge in Qijiang County，Chongqing caused the death of 40 people，the injuring of 14 people and the major economic loss about more than 6 million（Fig. 7.28）. This bridge was built in less than 3 years. After the accident，the experts made the quality identification for Rainbow Bridge and found many fatal quality problems：① The arch steel pipe welding had serious defects，and the welding was unqualified. ② The strength of steel pipe concrete was insufficient and the strength of the used concrete was less than one-third of the design strength. ③ The bridge structural design was unreasonable，which resulted in the severe corrosion of the strand cable，the anchor and the anchor piece connecting the arch，the bridge and the deck. ④ The pressure grouting was not dense.

(a) 垮塌前
Before the collapse

(b) 垮塌后
After the collapse

图 7.28　重庆市綦江县旧彩虹桥整体垮塌
Fig.7.28　Overall collapse of the old rainbow bridge in qijiang county, Chongqing

从技术方面分析,发生工程事故灾难的原因有:① 地质资料勘察严重失误,或根本没有进行勘察;② 地基承载力不够,同时基础设计又严重失误;③ 结构方案、结构计算或结构施工图有重大错误,或凭"经验"设计,无图施工;④ 材料和半成品的质量严重低劣,甚至采用假冒伪劣的产品和半成品;⑤ 施工和安装过程中偷工减料,粗制滥造;⑥ 施工的技术方案和措施有重大失误;⑦ 使用过程中盲目增加使用荷载,随意变更使用环境和使用状态;⑧ 任意对已建成工程打洞、拆墙、移柱、改扩建、夹层等。

从管理方面分析,发生工程事故灾难的原因有:① 由非相应资质的设计、施工单位进行设计、施工;② 建筑市场混乱无序,出现"六无"工程项目(无报建程序、无设计图纸、无勘察资料、无招投标、无资质施工、无质量监督);③ "层层分包"现象普遍,使设计、施工的管理处于严重失控状态;④ 企业经营思想不正,片面追求利润、产值,没有建立可靠的质量保障制度;⑤ 无固定技工队伍,技术工人和管理人员素质不高。

Analyzing from the technical aspects, the reasons of the engineering accident disasters are as follows: ① The geological exploration data are seriously wrong or no exploration at all. ② Bearing capacity of the foundation is not enough, and the basis design is also seriously wrong at the same time. ③ Structural program, structural calculation or structural drawing have major mistakes, and the design is made by experience. ④ The used materials and semi-finished products are seriously poor, or even belong to fake and shoddy product. ⑤ There are cutting corners and shoddy in the construction and installation process. ⑥ There are serious mistakes in the technology programs and measures. ⑦ The load is blindly increased during the production, and the environment and the used state are freely changed. ⑧ The completed project is freely changed by the holes, tearing down walls, column shift, expansion and mezzanine.

Analyzing from the management aspects, the reasons of the engineering accident disasters are as follow: ① The construction and the design are not finished by the corresponding construction units and the design units. ② Due to the confusion in building market, there exist "six none" projects, including no reported course, no design drawings, no survey data, no bidding, no qualified construction and no quality supervision. ③ "Sub-layers" is a widespread phenomenon that make the management of design and construction out of control. ④ The business's thinking is not correct, pursuit one-side profit and production value, and there is no reliable quality assurance system. ⑤ The construction company has no fixed mechanic team, and the quality of the skilled workers and managers are not high.

工程结构灾害检测与加固

Disaster detection and reinforcement of engineering structures

7.3.1 材料在灾害环境下的性能

工程结构受到地震、风、火、水、冰冻、腐蚀和施工不当引起的灾害,最终将落实到受灾后的结构检测与加固上。工程结构检测鉴定与加固涉及灾害材料学、灾害检测鉴定学、灾害修复和加固等领域。

在工程结构的灾后检测加固研究中,首先要关注的是材料受灾后的性能(如强度、弹性模量、本构关系等)变化,在这方面国内外都已做了许多研究,定性和定量地得出了一些结论,但是系统性还显不够,故在土木工程领域中,灾害材料学还未形成一个专门的学科。然而,在工程结构的加固设计、工程鉴定和工程咨询等实践中又必不可少地需要这方面的知识。

灾害材料学涉及土木工程材料的一般力学性能,如混凝土的内部裂缝和破坏机理、钢筋的内部结构破坏机理、砌体的一般破坏机理等;动力荷载对材料的影响,如钢筋的疲劳、冲击荷载对混凝土和钢筋的作用;火灾对材料性能的影响,如对混凝土或钢筋的影响、对混凝土与钢筋间黏结力的影响等;冰冻对材料性能的影响,如受冻混凝土的力学性能;腐蚀对材料性能的影响等。

7.3.1 Material performance in the disaster environment

Engineering structures under earthquake, wind, fire, water, freezing disasters, corrosion, and improper construction should be implemented to the detection and reinforcement of the affected structure finally. Detecting and reinforcement of the engineering structures involve disaster material science, disaster detection identification science, disaster restoration and reinforcement, etc.

In the study of the detection and reinforcement of engineering structures after a disaster, the primary concern is the changes of the material performance (such as strength, elastic modulus, the constitutive relation). Many researches have been done, and some conclusions have been obtained qualitatively and quantitatively, but these studies are not systematic. Therefore, in the field of civil engineering, disaster material science has not yet formed a specialized discipline.

Disaster material science involves the general mechanical properties of civil engineering materials. For example, the interior crack and failure mechanism of the concrete, the internal structure failure mechanism of the reinforcement and the failure mechanism of the brickwork; The influence of the dynamic loading on the material, such as fatigue of the reinforcement, the action of the impact load on concrete and steel bar; The influence of the fire on the properties of materials, such as the impact on concrete or steel, the influence of the bond between the concrete and reinforcement; The influence of freeze on the properties of materials, such as the mechanics performance of freezned concrete; the influence of corrosion on the properties of materials.

7.3.2 结构灾害检测与鉴定

工程结构检测与鉴定是采用各种检测方法对工程结构进行耐久性检测,并对其安全性、可靠性进行鉴定,得出鉴定等级和是否需要加固的结论的整个过程,其程序为:检测任务委托—调查—编制检测方案—现场检测—发出检测报告—进行鉴定评级和出具鉴定报告。

例如混凝土和砌体结构的检测过程,通常应先根据图纸数据进行理想结构的受力分析,找出关键构件,如关键的梁、柱或墙;然后在建筑现场对关键构件的关键部位去除表面装饰层和粉刷层,采用适当的检测方法确定混凝土碳化深度和弹性模量,或确定砌体结构砂浆和砌块的实际强度和弹性模量(如图7.29所示)。对部分关键混凝土构件可采用钻孔取芯得到真实的混凝土块,在实验室测得其真实混凝土强度(如图7.30所示)。将实测材料强度和弹性模量代入结构计算模型进行受力分析,分析结果可以评估结构是否安全并指出不安全的构件,作为结构加固的依据。

7.3.2 Structure disaster detection and identification

Engineering structure detection and identification is a complete process by using a variety of detection methods for durability testing of the engineering structure and identification of the security and reliability, then the identified level and the conclusion about whether the reinforcement is needed will be reached. The program includes detection commissioned, investigation, making detection plan, field testing, test report, rate identification and issuing reports.

For example, in the testing process of concrete and masonry structure, you should analyze the force on ideal structure and find out the key components such as beam, column or wall firstly, according to the drawing data. Secondly, at the scene of the construction, the surface decoration layer and paint layer are cleaned in the key components of the key parts, and concrete carbonation depth and elastic modulus, or the actual strength of the mortar and brick masonry structure and elastic modulus are determined using the appropriate test method (Fig. 7.29). As to some key concrete component, concrete blocks can be obtained by drilling core, then the real strength of concrete can be measured in the laboratory (Fig. 7.30). The measured material strength and modulus of elasticity are brought into structure calculation model. As the basis of structure reinforcement, the results of the analysis can evaluate whether the structure is safe and detect unsafe components.

图 7.29 原位压力机检测砌体强度
Fig.7.29 Press testing in masonry strength

图 7.30 钻孔取芯机取样
Fig.7.30 Sampling by core drilling machine

7.3.3 工程结构改造及加固

工程结构改造加固学是一门研究使受损的工程结构重新恢复使用功能或适应新的使用功能的学科。由于土木工程和现代科学技术的迅猛发展,这门学科的发展速度也异常迅速。工程结构需要加固和改造的原因通常有以下几种:

① 使用荷载增加。如桥梁车辆吨位增加、对建筑物进行加层或二次装修等,使原有结构增加负担。

② 抗震加固。现有结构达不到抗震设防指标,对结构的基础、柱、砌体墙、梁板进行加固改造,增加抗震能力(如图7.31 所示)。

③ 结构损害。建筑物年久失修或由于各种灾害导致结构损伤或破坏,不能满足目前的使用要求或安全度不足。

④ 纠正设计或施工失误。如纠正设计中配筋不足、构件截面太小,或施工中偷工减料等问题。

⑤ 历史性建筑的保护(如图 7.32

7.3.3 Engineering structural transformation and reinforcement

Engineering structure reinforcement learning is the study that restoring the using function of the damaged engineering structure or adapting to the new function. Due to the rapid development of civil engineering and the modern science and technology, this subject also developed rapidly. The reasons that building need to reinforce and reconstruct usually fell into the following kinds:

① The increase of using load. Adding bridge tonnage vehicles, layer or secondary decoration can increase the burden of the original structure.

② Seismic strengthening. The existing structures can not meet the seismic fortification target, then seismic capability should be increased for the structure of the foundations, columns, masonry walls, beams and slabs reinforcement renovation (Fig. 7.31).

③ The structural damage. The dilapidated and damaged structures caused by various disasters cannot meet current demand or insufficient degree of safety.

④ Correcting the design or construction mistakes.

⑤ The protection of the historic buildings(Fig. 7.32).

所示）。

图 7.31　中学教学楼基础隔震抗震加固
Fig.7.31　Building seismic strengthening

图 7.32　南京博物院老大殿整体顶升与隔震加固
Fig.7.32　The Integrally jack-up and isolation reinforcement

在发达国家和地区，工程改造加固已成为建筑业的重要组成部分。目前我国既有建筑物总量达 400 多亿平方米，建筑业也已开始由大规模新建时期迈向新建与维护并重时期。土木工程结构的加固方法主要有：加大截面法、外包钢加固法、预应力加固法、钢绞线网聚合物砂浆加固法、改变传力途径加固法、粘钢加固法、化学灌浆法、基础托换、基础纠偏等，如图 7.33 所示。

In the developed countries and regions, engineering reinforcement has become an important part of the construction. Currently, total quantity of existing buildings in China is more than 40 billion square meters, and the construction in China has also started from a large-scale construction period towards a building and maintenance period. There are many methods of the reinforcement applying to different engineering structures respectively (Fig. 7.33). For example, enlarging section method, the steel-encased reinforcement method, the prestressed strengthening method, the steel wire mesh-polymer mortar strengthening method, changing force's transfer road, stick steel strengthening method, chemical grouting method, and foundation underpinning, etc.

(a) 加大截面法
Enlarging section method

(b) 外包钢加固法
The steel-encased reinforcement method

土木工程防灾与减灾　Civil Engineering Disaster Prevention and Mitigation

(c) 预应力加固法
The prestressed strengthening method

(d) 钢绞线网聚合物砂浆加固法
Steel wire mesh-polymer mortar strengthening method

(e) 改变传力途径加固法
Changing force's transfer road

(f) 基础托换
Foundation underpinning

图 7.33　土木工程结构主要加固方法
Fig.7.33　Strengthening methods of civil engineering

7.4 工程防灾减灾的新成就与发展趋势

7.4.1 工程防灾减灾的新成就

城市化的加速,使得生命线系统工程防灾减灾成为研究的热点。近年来,我国学者对生命线工程系统减灾开展了基础研究并取得了重大的研究成果。

① 建立了生命线工程系统的场地危险性评估方法。

② 建立了地下管网等生命线工程系统在地震作用下的反应分析方法。

③ 建立了城市多灾害损失的评估模型。

7.4 The new achievements and trends of disaster prevention and mitigation engineering

7.4.1 New achievements of disaster prevention and mitigation

With the acceleration of urbanization, the lifeline systems engineering disaster prevention and mitigation has become a research hotspot. In recent years, Chinese scholars carried out many basic researches and made significant achievements in the lifeline engineering system mitigation.

① Established the site risk assessment method of lifeline engineering systems.

② Established the earthquake response analysis method of the underground pipe network system.

③ Established loss assessment model for city multi-hazard.

④ 提出了<u>基于性能的结构抗震设计方法</u>。

⑤ 研究了城市地震触发滑坡、岩溶塌陷、采空区塌陷以及地震火灾和渗水引发滑坡等灾害链现象，并提出了相应的评估方法。

⑥ 提出了大跨度建筑与桥梁结构的抗震抗风分析和隔震减振控制方法。

7.4.2 工程防灾减灾的发展趋势

自然灾害的预测、预报、预防和救助科学理论和技术，涉及自然科学领域的众多方面，而且大多数都是当前世界性的高科技前沿课题。防灾减灾科学研究处于自然科学、技术科学和社会科学的交汇点，体现了三者的相互渗透和结合。防灾减灾科学技术具备跨学科性，其成果具有广泛的社会应用性是它的重要特点。例如，防灾减灾工作中的综合减灾、灾害系统论、减灾模型、失误控制论、灾害哲学及文化、人为灾害、数字减灾系统等概念及评估方法都体现了上述特点。实践还表明，防灾减灾预测预防基础研究具有鲜明的超前性，突破途径的非常规性，某些重大发现的偶然性以及科学创新的艰难性。由于灾害涉及的学科较多，灾种间还有交叉，各地区所面临的重点有所不同，所以土木工程防灾减灾学的内容可以扩展，并不断地发展和完善。

① 开展自然灾害危险性评价和风险评估。美国和俄罗斯等国家重点对实际

④ Put forward <u>performance based seismic design method</u>.

⑤ Studied the disaster chain phcnomena including landslides, collapse caused by urban earthquakes, and landslides caused by seepage, then put forward the corresponding evaluation method.

⑥ Put forward the seismic and wind resistance analysis method as well as the vibration control method of long-span buildings and bridges.

7.4.2 New trends of disaster prevention and mitigation

Natural disaster prediction, forecasting, prevention and rescue involve many aspects of the natural science, and most of them are the current worldwide high-tech frontier. Disaster prevention and mitigation science research is the intersection of natural science, technology science and social science. It embodies the mutual infiltration and combination of all three. Disaster prevention and mitigation is interdisciplinary science and technology, and its achievements with extensive social application is its important characteristic. For example, the comprehensive disaster reduction, disaster system theory, disaster reduction model, disaster philosophy and culture, man-made disasters, digital disaster reduction system concept and evaluation method in the disaster prevention and mitigation work reflect the above characteristics.

① Carrying out the risk assessment for natural disasters. The United States, Russia and other countries are focusing on practical

风险和可承受风险进行评估。其成果已成为减灾应用基础研究的重要前沿领域之一,并为城市生命线工程的抗灾能力评估提供了依据。

② 开展承灾体易损性研究。美国等国家积极开展大城市的底层易损性研究、自然灾害社会易损性研究、积极易损性与社区易损性研究。这些研究在理论和方法上促进了承灾体易损性研究的深入,也表明承灾体易损性研究是综合科技减灾的前沿领域。

③ 灾害信息系统建设。美国、日本、加拿大和欧盟等国家,为进行灾害和紧急事物的管理,更好地沟通灾害信息,减轻灾害损失,均建立了灾害信息系统,实现灾害信息共享,以达到在灾害面前各方快速应急反应的要求。

④ 新材料、新技术的应用。传统的结构加固方法所用到的材料,如焊接、螺栓连接等会对原结构产生新的损伤和应力,而纤维增强复合材料是由高性能纤维材料与基体按一定比例并经过一定工艺复合形成的一种高性能新型材料,具有高强、轻质、耐腐蚀及施工方便等优点,并开始以各种形式应用于土木与建筑结构工程中。

and affordable risk assessment. The achievements become one of the important frontier fields in the basic research, and provide a basis for the evaluation of urban lifeline engineering disaster ability.

② Researching on vulnerability of hazard-affected bodies. The United States and other countries actively develop the underlying vulnerability research on big cities, social vulnerability research, positive vulnerability and community vulnerability research.

③ Disaster information system construction. America, Japan, Canada and some countries in the European Union have established the disaster information system, for the purpose of disaster management, better communication of disaster information and reducing disaster losses.

④ The application of new material and technology. Traditional structure reinforcement methods, such as welding, and bolt connection on the original structure produce new trauma and stress. The fiber reinforced polymer(FRP) is a high-performance new material which is composed by a certain proportion of high-performance fiber and matrix and through a certain process to form. Due to the advantages of high strength, light weight, corrosion resistance, and it begin to be applied in civil engineering structures.

注:本章图片均来源于网络。
Note:In this chapter, all pictures are from webs.

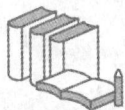

知识拓展
Learning More

相关链接　Related Links

（1）中国国家减灾网 http://www.jianzai.gov.cn/

（2）中国地震局 http://www.cea.gov.cn/

（3）太平洋地震工程研究中心 http://nisee2.berkeley.edu/

小贴士　Tips

我国《建筑抗震设计规范》（GB 50011—2010）明确提出设防要求："小震不坏，中震可修，大震不倒。"

The design requirements were put forward in *Code for Seismic Design of Buildings*（GB50011-2010）in China. No damage in small earthquake, repairable under moderate earthquake, no collapse under rare earthquake.

思考题　Review Questions

（1）土木工程灾害有哪些主要类型？

What are the main types of civil engineering disasters?

（2）如何减轻地震灾害？

How can earthquake disasters be mitigated?

（3）如何防止风灾？

How can wind disaster be prevented?

（4）如何防治地质灾害？

How to prevent and control geological disaster?

参考文献
References

［1］中华人民共和国建设部：《建筑抗震设计规范》（GB 50011—2010），中国建筑工业出版社，2011 年。

［2］中华人民共和国建设部：《建筑结构荷载规范》（GB 50009—2012），中国建筑工业出版社，2012 年。

［3］中华人民共和国建设部：《建筑防火设计规范》（GB 50016—2006），中国计划出版社，2006 年。

［4］张相庭：《结构风工程——理论·规范·实践》，中国建筑工业出版社，2006 年。

［5］江见鲸，徐志胜，等：《防灾减灾工程学》，机械工业出版社，2005 年。

［6］李爱群：《工程结构减振控制》，机械工业出版社，2007 年。

［7］Palanichamy M. S. Basic Civil Engineering. China Machine Press, 2005.

［8］Shen Zuyan. Introduction of Civil Engineering. China Architecture & Press, 2010.

［9］叶志明:《土木工程概论》(第 3 版),高等教育出版社,2009 年。

［10］Williams A. Seismic Design of Buildings and Bridges. China Waterpower Press, 2002.

［11］周新刚:《土木工程概论》,中国建筑工业出版社,2011 年。

［12］周云:《土木工程防灾减灾学》,华南理工大出版社,2002 年。

［13］董羡,黄林青:《土木工程概论》,中国水利水电出版社,2011 年。

［14］袁海军,姜红:《建筑结构检测鉴定与加固手册》,中国建筑工业出版社,2006 年。

［15］任建喜:《土木工程概论》,机械工业出版社,2011 年。

［16］Emil Simiu, Robert H. Scanlan. Wind Effects on Structures: Fundamentals and Applications to Design. 3rd ed. John Willey & Sons, INC., 1996.

［17］王茹:《土木工程防灾减灾学》,中国建材工业出版社,2008 年。

［18］李文虎,代国忠:《土木工程概论》,化学工业出版社,2011 年。

土木工程导论

第8章 桥梁与隧道工程

Chapter 8　Bridge and Tunnel Engineering

在日常生活中,很少有人关注桥梁和隧道。但是,作为基础建设中的重要组成部分,桥梁和隧道穿越不同的障碍,开辟出新的道路,给人们带来诸多便利。20世纪80年代以来,在桥梁跨度和隧道长度方面,世界纪录不断被刷新。本章在桥梁和隧道基本概念及发展历史的基础上,对桥梁和隧道的常见结构形式进行了简单介绍,并对当今时代桥梁和隧道发展的前沿进行了展望。

In daily life, few people concern about bridges and tunnels. However, as a basis of the important part in construction, bridges and tunnels bring convenience to people. Since the 1980s, the world record of the bridge span of and tunnel length has been constantly renewed. In this chapter, we will give a brief introduction to the common forms of structure profile on the basic of the concepts, the historical development of bridges and tunnels, and the future for the development of bridges and tunnels.

8.1　桥梁

8.1　Bridge

8.1.1　桥梁概念及现状

（1）桥梁的基本概念

桥梁是为跨越天然或人工障碍物而修建的建筑物。

（2）桥梁的发展历史和现状

从工程技术的角度来看,桥梁的发展按时间可划分为古代桥梁和现代桥梁。

8.1.1　The concept and status of bridges

（1）The basic concept of bridges

Bridges refers to constructions that were constructed to across the natural or artificial barriers.

（2）The development history and current status of bridges

From the engineering point of view, the development of the bridges can be divided

1）古代桥梁

人类在原始时代，只能利用自然倒下的树木，自然形成的石梁或石拱，溪涧突出的石块，谷岸生长的藤萝等，跨越水道和峡谷。古代桥梁在 17 世纪以前，一般是用木、石材料建造的，并可按建桥材料把桥分为石桥和木桥。天下闻名的赵州桥是世界上现存最早，保存最好的石拱桥，距今已有 1 400 年的历史，如图 8.1 所示。苏州宝带桥，是驰名中外的多孔石桥，如图 8.2 所示。目前世界上最长的石梁桥，是福建泉州安平桥，如图 8.3 所示。扬州五亭桥建于 1757 年，桥上建有 5 座亭子，故又称莲花桥，如图 8.4 所示。

1) Ancient bridges

In ancient times, people crossed waterways and valleys using the trees fell down naturally, stone beam or stone arch formed naturally, prominent rocks in the stream, vines and so on. Before the 17th century, ancient bridge is generally constructed with wood and stone materials, and can be classified into stone bridge and wooden bridge according to materials. The famous Zhaozhou Bridge is the oldest and best preserved stone arch bridge in the world, and it has a history of more than 1 400 years, as shown in Fig. 8. 1. Suzhou Precious Belt Bridge is the famous stone bridge with many holes, as shown in Fig. 8. 2. Currently, the longest stone beam bridge in the world is Anping bridge located in Quanzhou, as shown in Fig. 8. 3. Five Pavilion Bridge in Yangzhou was built in 1757. It has five pavilions, also called Lotus Bridge, as shown in Fig. 8. 4.

图 8.1　赵州桥
Fig.8.1　Zhaozhou Bridge

图 8.2　苏州宝带桥
Fig.8.2　Suzhou Precious Belt Bridge

图 8.3　泉州安平桥
Fig.8.3　Anping Bridge

图 8.4　扬州莲花桥
Fig.8.4　Five Pavilion Bridge

在一些欧洲国家,也有许多著名的古桥。

在瑞士卢塞恩,最引人注目的是斜跨在罗伊河面的木制长桥——卡贝尔桥,如图8.5所示。法国嘉德水道桥,最大跨度为24 m,如图8.6所示。

In some European countries, there are also many famous bridges.

In Lucerne, Switzerland, the most striking view is the wooden long bridge, named as Chapel Bridge over Roy river, as shown in Fig. 8.5. Guardian Aqueduct in France with the maximum span of 24 m is shown in Fig. 8.6.

图8.5 卡贝尔桥
Fig. 8.5 Chapel Bridge

图8.6 法国嘉德水道桥
Fig. 8.6 Guardian Aqueduct in France

2) 现代桥梁

现代桥梁指18世纪中期以来,通过运用现代技术及知识所建的桥梁。

世界跨度第一的悬索桥——日本明石海峡桥,如图8.7所示,横跨日本内海,使日本神户与淡路岛紧紧相连。世界上最长的跨海大桥——青海海湾大桥,如图8.8所示。位于中国江苏省南通市的苏通大桥,横跨长江,大桥于2008年7月1日正式通车,总长1 088 m,超过了长为809 m的日本的Tatara大桥,成为世界最长的斜拉桥,如图8.9所示。东海大桥是世界上最长的跨海大桥之一,长约19 km,建成于2005年,如图8.10所示。

2) Modern bridge

Modern bridge is concerned with the bridges which were built through technology and knowledge since the mid-18th century.

The longest suspension bridge in the world is Akashi Strait Bridge in Japan, as shown in Fig. 8.7. It crosses the sea of Japan and linked the Japanese Kobe with Awaji Island closely. The longest cross-sea bridge in the world is Qingdao Bay Bridge, as shown in Fig. 8.8. Sutong Bridge which located in Nantong, Jiangsu Province, crossing the Yangtze River. It officially opened on July 1st, 2008, with the length of 1 088 m. It is longer than the Tatara Bridge in Japan, with the length of 809 m, and it becomes the longest cable-stayed bridge in the world, as shown in Fig. 8.9. Donghai Bridge was built in 2005, and it is one of the longest cross-sea bridge in the world. It is about 19 kilometers, as shown in Fig. 8.10.

图 8.7　日本明石海峡桥
Fig. 8.7　Akashi Strait Bridge in Japan

图 8.8　青岛海湾大桥
Fig. 8.8　Qingdao Bay Bridge

图 8.9　苏通大桥
Fig. 8.9　Sutong bridge

图 8.10　东海大桥
Fig. 8.10　Donghai Bridge

① 梁桥

梁桥构造简单,施工方便,工期短,造价低且维修容易,应用非常广泛。

目前世界上跨度最大的钢梁桥是加拿大的魁北克桥,桥的主跨为 548.6 m,如图 8.11 所示。日本于 1976 年建成了世界上跨径最大的连续钢架构桥——浜名大桥。1980 年,在菲律宾的帕罗斯岛,建成了中跨为 240.8 m 的科勒巴贝尔塞浦桥,从而使浜名大桥退居第二。目前世界上跨度最大的混凝土梁桥是我国重庆的长江大桥复线桥,如图 8.12 所示。

① Beam bridge

Beam bridge has the characteristics of simple structure, convenient construction, short construction period, low cost and easy maintenance, and it is widely used now.

Currently, the world's longest beam bridge is Quebec Bridge in Canada, and the main span of the bridge is 548.6 m, as shown in Fig. 8.11. The world's largest span continuous steel structure bridge, named as Hamana Bridge, was built in Japan in 1976. In 1980, Philippines built the Kohler Babel Cyprus Bridge in the Paros Island with the mid-span of 240.8 m, and Hamana bridge was relegated to the second place. Currently, the world's longest span of concrete girder bridge is Yangtze River Bridge of double-track bridge in Chongqing, as shown in Fig. 8.12.

图 8.11　魁北克桥
Fig. 8.11　Quebec Bridge

图 8.12　重庆长江大桥复线桥
Fig. 8.12　Yangtze River Bridge of double-track bridge

② 拱桥

古今中外拱桥遍布各地,在桥梁工程中占有重要地位。又因为其造型美观,也常用于城市、风景区的桥梁建筑。

1873 年,法国人首创建成了一座拱式人行桥,它代表着钢筋混凝土拱桥的崛起。从 19 世纪末到 20 世纪 50 年代,钢筋混凝土拱桥无论是在跨越能力,还是结构体系等方面均有很大的发展。2009 年建

② Arch bridge

Arch bridges are very popular, and they occupy an important position in the bridge engineering. It also used commonly in cities and scenic bridge constructions because of its attractive appearance.

In 1873, the Frenchman built a pedestrian arch bridge firstly which represents the rise of the reinforced concrete arch bridge. From the end of 19th century to the 1950s, the reinforced concrete arch bridge has developed in terms of spanning capacity, structural system, etc. Chaotianmen Bridge which has been open to traffic in 2009 is the

成通车的重庆朝天门大桥,是世界上跨度最大的拱桥,如图 8.13 所示。主跨超过 550 m 的上海卢浦大桥,主跨 540 m 的韩国傍花大桥,以及主跨 518 m 的美国的新河谷大桥,都是世界大跨度拱桥的典范(如图 8.14,图 8.15,图 8.16 所示)。

world's longest arch bridge, as shown in Fig. 8.13. The main span of the Lupu Bridge is over 550 m, as shown in Fig. 8.14. The span of the Korean Bangwha bridge is 540 m, as shown in Fig. 8.15. The New River Gorge bridge in USA, with span of 518 m, is shown in Fig. 8.16. They are all the models of the world's largest span arch bridge.

图 8.13 重庆朝天门大桥
Fig. 8.13 Chaotianmen Bridge

图 8.14 上海卢浦大桥
Fig. 8.14 Lupu Bridge

图 8.15 韩国傍花大桥
Fig. 8.15 Korean Bangwha Bridge

图 8.16 美国新河谷大桥
Fig. 8.16 New River Gorge Bridge

③ 斜拉桥

斜拉桥,又称斜张桥,是用锚在桥塔上的多根斜向钢缆索吊住主梁的缆索承重桥。

现代斜拉桥起源于德国。1938 年,德国工程师迪辛格尔在设计汉堡附近跨越易北河的一座悬索桥时,明确提出了在悬

③ Cable-stayed bridge

Cable-stayed bridge, also known as stayed bridge, is a kind of supported bridge that anchoring the girder on the bridge tower by a plurality of diagonal cables.

Modern cable-stayed bridge originated from Germany. In 1938, German engineer Singhal designed a suspension bridge across the Elbe near Hamburg. He clearly presen-

索桥体系中结合使用斜拉桥,可大大地减少荷载引起的挠度。近年来,斜拉桥这种结构形式已日益广泛地在世界各地得到应用。2009 年建成的苏通大桥为世界上跨径最大的斜拉桥,超过了主跨 856 m 的法国诺曼底大桥(如图 8.17 所示),以及主跨 890 m 的日本多多罗大桥(如图8.18 所示)。

ted the use of cable-stayed bridge in conjunction with cable-stayed suspension bridge system, which can greatly reduce the deflection caused by the load. In recent years, the structure of cable-stayed bridge has been applied widely in the world. The construction of Sutong Bridge in 2009 set the record of the world's largest cable-stayed span. The span of Sutong Bridge surpasses the French Normandy bridge with its 856 m main-span (Fig. 8.17), and the Tatara bridge in Japan, with its span of 890 m(Fig. 8.18).

图 8.17　法国诺曼底大桥
Fig. 8.17　French Normandy Bridge

图 8.18　日本多多罗大桥
Fig. 8.18　Tatara Bridge in Japan

目前世界上跨度最大的公铁两用斜拉桥是我国的武汉天兴洲大桥,它也是世界上铁路列车通行速度最快的桥梁,如图 8.19所示。

Currently, the world's longest rail-cum-span cable-stayed bridge is Tianxingzhou Bridge in Wuhan province, and it is also the world's fastest train traffic bridge, as shown in Fig. 8.19.

图 8.19　武汉天兴洲大桥
Fig. 8.19　Tianxingzhou bridge in Wuhan

④ 悬索桥

悬索桥，又称吊桥，是一种古老的桥型，是以缆索为主要的承重结构，并与桥塔、吊杆、锚定和桥面结构共同组成缆索的承重桥。现代悬索桥是特大跨度桥梁的主要形式之一，特别是在跨越大河、深谷等不易修筑桥墩的地区，悬索桥具有独特的优势。

1883年，著名的布鲁克林悬索桥在纽约东河上建成，如图8.20所示，它是当时世界上最长的大桥。之后该纪录被旧金山金门大桥刷新，如图8.21所示。

图8.20　布鲁克林悬索桥
Fig. 8.20　Brooklyn suspension bridge

目前，世界上跨度最大的悬索桥是日本的明石海峡大桥，它也是世界上跨径最大的桥梁，如图8.7所示。我国现代悬索桥的建造虽然起步较晚，但是从20世纪90年代开始取得非常大的进步，修建了不少大跨度悬索桥。2009年建成通车的西堠门公路大桥，成为我国跨度最大的桥梁，如图8.22所示。香港青马大桥是一座公铁两用悬索桥，于1997年建成通车，是

④ Suspension

Suspension bridge, also known as draw-bridge, is an ancient bridge type, which uses the rope as the main load-bearing structure. It forms the cable bearing bridges with ropes, pylon, boom, anchoring and cable. Modern suspension bridge is one of the main forms of long-span bridges, and it has a unique advantage, especially when it crosses rivers, valleys and other areas.

In 1883, the famous Brooklyn suspension bridge was built on the East River in New York, as shown in Fig. 8. 20, which was the world's longest bridge at that time. The San Francisco golden gate bridge rewrote record shortly (Fig. 8. 21).

图8.21　旧金山金门大桥
Fig. 8.21　The San Francisco golden gate bridge

Currently, the suspension bridge with the largest span is Akashi Kaikyo Bridge in Japan, and it is also the largest span bridge, as shown in Fig. 8. 7. Despite a late start of the construction of modern suspension bridge, a very great progress has been made in China since 1990s' and a number of large-span suspension bridges have been built. Xihoumen Bridge, opened to traffic in 2009, has become the largest span bridge in China, as shown in Fig. 8. 22. Tsing Ma Bridge in Hong Kong is a rail-cum-bridge. It opened to traffic in 1997, and it is the largest span rail-

当今世界上跨度最大的两用桥，如图 8.23 所示。

图 8.22　西堠门公路大桥
Fig. 8.22　Xihoumen Bridge

cum-bridge in the world, as shown in Fig. 8.23.

图 8.23　香港青马大桥
Fig. 8.23　Tsing Ma Bridge in Hong Kong

8.1.2　桥梁的组成和类型

（1）桥梁的组成

桥梁有 5 个"大部件"（如图 8.24 所示）和 5 个"小部件"组成。这五大部件介绍如下。

① 桥跨结构（或称桥孔结构）。桥跨结构是线路遇到障碍（如江河、山谷等）时，跨越这类障碍的主要承载结构。

② 支座系统。支座系统支承上部结构并传递荷载于桥梁墩台上，其应满足上部结构在荷载、温度或其他因素作用下所预计的位移功能。

③ 桥墩。桥墩是支承两侧桥跨上部结构的建筑物。

④ 桥台。桥台位于河道两岸，一端与路堤相接，防止路堤滑塌；另一端支承桥跨上部结构。

⑤ 基础。基础是保证墩台安全并将荷载传至地基的结构部分。基础工程在整个桥梁工程施工中是比较困难的部分，

8.1.2　The composition of bridge

（1）The composition of bridge

The bridge has five "big parts" (Fig. 8.24) and five "widgets". The five big parts include:

① Bridge span structure (also known as the bridge opening structure). It is the main load-bearing structure when crossing line obstacles(such as rivers, valleys or other routes).

② Support system. It is used to support the upper structure and pass loads to bridge piers, and it should meet the displacement function under load, temperature, or other factors that are expected of the supper structure.

③ Bridge pier. It is the structure supporting the superstructure on both sides of a bridge.

④ Bridge abutment. It is located on both sides of the river , and one side connects with the embankment to prevent embankment slumping; the other side supports the superstructure of the bridge.

⑤ Foundation. It is a structure that ensures the safety of abutment and transmit

而且常常需要在水下施工,因而遇到的问题复杂得多。

loads to subgrade. Foundation engineering is a difficult part across the bridge construction, and it often requires the underwater construction, thus its problems are very complicated.

桥跨结构和支座系统是桥梁的上部结构,桥墩、桥台和基础为桥梁下部结构。一般在路堤与桥台衔接处,或者桥台的两侧设置锥形护坡,以保证迎水部分路堤边坡的稳定。

Bridge span structures and support systems are called the bridge superstructure, and piers, abutments and foundation are called the bridge substructure. On the connection of the embankment and the bridge abutment, there is a tapered abutment slope on both sides of the abutment. It can ensure the stability of the embankment slope.

五小部件是:桥面铺装(或称行车道铺装)、排水防水系统、栏杆(或防撞护栏)、伸缩缝和灯光照明。这五小部件均为与桥梁服务功能有关的部件,总称为桥面构造。

Five widgets are bridge deck pavement (or called carriageway pavement), drainage waterproofing system, railing (or crash barrier), expansion joints and lighting. These five widgets are all related to the service functions of a bridge, collectively known as the bridge deck structure.

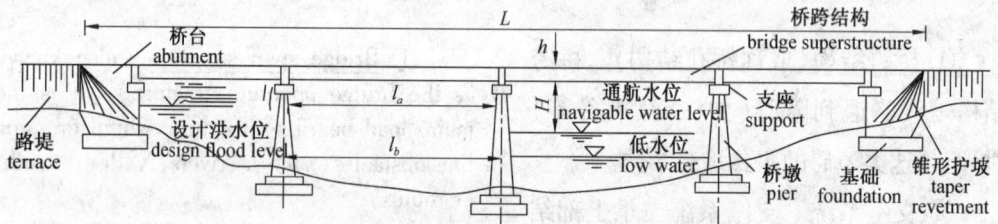

图 8.24 桥梁的基本组成
Fig. 8.24 The composition of a bridge

(2) 桥梁的类型

桥梁的分类方法多种多样,主要有以下几种:

① 按桥梁的长度规模,分为特大桥、大桥、中桥、小桥等。

② 按桥梁主体结构材料分类,分为钢桥、混凝土桥、石桥、木桥。

③ 按桥梁的用途,分为铁路桥、公路桥、公铁两用桥、人行桥、输水桥等。

(2) The types of bridges

The classification of bridges is varied, mainly includes the following aspects:

① According to the length, bridges are divided into grand bridge, major bridge, medium bridge, and so on.

② According to the materials of the main structure, bridges are divided into steel bridge, concrete bridge, stone bridge, and wooden bridge.

③ According to the function, bridges

④ 按桥梁的跨越对象,分为跨河桥、跨谷桥、跨线桥、立交桥等。

⑤ 按行车道在桥跨的位置,分为上承式桥、中承式桥、下承式桥。

⑥ 按桥梁的平面布置,分为正桥、斜桥、弯桥等。

⑦ 按桥梁结构体系,分为梁桥、拱桥、悬索桥3种基本体系和组合体系。

8.1.3 梁式桥

梁式桥是古老的结构体系之一,是一种在竖向荷载下不产生水平反力的结构,包括简支梁、连续梁和悬臂梁,如图8.25所示。

are divided into railway bridge, road bridge, combined rail-cum-road bridge, footbridge, water bridge and so on.

④ According to the across object, bridges are divided into bridge crossing the river, bridge crossing the valley, overpass bridge, flyover and so on.

⑤ According to the location of the carriageway on the bridge span, bridges are divided into deck bridge, half-through bridge and through bridge.

⑥ According to the layout, bridges are divided into main bridge, skew bridge and curved bridge.

⑦ According to the structural system, bridges are divided into beam bridge, arch bridge, suspension system and a combination of three basic systems.

8.1.3　Beam bridge

Beam bridge is one of the ancient structures, and it is a kind of structure without producing the horizontal reaction force by the vertical load. Beam bridge can be divided into simply supported beam, continuous beam and cantilever beam, as shown in Fig. 8.25.

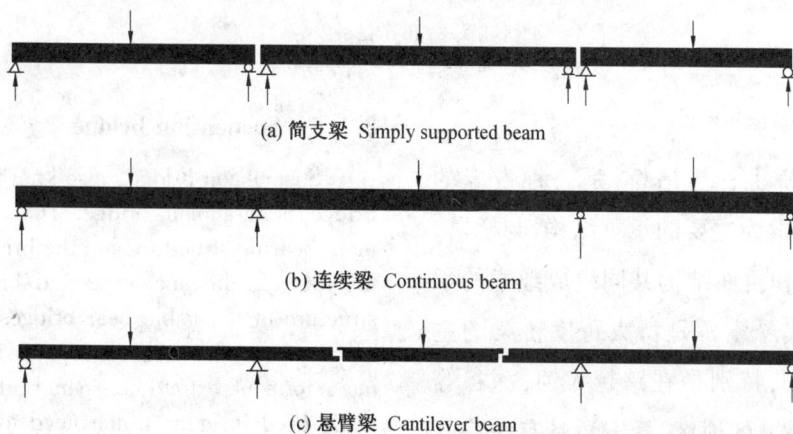

(a) 简支梁　Simply supported beam

(b) 连续梁　Continuous beam

(c) 悬臂梁　Cantilever beam

图 8.25　梁式桥
Fig. 8.25　Beam bridge

8.1.4　拱桥

拱桥的主要承重结构是具有曲线外

8.1.4　Arch bridge

The main load-bearing structure of the

形的拱圈,拱圈主要承受轴向压力,但也具有较大的水平反力,其在很大程度上抵消拱圈内的弯矩作用(如图 8.26 所示)。拱桥具有以下的优势:

① 拱弯矩内力小;

② 由冲击力影响的振动较小;

③ 外形美观。

arch bridge is the arch which has a curved shape. Arch mainly bears axial pressure, but it also has a large horizontal reaction force which largely offsets the bending moment within the arch. Arch bridge has the following advantages:

① The arch has a smaller moment value than beam with the same span.

② The shock effect by the impact is lower.

③ The appearance is attractive.

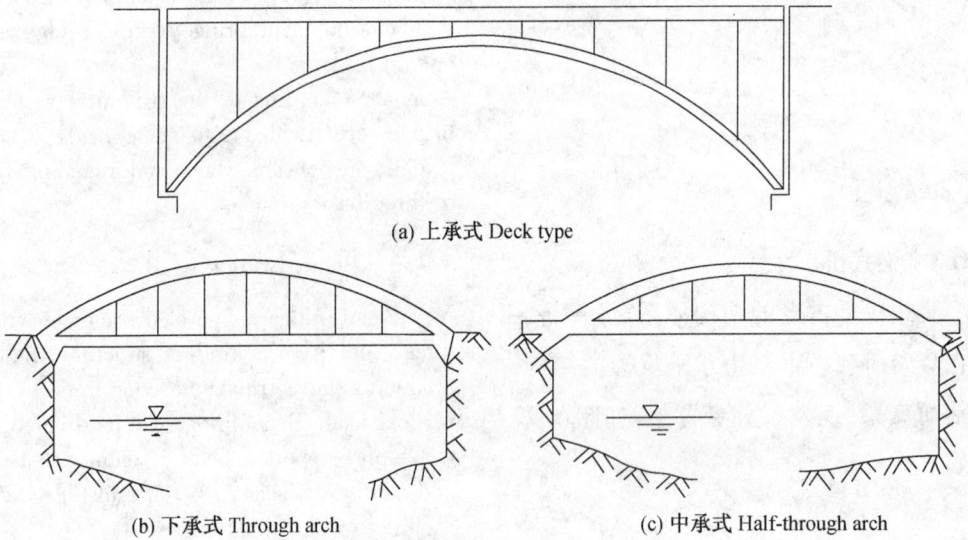

(a) 上承式 Deck type

(b) 下承式 Through arch

(c) 中承式 Half-through arch

图 8.26　拱桥
Fig. 8.26　Arch bridge

8.1.5　悬索桥

悬索桥也称为吊桥,是一种古老的桥,是以缆索为主要的承重结构,与桥塔、吊杆、锚定和桥面结构共同组成缆索的承重桥。现代悬索桥是特大跨度桥梁的主要形式之一,特别是在跨越大河,深谷等不易修筑桥墩的地区,悬索桥具有独特的优势(如图8.27 所示)。

8.1.5　Suspension bridge

Suspension bridge, also known as drawbridge, is an ancient bridge. The cable is the main bearing structure, and the bridge tower, the derrick, the anchoring and bridge deck structure of the cable bear bridges together. The modern suspension bridge is one of the main form of extra long-span bridges, especially used in areas which need to cross the great river, barranca and so on where are difficult to build bridge area. Suspension bridge has a unique advantage(Fig. 8.27).

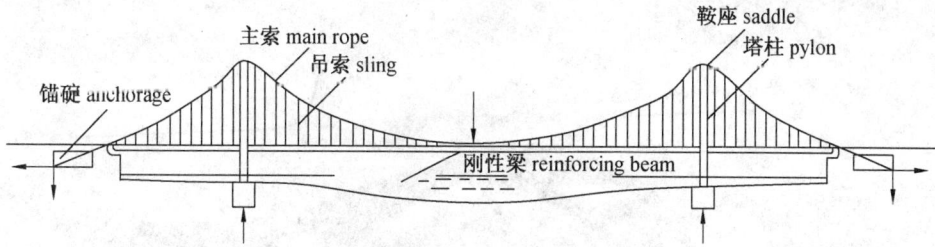

图 8.27 悬索桥
Fig. 8.27 Suspension bridge

8.1.6 组合体系桥梁

组合体系桥梁指承重结构采用两种基本体系,或者一种基本体系与某些构件组合在一起的桥梁。

(1) 斜拉桥

斜拉桥是梁与塔、斜索组成的组合体系,如图 8.28 所示。

8.1.6 Combination system bridge

The bearing structure of the combination system bridge has two kinds of basic system, or a kind of basic system together with some components of the bridge.

(1) Cable-stayed bridge

Cable-stayed bridge is composed of beams and columns, and it is a combination system, as shown in Fig. 8.28.

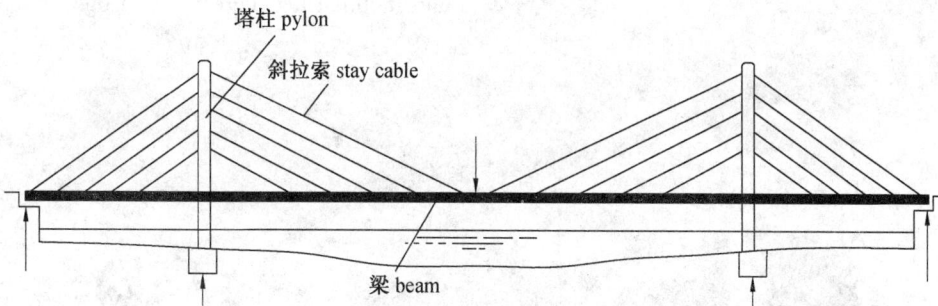

图 8.28 斜拉桥
Fig. 8.28 Cable-stayed bridge

(2) 梁拱组合体系

梁拱组合体系同时具备梁的受弯和拱的承压特点,梁和拱都是主要承重结构,如图 8.29 所示。

(2) Beam arch combination system

Beam arch combination system has the combined characteristics of bendy beams and pressure-bearing of the arch. Beams and arches are the main load-bearing structures, as shown in Fig. 8.29.

桥梁与隧道工程 Bridge and Tunnel Engineering

图 8.29　梁拱组合体系
Fig. 8.29　Beam arch combination system

（3）刚架桥

刚架桥的主要承重结构是梁与立柱组合的体系，如图8.30所示。其分为门式刚架桥和斜腿刚架桥。

（3）Rigid frame bridge

The main bearing structure of a rigid frame bridge is the combined system of beams and columns, as shown in Fig. 8. 30. It is divided into gate-type rigid frame bridge and inclined leg rigid frame bridge.

图 8.30　刚架桥
Fig. 8.30　Rigid frame bridge

（4）其他组合体系

除了上述基本体系和组合体系之外，国内外桥梁工程师还一直在研究其他各种体系，包括斜拉体系与梁、拱、索的组

（4）Other combination system

In addition to the above basic system and the combination system, bridge engineers have studied a variety of other systems at home and abroad. They include the suspen-

合，如马来西亚的 Seri Saujana 桥和我国的湘潭莲城大桥，如图 8.31、图 8.32 所示。

图 8.31　马来西亚的 Seri Saujana 桥
Fig. 8.31　Seri Saujana Bridge in Malaysia

图 8.32　中国湘潭莲城大桥
Fig. 8.32　Xiangtan Liancheng Bridge in China

8.1.7　桥梁的施工技术

桥梁施工技术，特别是大跨度桥梁施工技术要求较高，施工难度较大。随着科学技术的进步，施工机具、设备和土木工程材料的发展，桥梁施工技术也得到不断地改进与提高，逐步发展和丰富。目前，桥梁施工技术的发展主要体现在大型深水基础施工技术、无支架施工技术（如图 8.33 所示）、大型施工机具、大型构件的高精度制造技术和施工全过程的控制技术等方面。

sion system and the combination of the beam, arch and cable, such as Seri Saujana Bridge in Malaysia and Xiangtan Liancheng Bridge in China(Fig. 8.31 and Fig. 8.32).

8.1.7　**Bridge construction technology**

Bridge construction technology, especially the long-span bridge has a higher demand for construction technology, and the construction process is difficult. With the development of science and technology, construction machinery, equipment and the development of civil engineering materials, bridge construction technologies also get continuous improvement, and they have been gradually developed and enriched. At present, the development of bridge construction technology is mainly manifested in the large-scale deep water foundation construction technology, no bracket construction technology(Fig. 8. 33), large-scale construction machinery, precision manufacturing technology of large components and construction process control technology, etc.

图 8.33 拱的无支架施工
Fig. 8.33 Arch of non-bracket construction

(1)桥梁基础施工

桥梁基础是桥梁结构物直接与地基接触的最下部分,是桥梁下部结构的重要组成部分。桥梁基础的施工质量直接决定着桥梁的强度、刚度、稳定性、耐久性和安全性。

基础施工的形式有以下几种:

① 明挖扩大基础施工,如图 8.34 所示;

② 沉入桩基础施工;

③ 钻孔桩基础施工,如图 8.35 所示;

④ 沉井基础施工,如图 8.36 所示;

⑤ 地下连续墙基础施工,如图 8.37 所示;

⑥ 组合式基础施工。

(1) The foundation construction of bridge

The foundation of the bridge is the bottom part in the direct contact with the ground. It is the important part of the lower part of bridge structure. The construction quality of the bridge foundation directly decides the strength, stiffness, stability, durability and safety of the bridge.

There are several forms of the foundation construction.

① Cutting and expanding the foundation construction, as shown in Fig. 8.34.

② Sinking into the pile foundation construction.

③ Bored pile foundation construction, as shown in Fig. 8.35.

④ Open caisson foundation construction, as shown in Fig. 8.36.

⑤ Foundation construction of underground continuous wall, as shown in Fig. 8.37.

⑥ Composite foundation construction.

图 8.34　明挖扩大基础施工
Fig. 8.34　Cutting and expanding the foundation construction

图 8.35　钻孔桩基础施工
Fig. 8.35　Bored pile foundation construction

图 8.36　沉井基础施工
Fig. 8.36　Open caisson foundation construction

图 8.37　地下连续墙基础施工
Fig. 8.37　Foundation construction of underground continuous wall

（2）桥梁墩台施工

桥墩是桥梁的重要结构,起着支承桥梁上部结构荷载的作用,并且将它传递给地基基础。桥墩水下施工如图 8.38 所示。

(2) Construction of the bridge pier

Pier is an important structure of bridge construction. It bears superstructure loads of the bridge, and passes it to the foundation. Underwater pipers construction is shown in Fig. 8.38.

图 8.38　桥墩水下施工
Fig. 8.38　Underwater piers construction

8.2.1 隧道的基本概念

隧道,是修筑在岩体、土体或水底的两端有出入口的通道,它是交通运输线路穿越山岭、丘陵、土层、水域等天然障碍最有效的途径。1970 年经合组织(OECD)的隧道会议对隧道所下的定义为:以某种用途,在地面下用任何方法按规定形状和尺寸,修筑的断面积大于 2 m 的洞室。隧道除可应用于铁路、公路交通和水力发电、灌溉等领域,也用于上下水道、输电线路等大型管路的通道。

(1) 隧道的历史

人类在很早以前就知道利用自然洞穴作为栖息之所。当社会发展到能制造挖掘工具时,就出现了人类挖掘的隧道。

在其他古代文明地区有很多著名的古隧道,如公元前 2180—2160 年,在古巴比伦城幼发拉底河下修筑的人行隧道为砖砌构造物,是迄今为止最早用于交通的隧道。

近代隧道兴起于运河时期,从 17 世纪开始,欧洲陆续修建了许多运河隧道。法国的兰葵达克运河隧道,建于 1666—1681 年,长 157 m,它可能是最早用火药开凿的隧道。1830 年前后,铁路成为新的运输手段。随着铁路运输事业的发展,隧道也越

8.2.1 The basic concept of the tunnel

Tunnel is a channel that is built on rock, soil or under water, with an exit channel and an entrance channel at both ends. It is the most effective way for transportation route to across the mountains and hills, soil, water and other natural barriers. In 1970, the tunnel meeting of Organisation for Economic Co-operation and Development (OECD) defined tunnel as the following: it is a cavity with more than 2 m broken area, which was constructed under the ground by any shape and size with a certain purpose. Tunnel can be used in railway, highway transportation, and hydraulic power generation, and irrigation. Besides, it can be also used in large pipe channels, such as sewer and transmission lines.

(1) The history of the tunnel

Human beings used natural caves as a residence long time ago. When human beings can make mining tools, the mining tunnel appears.

There are many famous ancient tunnels in other ancient civilization areas, such as the ancient city construction of the pedestrian tunnel built in 2180-2160 B. C., under the Euphrates river. It is the brick structure which is earliest used in traffic tunnel so far.

Modern tunnel arose in the period of the canal. Since the 17th century, many canal tunnels have been built in Europe. Languedoc canal tunnel in France, which was built in 1666-1681, was 157 meters long. It is probably the earliest tunnel constructed by gunpowder. Around 1830, railway became a new transport way. With the development of

来越多。1895—1906 年已出现了长 19.73 km 穿越阿尔卑斯山脉的最长铁路隧道。目前,世界最长的铁路隧道已达 53.85 km。较为完善的水底通道隧道建于 1927 年,是位于纽约哈德逊河底的 Holland 隧道。现在,世界上的长大道路隧道(2 km 以上)和长大水底隧道(0.5 ~ 2.0 km)已超过百条。

(2) 典型隧道

直布罗陀海峡跨海通道由联结欧洲和非洲的海底隧道及部分海上桥梁构成,目前尚由西班牙、英国、摩洛哥等拥有海峡主权的国家共同规划。根据英国广播公司报道,其兴建费用将达 100 亿美元。

挪威的洛达尔隧道,是世界最长的公路隧道,全长 24.5 km(如图 8.39 所示)。

瑞士的圣哥达隧道,是世界第三长的公路隧道,全长 16.32 km,连接瑞士的乌里州和提契诺州。

英法海底隧道,是世界第二长的铁路隧道,全长 50.5 km,海底长度 38.9 km,也是世界海底长度最长的海底隧道。它跨越英吉利海峡,连接英国和法国(如图 8.40 所示)。

railway transportation enterprises, the tunnel became long and long. In 1895-1906, a 19.73 km long railway which is the longest tunnel crossing the Alps appeared. Currently, the longest railway tunnel in the world amount to 53.85 km. Relatively, perfect underwater channel tunnel was built in 1927, which is the Holland Tunnel located in New York's Hudson River. The number of the long and large road tunnel (2 km) and the long and large underwater tunnel (0.5 ~ 2.0 km) has over one hundred in the world now.

(2) Typical tunnel

Gibraltar sea-crossing channel is composed of undersea tunnel linking Europe and Africa and parts of the sea bridge. Currently, this channel was owned by Spain, England, Morocco. According to BBC reports, the construction cost is estimated at 10 billion dollars.

Los Dahl Tunnel in Norway, as shown in Fig. 8.39, is the world's longest road tunnel with the length of 24.5 km.

Gotthard tunnel in Swiss is the world's third longest highway tunnel with the length of 16.32 km, connected the Canton of Uri with Ticino.

Britain and France Subsea Tunnel, as shown in Fig. 8.40, the world's second longest rail tunnel with the length of 50.5 km, and its submarine length is 38.9 km. It is the underwater tunnel with world's longest submarine length. It cross the English Channel connected UK and France.

图 8.39　洛达尔隧道
Fig. 8.39　Los dahl tunnel

图 8.40　英法海底隧道
Fig. 8.40　Britain and France subsea tunnel

香港海底隧道是世界上最繁忙的行车隧道之一,全长 1.8 km,平均每日行车量达 121 700 辆,跨越维多利亚港连接九龙半岛和香港岛(如图 8.41 所示)。

日本青函隧道是目前世界上最长的铁路隧道,全长 53.9 km,海底长度 23.3 km(如图 8.42 所示)。此隧道跨越津轻海峡,连接日本的北海道和本州。

Subsea Tunnel in Hong Kong, as shown in Fig. 8.41, is one of the world's busiest vehicular tunnel with the total length of 1.8 km. The average daily traffic volume reached 121 700 units, and it is across Victoria harbour connected Hong Kong island with Kowloon Peninsula.

Seikan Tunnel, as shown in Fig. 8.42, is the world's longest rail tunnel currently with the total length of 53.9 km, and its submarine length is 23.3 km. The tunnel crossing the Tsugaru connects Hokkaido and Honshu in Japan.

图 8.41　香港海底隧道
Fig. 8.41　Subsea Tunnel in Hong Kong

图 8.42　日本青函隧道
Fig. 8.42　Seikan Tunnel in Japan

秦岭—终南山特长公路隧道是中国乃至亚洲最长的公路隧道,也是世界最长的双孔公路隧道,全长 18.02 km,2006 年完工后已超过圣哥达隧道成为世界第二

Qinling-Zhongnan Mountains extra-long highway tunnel is the longest road tunnel in Asia, which is the world's longest double holes road tunnel with the length of 18.02 km. It was the world's second longest road

长的公路隧道。

美国的德拉瓦隧道是世界最长的输水隧道，全长 169 km。

美国纽约的林肯隧道跨越哈德逊河，连接纽约市和新泽西州，是世界最繁忙的公路隧道之一，全长 2.4 km。

8.2.2 隧道的组成

隧道的结构包括主体建筑物和附属设备两部分。主体建筑物由洞身和洞门组成，附属设备包括避车洞、消防设施、应急通信和防排水设施。长大隧道还有专门的通风和照明设备。

高速铁路隧道内不设置供养护维修人员待避的洞室，但应考虑设置存放维修工具和其他业务部门需要的专用洞室。

(1) 洞门构造

洞门是隧道结构物的一部分，其作用在于支挡正面仰坡及路堑边坡，拦截仰坡小量的剥落土石，并将仰坡上的汇水引离隧道，以稳固洞口，保证线路行车安全。

洞门类型主要有：环框式洞门，如图8.43 所示；端墙式洞门，如图 8.44 与图8.45 所示；柱式洞口、翼墙式洞门，如图8.46 与图 8.47 所示；台阶式洞门，如图8.48 所示。

tunnel in 2006.

The Delaware Tunnel in America is the world's longest water conveyance tunnel with the total length of 169 km.

Lincoln Tunnel in New York crossing the Hudson River connected New York and New Jersey. It is one of the world's busiest highway tunnel with the length of 2.4 km.

8.2.2 The composition of the tunnel

The structure of the tunnel consists of a main building and ancillary equipment. Main building is composed of the tunnel trunk and the tunnel portal. Ancillary equipment includes refuge hole, fire control facilities, emergency communications and waterproof and drainage facilities. The long and large tunnel has special ventilation and lighting equipment.

Cavity for maintenance person is not set in high-speed railway tunnel, but special chambers which can be used to store maintenance tools and meet the needs of other department should be set up.

(1) Tunnel portal structure

Tunnel portal is the part of the tunnel structure, and its role is positive upward slope retaining and cutting slope, interception of upward slope small spalling jades, and will be back on the slope catchment away from the tunnel, in order to stabilize the hole of the cave and ensure driving safety.

The types of tunnel portal are as follows: the ring frame of the tunnel portal is shown in Fig. 8.43; the end wall type tunnel portal is shown in Fig. 8.44 and Fig. 8.45; the column type tunnel trunk and the wing wall type tunnel portal are as shown in Fig. 8.46 and Fig. 8.47. The setback type tunnel portal is shown in Fig. 8.48.

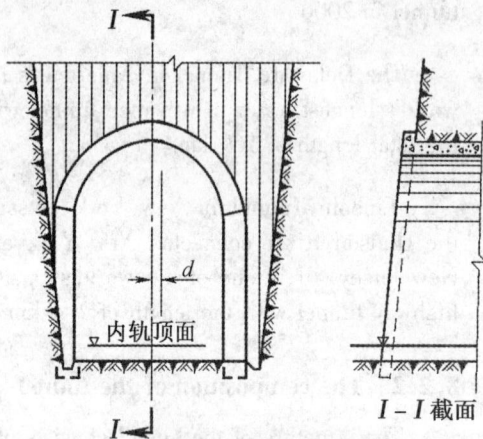

图 8.43　环框式洞门
Fig. 8.43　Ring frame of the tunnel portal

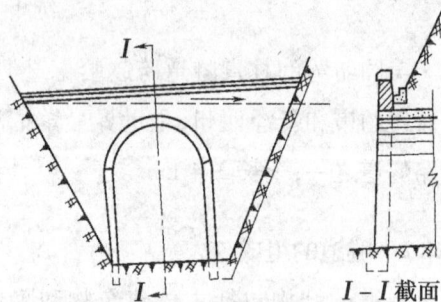

图 8.44　端墙式洞门
Fig. 8.44　End wall type tunnel portal

图 8.45　中部抬高端墙式洞门
Fig. 8.45　The central drive up side wall tunnel portal

图 8.46　翼墙式洞门
Fig. 8.46　Wing wall tunnel portal

图 8.47　带耳墙的翼墙式洞门
Fig. 8.47　Wing wall tunnel portal with the ear wall

图 8.48　台阶式洞门
Fig. 8.48　Sidestep tunnel portal

当地形等高线与路斜交，采用正交洞门将出现外侧岩壁太薄或洞门墙后露空，而靠山侧边，仰坡开挖过高，不能保证安全，或靠河侧岩壁陡峻，基础不好处理时，

When topographic contour and road are heterotropic and adopting orthogonal tunnel portal will make the lateral wall too thin, or the back of wall empty of tunnel portal, and yet beside mountain broadside, excavate overtop of the upward slope cannot guarantee

可以采用斜交洞门。斜交洞门其端墙与线路中线的交角不应小于 45°，一般采用 45°和 60°。

security, or because of cliff steep beside the river side, foundation is not easy to handle, we can adopt oblique tunnel portal. The angle of intersection of the side wall of heterotropic tunnel portal and the angle of center line should not be less than 45°, general use 45° and 60°.

（2）洞内附属设备

洞内附属设备有避车洞、排水沟、供电照明设备、运营通风设备等，下面介绍避车洞。

（2）Ancillary service inside the hole

Ancillary equipment has refuge hole, drainage, power supply, lighting equipment, operation ventilation equipment, etc., and refuge hole is described below.

① 避车洞的作用。

隧道内应设置小避车洞，以便当列车通过时，隧道内的养护工作人员与行人进入其中避让。另外，隧道内还需设置大避车洞，以便于工务与大修的轻型车辆、小车等避让列车。

① The role of refuge hole

Small refuge hole should be set inside the tunnel, so that maintenance staff and pedestrians can enter the tunnel when the train goes through the tunnel. Moreover, large refuge holes need to be set inside the tunnel.

② 避车洞的间距和尺寸见表 8.1。

② The space and size of the refuge hole are shown in Table 8.1.

表 8.1 避车洞的间距尺寸
Table 8.1 The door distance size of the refuge hole

名称 Description	一侧间距/m One side separation distance		尺寸/m Size		
			宽 Width	深 Depth	中心高 Height of center
大避车洞 big refuge hole	碎石道床 ballast bed	300	4.0	2.5	2.8
	混凝土宽枕道床或整体道床 concrete wide pillow ballast bed or integrated ballast bed	420			
小避车洞 small refuge hole	碎石道床 ballast bed	60	2.0	1.0	2.2
	混凝土道床或整体道床 concrete ballast bed or integrated ballast bed				

8.2.3 隧道的分类

当前隧道除仍用于铁路、公路交通（如图8.49、图8.50所示）和水力发电、灌溉等水工隧洞外，也用于上下水道、输电线路等大型管路的通道。另外，现已将过去理解为"地下通路"的隧道概念扩大到地下空间的利用方面，包括诸如地下发电变电所、地下汽车停车场、大型地下车站、地下街道等适用隧道工程技术的建筑物。

8.2.3 The classification of tunnel

Currently, tunnel is used in railway, highway transportation(Fig. 8.49 and Fig. 8. 50), and the hydroelectricity, irrigation. Besides, it can also be used in sewer, transmission lines and other large pipe channels. Moreover, the past concepts of underground channel tunnel has extended to the utilizing of underground space, such as the underground power substation, underground car parking lot, large underground station, underground streets for tunnel engineering of buildings.

图 8.49 北京八达岭公路隧道
Fig. 8.49 Beijing Badaling highway tunnel

图 8.50 青藏线锡铁山分水岭隧
Fig. 8.50 Xitie mountain watershed tunnel along the qinghai-tibet railway

隧道除按上述用途分类外，从地质上还可按开挖对象划分为岩石隧道和土砂隧道，并可根据施工场所的不同区分为公路隧道、城市隧道和水下隧道等。此外，过去视为特殊施工方法的盾构法或沉管法，现今已经普及，因此也可以将这些方法包括在内，而按施工方式、方法进行详细的分类。

In addition to the classification above, tunnel can be divided into rock tunnel and soil tunnel according to the excavation objects from the geological perspective. It can be distinguished into highway tunnel construction site, urban tunnel and underwater tunnel, etc. according to different construction site. In addition, shield method or immersed tube method which is considered as special methed become popular, so these methods can be included, and to be classified in detail according to the construction method.

8.2.4 隧道施工

根据隧道穿越地层的不同情况和目前隧道施工方法的发展，隧道施工方法可

8.2.4 Tunnel construction

According to the different situations of the tunnel through the formation and the development of tunnel construction method,

以按以下方式分类：

①	山岭隧道施工法，包括矿山法、掘进机法，其中矿山法又分为传统矿山法和新奥法。

②	浅埋及软土隧道施工方法，包括明挖法、地下连续墙法、盖挖法、浅埋暗挖法和盾构法。

③	水底隧道施工法，包括沉埋法和盾构法。

8.3 桥梁与隧道的展望

随着经济的发展和综合国力的增强，我国的建筑材料、设备、建筑技术都有了较快的发展，为广大工程技术人员提供了方便、快捷的计算分析手段。未来桥梁隧道工程发展的将着重体现在以下几个方面：

①	桥梁隧道结构形式的创新；

②	桥梁隧道的全寿命概念设计、性能设计；

③	设计分析软件工具的多元化（BIM，MIDAS 等）；

④	高性能材料（CFRP、聚脲技术、瓦克化学产品）在桥隧工程中的应用研究；

⑤	考虑极端环境因素对桥隧工程传统强度、耐久性设计思想的冲击；

⑥	在建、既有桥梁全过程施工监控、性能检测及寿命评估技术的发展；

⑦	长大跨桥梁、深海隧道的施工关键

tunnel construction methods can be classified as follows：

①	Mountain tunnel construction includes the mining method and the tunnel-boring machine method, and the mining method is divided into the traditional mining method and the new Austrian Tunnelling method.

②	Shallow buried and soft soil tunnel construction method includes the open cut method, the underground wall method, the cover-excavation method, the shallow tunneling method and the shield method.

③	The underwater tunnel construction method includes the immersed tunneling method and the shield method.

8.3 Bridge and tunnel vision

With the development of economy and the enhancement of comprehensive national strength, construction materials, equipment, construction technology in China has developed rapidly. It provides convenient and efficient means of calculation and analysis for the general engineering and technical personnel. Bridge and tunnel engineering in the future will focus on the following aspects：

①	Innovation of the bridge form and tunnel structure；

②	The full life concept design, function design of the tunnel and bridge.

③	The diversity of the design analysis software tools(BIM, MIDAS, etc.).

④	The application of high performance materials(CFRP, polyurea technology, wacker chemical products) in bridges and tunnels.

⑤	The extreme environmental factors impacted on traditional strength, durability design thought of the bridge and tunnel engineering need to be considered.

⑥	The development of the whole monitoring process of construction of being built

技术的相关研究成为研究热点——要能解决实际工程问题。

and existing bridge, performance test and life assessment technology development.

⑦ Key construction technology research of the long and large span bridges and deep sea tunnels become a hotspot so as to solve practical engineering problems.

注:本章图片均来自网络。
Note: In this chapter, all pictures are from webs.

知识拓展
Learning More

相关链接 Related Links

如果想了解桥梁与隧道工程的详细知识、最新发展态势及相关政策,可访问中国桥梁工程网和中国隧道工程网。

If you want to acquire the detailed knowledge of bridge and the tunnel engineering, the latest development trend and related policies, please get Chinese bridge project network and the tunnel engineering network.

小贴士 Tips

桥梁与隧道工程的注册师制度

国内和桥梁与隧道工程有关的注册师有:注册结构师、注册岩土师、注册建造师、注册造价工程师、注册监理工程师等,通过考试后可从事相应市政工程专业的设计、施工、管理等工作。

Bridge and Tunnel Engineering of the System registered polices

Domestic registered architect is associated with bridge and tunnel engineering are as follows: registered structure certificate, geotechnical engineers, national certified architect, certified cost engineer etc., certified supervision engineer. After passing the exam, you can engaged in relevant municipal engineering design, construction and management.

思考题 Review Questions

(1) 桥梁与隧道工程包括哪些内容?

What is the content of bridge and tunnel engineering?

(2) 桥梁可分为哪些类型?

What are the types of bridges?

(3) 隧道的施工方法有哪些?

What are the tunnel's construction methods?

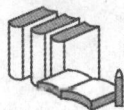

参考文献
References

［1］徐伟:《桥梁工程》,人民交通出版社,2008 年。

［2］夏禾:《桥梁工程》,高等教育出版社,2011 年。

［3］姚玲森:《桥梁工程》,人民交通出版社,2008 年。

［4］顾安邦,向中富:《桥梁工程》,人民交通出版社,2011 年。

［5］戴俊:《隧道工程》,机械工业出版社,2012 年。

［6］彭立敏,刘小兵:《隧道工程》,中南大学出版社,2009 年。

［7］徐伟:《桥梁施工》,人民交通出版社,2008 年。

［8］于书翰,杜谟远:《隧道施工》,人民交通出版社,1999 年。

第9章 工程项目管理

Chapter 9 Construction Project Management

工程项目管理是对建设工程的全过程实施的计划、协调、控制等工作的总称，是保证工程按期交付并实现预期目标的重要工作。本章介绍工程项目管理的基本概念和工作框架、建设工程的主要实施模式、相关执业认证制度以及工程项目管理的发展趋势。

Construction project management (CPM) or construction management is the overall planning, coordination, and control of a project from the beginning to the end. The purpose of construction project management is to meet a client's requirements, so as to produce a viable project. This chapter introduces concepts and framework of construction project management, construction models, professional qualification systems and the tendency of construction project management.

9.1 工程项目管理总述

9.1 Overview of construction project management

管理是为了实现特定目标，对管理对象进行的决策、计划、组织、协调、控制等一系列工作的总称。工程项目管理是为了使工程项目在一定的约束条件下取得成功，对项目的活动实施决策与计划、组织与指挥、控制与协调等一系列工作的总称。它以工程项目为管理对象，以工程项目的实施工作和项目相关各方的管理为

Management is a series of work to achieve certain goals, such as overall planning, coordination, and control of a project from beginning to completion. Construction project management is a series of work to achieving the goal of a construction project. The work include making decision, planning, organizing, coordinating, controlling, etc. Construction project management aims at meeting a client's requirements in order to produce a viable project. The activities and participates in this process are the key fac-

重点。

工程项目管理是项目管理的分支之一。现代项目管理理论是在现代科学技术知识,特别是信息论、控制论、系统论、行为科学等基础上产生和发展起来的。变革管理、危机管理、集成管理、知识管理等现代管理理论为工程项目管理提供了科学的理论支撑;预测技术、决策技术、网络技术等现代项目管理技术为工程项目管理提供了有力的技术支撑。

随着现代工程项目的复杂化和社会分工的要求,工程项目管理日益体现出专业的发展趋势。社会中出现了专业化的工程项目管理公司,它们专门承接工程项目管理业务,为业主和投资者提供全过程的专业化咨询和管理服务。因此,现代的工程项目管理不仅是一门学科,而且成为一种职业,专业化的工程项目管理已成为一个新兴产业。

工程项目实施中,不同层次和角色的参与者都有各自的项目管理职责(见表9.1)和相应的项目管理组织,并在不同阶段承担不同范围和内容的工作任务。因此,工程项目管理是分层次、多角度展开的。

tors of construction project management.

Construction project management is a branch of modern project management. Modern project management develops on the basis of information theory, cybernetics, system theory, behavior science, etc. Change management, crisis management, integrated management, knowledge management, etc. are theoretical supports for construction project management. Forecasting technique, decision-making technique, network technique, etc. are technical support for construction project management.

Construction project management is becoming more and more specialized, with the complication of modern construction project management and division of labor in society. For example, a project management company provides a series of specialized management and consultancy services which can cover all the processes of construction project management. Therefore, construction project management is not only a kind of vocation, but also a new industry.

Different participators of construction project usually have different responsibilities (Table 9. 1) and construction organizations correspondingly. Therefore, construction project management is a multi-level and multi-dimensional management.

表9.1 不同工程项目管理者的职责
Table 9.1 Responsibilities of different construction project participators

管理角色 Management role	工作职责 Responsibilities
投资者 investor	筹措资金,对投资方向、投资的分配、投资计划、项目的规模、管理模式等重大、宏观问题进行决策和控制 financing, making decision, controlling important and strategical factors, such as investment orientation, project scale, management mode
业主 owner	以项目所有者的身份开展工作,主要承担项目的宏观管理以及与项目有关的外部事务,居于项目组织最高层 The owner is on the highest level of the project organization macro-manage the project, deal with the external business
项目管理公司 project management company	受业主委托,提供工程项目管理服务,进行合同管理、投资管理、质量管理、进度控制、信息管理等,协调与业主签订合同的各设计单位、承包商、供应商的关系,并为业主承担项目中的事务性管理工作和决策咨询工作等 provide the management on the behalf of the owner such as contract management, investment management, quality management, schedule management, information management
承包商 contractor	承担具体的施工活动和相关管理任务 doing the specific management activities of the construction
政府 government	履行社会管理职能,依据法律和法规对项目进行行政管理,提供服务和开展监督工作 administration, manage the construction market and the project, provide services
其他 others	项目管理组织中,不同的管理人员在不同的岗位上承担不同的项目管理工作任务,如项目经理、计划管理人员、成本管理人员等 Different managers in different positions hold their own task, such as project manager, schedule and plan manager, cost control manager, etc.

9.2 工程项目管理的任务

工程项目管理面向工程的建设全过程,包括预测、决策、计划、控制、反馈等一系列管理过程和活动。实施中,需要涉及投资管理、质量管理、进度管理、费用管理、合同管理、组织管理、安全管理等众多管理对象和任务(见表9.2)。工程项目本身也是一个复杂系统,可分解为若干子项

9.2 Tasks of construction project management

Construction project management covers the whole process of a project. It includes a series of management tasks, such as investment management, quality management, schedule management, management, cost management, contract management organization management, and safety management (Table 9.2). Besides, a construction project can be divided into some sub-projects, work packages, etc. The framework of con-

目、项目单元、任务、活动等,各管理过程和任务都要落实到不同的系统组成成分上。因此,工程项目管理的工作架构是一个集成化的三维结构(如图9.1所示)。

struction project management is a 3D integrated structure (Fig. 9.1).

表 9.2　工程项目管理的职能和任务

Table 9.2　Functions and main tasks of construction project management

职能 Function	主要任务 Main task
质量管理 quality management	建立工程质量保证体系 establishing the quality guarantee system
	对材料和设备的质量进行检查、验收 quality inspection and acceptance of materials and equipment
	对施工过程的质量进行全程监督和中间检查 quality inspection during construction
	对已完工程进行验收 quality acceptance of finished projects
	处置不符合要求的材料和工艺 dealing with defective materials and processes
	组织竣工验收,安装调试和移交 installation, debugging and completion acceptance
进度管理 schedule management	确定工程的总持续时间和活动之间的逻辑关系 determine the duration of project, logical relationship of activities
	确定总工期并安排各阶段工程的工期 make out the aggregate schedule of the project and scheduled plan in each step according to the total time limit of the project
	论证、审核工程的实施方案和进度计划 plan demonstration
	监督参与各方按计划开始和完成工作 supervision of performance
	要求承包商修改进度计划,指令暂停工程或指令加速 order contractors accelerate, pause and change plan
	处理工期索赔要求等 dealing with time claim, etc.

职能 Function	主要任务 Main task
费用管理 cost management	投资预测和计划,包括工程投资的估算、概算和预算 make a investment plan, including estimate, general estimate and budget
	对工程编制标底和报价,以及施工中对工程变更进行估价 put forward the offer in a suitable way after estimating the cost of construction on the basis of the stated construction plan
	制订支付计划、收款计划、资金计划和融资计划 make a payment scheme, collection plan, fund plan and financing plan
	对已完工程进行量方,指令各种形式的工程变更,处理费用索赔 calculation of constructed project, dealing with cost claim
	审查、批准进度付款 inspect and approve a payment
	审查监督成本支出,成本跟踪和诊断 trace and diagnose cost
	实施竣工结算及决算,提出结算报告等 make a completion settlement and report
合同管理 contract management	采购计划和采购工作安排的制订 make procurement plans
	招标投标管理,包括合同策划、招标准备工作、起草招标文件、合同审查 bidding management including contract planning, bidding preparation, drafting the document
	合同实施控制,解释合同,确保项目人员了解合同,遵守合同 controlling of contract execution to ensure the implementation of the contract
	监督合同实施,审查承包商的分包合同,批准分包单位等 supervision of contract execution and subcontract
	合同变更管理,索赔管理,解决合同争执等 change management, claim management, dispute management
	合同风险管理 risk management
组织管理 organization management	建立项目组织机构,明确责权利关系 establishing of project organization
	制定项目管理工作流程和规则 creating work flow and rules
	解决组织间的冲突和争执 dealing with conflict and dispute among organizations
	处理内部与外部关系,沟通、协调项目参与各方等 dealing with internal and external relationship

职能 Function	主要任务 Main task
信息管理 information management	建立管理信息系统,确定组织成员之间的信息形式、信息流 establishing management information system, creating information flow
	收集和保存工程过程中的各种信息 collecting and saving information
	起草各种文件,向承包商发布图纸、指令 making documents
	向业主、企业和其他相关各方提交各种报告 making and submitting reports

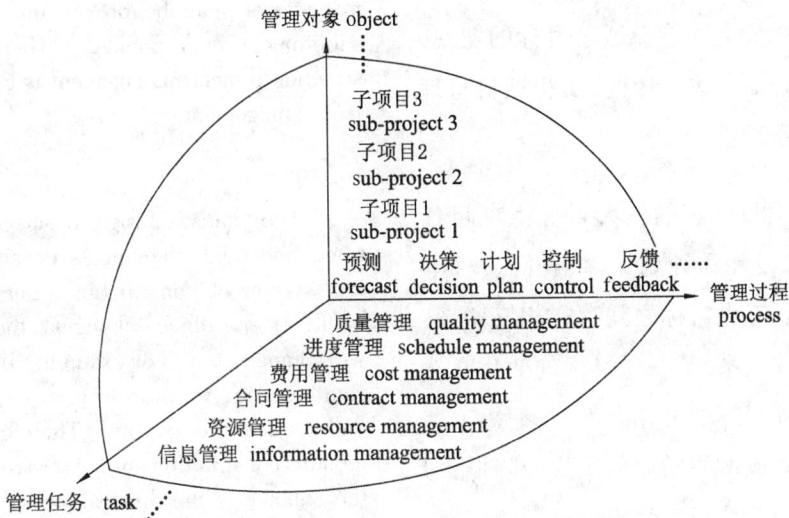

图 9.1　工程项目管理的工作架构
Fig. 9.1　Framework of construction project management

9.3　工程项目管理的目标

工程项目管理的目标包含 3 个方面,共同构成了工程项目管理的目标体系,如图 9.2 所示。

① 质量目标,如使用寿命、抗震等级、功能性能等;

② 进度目标,如工程的总工期、里程

9.3　Objects of construction project management

The object system of construction project management(Fig. 9.2) includes:

① Quality, such as service life, antiseismic level, and functional features.

② Time, such as duration of project, and date of milestone events.

③ Cost, such as total investment, labor costs, and material costs.

碑事件的日期等；

③ 费用目标，如总投资、劳动力成本、材料成本等。

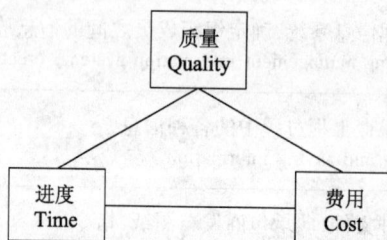

图 9.2　工程项目管理的目标体系

Fig. 9.2　Object system of construction project management

工程项目管理的目标必须分解落实到具体的各个项目单元(子项目、活动)和项目组织单元上才能保证总目标的实现，形成一个控制体系。因此，工程项目管理也是目标管理。

工程项目管理的三大目标由项目任务书、技术设计和计划文件、合同文件等具体定义。三大目标之间存在相互联系、相互影响的关系，某一项目标的变化必然引起另两项目标的变化。例如，缩短工期有可能带来成本的增加和质量的降低，即进度目标会影响费用和质量目标的实现。因此，成功的工程项目管理应追求三者之间的均衡性和合理性。

现代社会中，工程对推动社会进步和人类可持续发展的效用日益显著，工程项目管理的目标也不断得到扩展。例如，环境友好、全要素安全、工程各参与方全面满意等。

In order to realize the general target, the object system of construction project management can be broken down into each subproject, work package. Therefore, construction project management is a kind of object management.

The plans, design descriptions, contracts and other documents describe the object system of construction project management. If one object changes, the other one will change, too. For example, if we want to shorten time, we have to increase investment and reduce the quality. Therefore, a successful construction project management is the balance of the three objects.

The object system of construction project management develops and extends in modern world with the development of society. There are some new objects, such as friendly environment, safety and security of all factors, and satisfaction of all participators.

9.4 工程项目的建设模式

工程项目的建设模式也称采购模式或承发包模式。业主通过发包和合同委托项目任务,形成工程的合同体系,并通过合同实现对工程项目目标的控制。承包商依据所承担的合同内容,开展相应的工作。因此,工程项目的建设模式决定了与业主签约的承包商的数量和彼此间的关系,决定着项目组织结构的基本形式及管理模式。工程项目建设模式的发展历程如图9.3所示。

Construction model of the project is also called procurement model or contracting and awarding model. The owner builds the contract system by selecting contractors and awarding contracts. The owner can control the objects of project by using contracts. The performances of contractors are consistent with contracts. It means the construction model determines the relationship between the owner and contractors, the organization structure and the management model. The development of construction model is shown in Fig. 9.3.

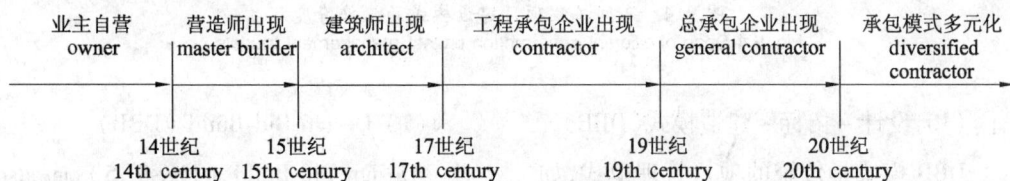

图9.3 工程项目建设模式的演变
Fig. 9.3 Development of construction project management model

目前,普遍采用的工程建设模式有:设计-招标-建造模式(DBB)、设计-建造模式(DB)、建设管理模式(CM)、设计-采购-建设模式(EPC)、项目管理模式(PM模式)、管理承包模式(MC)、项目融资模式(BOT)、项目伙伴模式(Project Partnering)等。这几种模式的工作范围跨度如图9.4所示。

The popular construction models in the world include Design-Bid-Build (DBB), Design-Build (DB), Construction Management (CM), Engineering Procurement Construction (EPC), Project Management (PM), Management Contracting (PM), Build-Operate-Transfer (BOT), Project Partnering, etc. Their span is shown in Fig. 9.4.

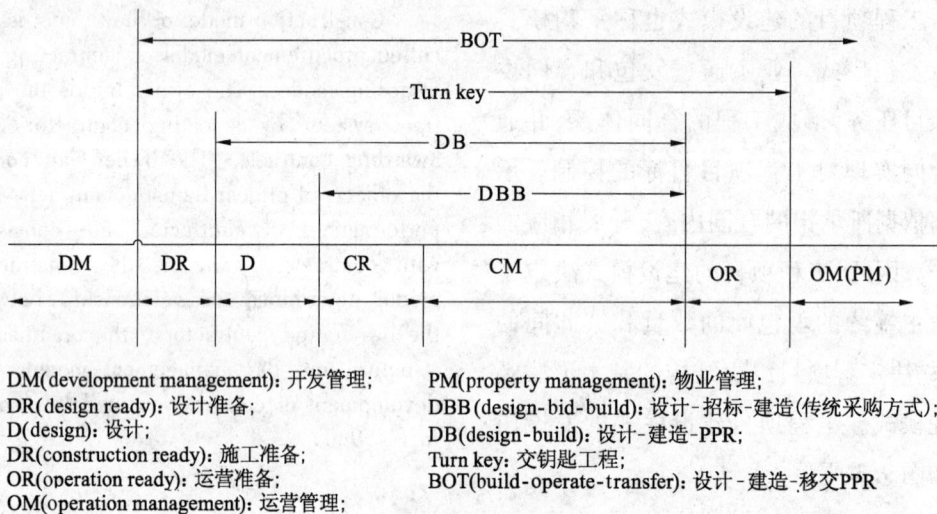

DM(development management): 开发管理；
DR(design ready): 设计准备；
D(design): 设计；
DR(construction ready): 施工准备；
OR(operation ready): 运营准备；
OM(operation management): 运营管理；

PM(property management): 物业管理；
DBB(design-bid-build): 设计-招标-建造(传统采购方式)；
DB(design-build): 设计-建造-PPR；
Turn key: 交钥匙工程；
BOT(build-operate-transfer): 设计-建造-移交PPR

图 9.4 不同工程项目建设模式的工作跨度
Fig. 9.4 Span of different construction project management models

（1）设计－招标－建造模式（DBB）

DBB 模式是传统的项目管理模式，如图9.5所示。采用 DBB 模式时，业主与设计商（建筑师/工程师）签订专业服务合同，建筑师/工程师负责提供项目的设计和合同文件。在设计商的协助下，通过竞争性招标将工程施工任务交给报价和质量都满足要求且/或最具资质的投标人（承包商）来完成。在施工阶段，设计专业人员通常担任重要的监督角色，并且是业主与承包商沟通的桥梁。

（1）Design-Bid-Build（DBB）

Design-Bid-Build（Fig. 9.5）is also known as Design-tender traditional method, and it is a project delivery method in which the agency or the owner contracts with separate entities for both the design and the construction of a project. The designer provides plan and blueprint of the project. The owner selects contractor who can meet the quality and financial requirements best. The designer also plays as a superintendent during the construction period. It is the bridge between the owner and contractor. DBB is the traditional method for project delivery and differs in several substantial aspects from design-build.

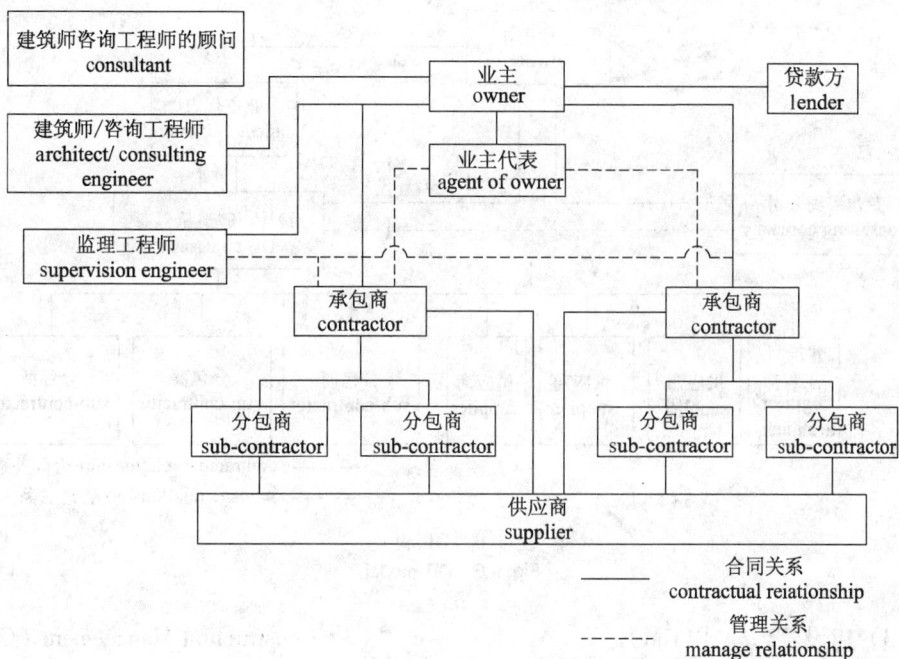

图 9.5　DBB 模式
Fig. 9.5　DBB model

（2）设计－建造模式（DB）

DB 模式是近年来常用的现代工程项目管理模式，它又被称为设计和施工（Design-Construction）、交钥匙工程（Turn-key）或者是一揽子工程（Package Deal），如图 9.6 所示。DB 模式下，设计－建设承包商承担了全部的工程设计和建设工作。业主与承包商之间只有一份合同，所以业主承担了最小的风险。而设计－建设一体的模式也能够最大限度地缩小两者直接的界面和摩擦，保证工程质量和按期交付。DB 的承包商承担了工程建设的大部分职能和风险，并要对工程中的失误负责。

（2）Design-Build（DB）

Design-Build is a modern project delivery system used in the construction industry (Fig. 9.6). It is also known as Design-Construction, Turn-key or Package Deal. It is a method to deliver a project in which the design and construction services are contracted by a single entity known as the design-builder or design-build contractor. Contrasted with DBB, DB relies on a single point of responsibility contract, and it is used to minimize risks for the project owner and to reduce the delivery schedule by over lapping the design phase and construction phase of a project. DB contractor will be responsible for all of the work on the project, regardless of the nature of the fault.

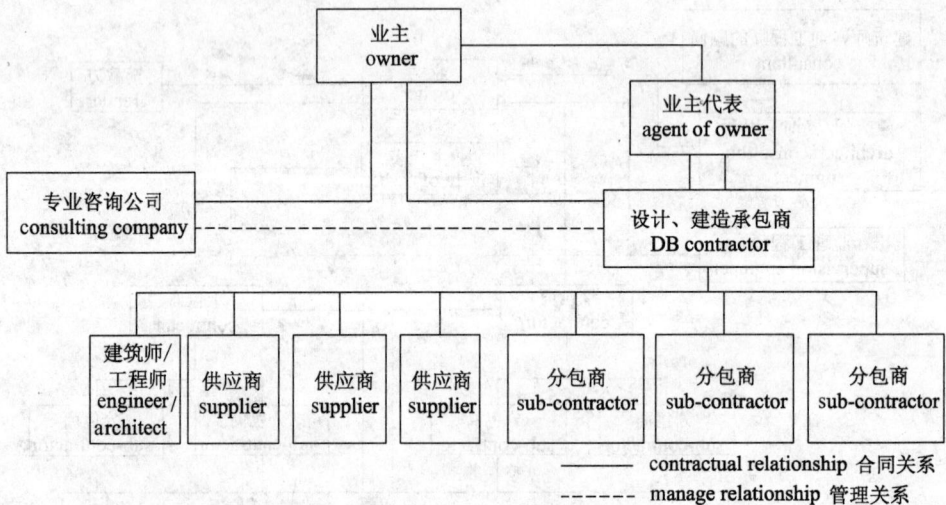

图 9.6　DB 模式
Fig. 9.6　DB model

（3）建设管理模式（CM）

CM 模式从项目开始阶段就雇用具有施工经验的 CM 单位参与到工程项目实施过程中来,以便为设计师提供施工方面的建议,并且负责管理施工过程,如图 9.7 所示。CM 模式改变了过去设计完成后才进行招标的传统模式,而采取分阶段招标,由业主、CM 单位和设计商组成一个联合小组,共同负责组织和管理工程的规划、设计和施工,CM 单位负责工程的监督、协调及管理工作,在施工阶段定期与承包商交流,对成本、质量和进度进行监督,并预测和监控成本和进度的变化。

（3）Construction Management（CM）

In the CM model（Fig. 9.7）, experienced contractors and the CM join the project at the first stage, and they can provide some advice to the designer and take on construction management. CM changes the traditional bidding model to multi-stage bidding model. The owner, CM and designer constitute a group which is responsible for plan, design and construction. The CM is responsible for supervision and coordination and management. It also communicates with the contractor about the cost, quality and schedule.

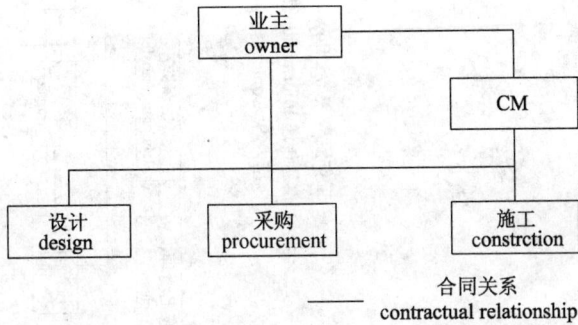

业主
owner

CM

设计
design

采购
procurement

施工
constrction

—— 合同关系
contractual relationship

图 9.7　CM 模式
Fig. 9.7　CM model

（4）设计 – 采购 – 建设模式（EPC）

（4）Engineering Procurement Construction（EPC）

EPC 模式是一种简练的工程项目管理模式,是一种具有特殊性的设计 – 建造方式,即由承包商为业主提供包括项目科研、融资、土地购买、设计、施工直到竣工移交给业主的全套服务,如图 9.8 所示。采用此模式,在工程项目确定之后,业主只需选定负责项目的设计与施工的实体——交钥匙的承包商,该承包商对设计、施工及项目完工后试运行全部合格的成本负责。项目的供应商与分包商仍须在业主的监督下采取竞标的方式产生。

EPC（Fig. 9. 8）is an acronym that stands for engineering, procurement and construction. It is a common form of contracting arrangement within the construction industry. Under an EPC contract, the contractor designs the installation, procures the necessary materials and builds the project, either directly or by subcontracting part of the work. In some cases, the contractor carries the project risk for schedule as well as budgets in return for a fixed price, called lump sum or LSTK depending on the agreed scope of work.

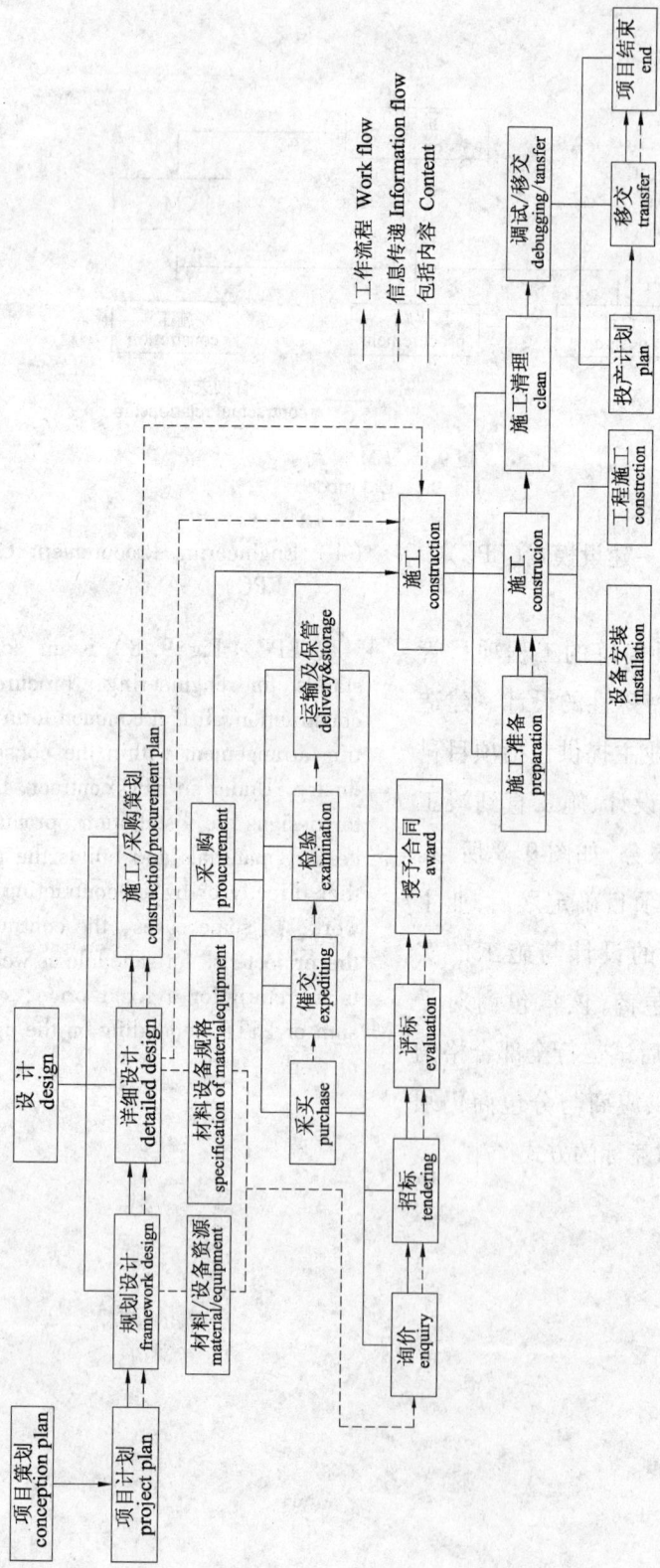

图 9.8　EPC 模式
Fig.9.8　EPC model

（5）建造 – 运营 – 移交模式（BOT）

BOT 模式是一种将基础设施建设项目依靠私人资本的一种融资、建造项目管理方式（如图9.9所示），即基础设施国有项目民营化。政府开放本国基础设施建设和运营市场，授权项目公司负责筹资和组织建设，建成后负责运营及偿还贷款，协议期满后，再无偿移交给政府。BOT 模式的基本思路是：由项目所在国政府或所属机构为项目的建设和经营提供一种特许权协议作为项目融资的基础，由本国公司或者外国公司作为项目的投资者和经营者安排融资，承担风险，开发建设项目，并在有限的时间内经营项目获取商业利润，最后根据协议将该项目转让给相应的政府机构。

（5）Build-Operate-Transfer（BOT）

Build-Operate-Transfer（Fig. 9.9）is a form of project financing, wherein a private entity receives a concession from the private or public sector to finance, design, construct, and operate a facility stated in the concession contract. This enables the project proponent to recover its investment, operating and maintenance expenses in the project. BOT finds extensive application in the infrastructure projects and in public-private partnership. In the BOT framework, a third party, for example the public administration, delegates to a private sector entity to design and build infrastructure and to operate and maintain facilities for a certain period. During this period, the private party has the responsibility to raise the finance for the project and is entitled to retain all revenues generated by the project and is the owner of the regarded facility. The facility will be then transferred to the public administration at the end of the concession agreement, without any remuneration of the private entity involved.

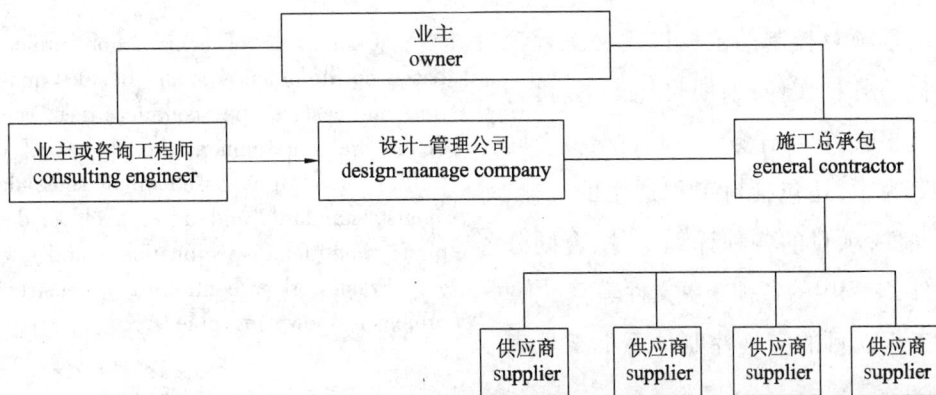

(a) 实现形式 1 The first realization form

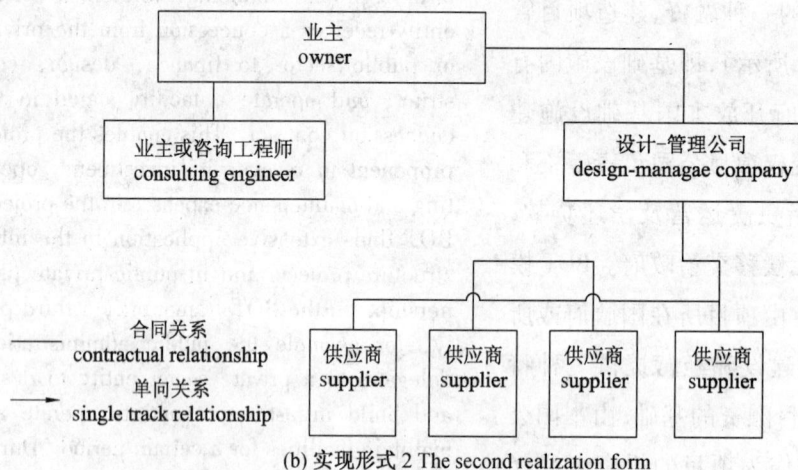

業主
owner

業主或咨询工程师
consulting engineer

设计-管理公司
design-managae company

合同关系
contractual relationship

单向关系
single track relationship

供应商
supplier

供应商
supplier

供应商
supplier

供应商
supplier

(b) 实现形式 2 The second realization form

图 9.9 BOT 模式
Fig. 9.9 BOT model

9.5 工程项目的实施控制

工程项目的实施控制是工程项目管理的重要职能。它通过与预定标准进行对比,及时发现存在的偏差,并采取纠偏措施,从而减少最终的误差。

工程项目控制的依据以定义工程项目目标的各种文件,如项目建议书、可行性研究报告、项目任务书、设计文件、合同文件等为主,还包括对工程适用的法律、法规文件,项目的各种计划文件、合同分析文件,工程中的各种变更文件等。

工程项目的实施控制体系框架见表9.3。

9.5 Controlling of construction project

Controlling is an important function of construction project management, because it helps to check the errors and to take the corrective action, so that deviation from standards is minimized and stated goals of the organization are achieved in a desired manner.

Controlling of construction project is based on the object system. Besides quality, time and cost, scope, contract, risk, safety, health and environment are objects of controlling, too. It is based on to some documents, standards and laws, such as design plans, contracts, specifications, and laws.

Framework of controlling of construction project is shown in Table 9.3.

表 9.3 工程项目的实施控制体系框架
Table 9.3 Framework of controlling of construction project

对象 Object	内容 Content	目标 Goal	依据 Basis
范围控制 scope control	保证按任务书(或设计文件、合同)规定的数量完成工程 complete the whole project and sub-project according to the plans	范围定义 scope definitions	范围规划和定义文件(项目任务书、设计文件、工程量表等) scope plans, definition documents, etc.
成本控制 cost control	保证按计划成本完成工程,防止成本超支和费用增加,达到盈利目的 control cost, avoiding overspend, cost plan supervision	计划成本 projected costs	各分项工程、分部工程、总工程计划成本、人力、材料、资金计划、计划成本曲线等 cost plan, labor cost plan, etc.
质量控制 quality control	保证按任务书(或设计文件、合同)规定的质量完成工程,使工程顺利通过验收,交付使用,实现使用功能 complete the project with all functions and be up to standard	规定的质量标准 quality standards	各种技术标准、规范、工程说明、图纸、工程项目定义、任务书、批准文件 technical standard, definition, plan, drawing, etc.
进度控制 schedule control	按预定进度计划实施工程,按期交付工程,防止工程拖延 complete the project in time	任务书(或合同)规定的工期 contract period	工期定额规定的总工期计划、批准的详细的施工进度计划、网络图、横道图等 schedule plan, Gantt chart, PERT, etc.
合同控制 contract control	按合同规定全面完成自己的义务,防止违约 supervision of performance, against default	合同规定的各项义务、责任 responsibilities in contract	合同范围内的各种文件、合同分析资料 contract documents, contract analysis, claim documents, etc.
风险控制 risk control	防止和减低风险的不利影响 reduce risk	风险责任 risk responsibility	风险分析和风险应对计划 risk analysis, risk response plan, etc.
安全、健康、环境控制 safety, health and environment control	保证项目的实施过程、运营过程和产品(或服务)符合安全、健康和环境保护要求 assure the safety, health and environment of standards	法律、合同和规范 laws, contracts, standard	法律、合同文件和规范文件 laws, contracts, standard, etc.

9.6 工程项目管理的执业资格认证制度

9.6 Professional qualification systems of construction project management in China

20 世纪 80 年代以来,通过借鉴国际

A series of professional qualification

惯例,我国在建设工程领域建立了执业资格认证制度,见表9.4。其中,直接与工程项目管理密切相关的有注册建造师、注册监理工程师、注册造价工程师、注册咨询工程师等。

systems of construction project management (Table 9.4) has been set up according to the international tradition in China since 1980s. It includes registered architect, registered structural engineer, registered constructor, certified supervision engineer, certified cost engineer, certified consulting engineer, etc.

表 9.4　工程项目管理的执业资格认证制度
Table 9.4　Professional qualification systems of construction project management in China

名　称 Title	执业范围 Scope of professional service
注册建造师 registered constructor	可以受聘担任建设工程施工的项目经理或从事其他施工活动的管理工作,如质量监督、工程管理咨询,以及法律、行政法规或国务院建设行政主管部门规定的其他业务
	being a manager of a construction project or doing works related to construction project management
注册监理工程师 certified supervision engineer	可从事工程监理、工程经济与技术咨询、工程招标与采购咨询、工程项目管理服务以及国务院有关部门规定的其他业务
	supervision of a construction project, economic and technical consulting, consulting of bidding and procurement, management service, etc.
注册造价工程师 certified cost engineer	从事工程项目投资估算的编制、审核及项目经济评价;工程概算、预算、结(决)算、标底价、投标报价的编审;工程变更及合同价款的调整和索赔费用的计算;工程项目各阶段工程造价控制;工程经济纠纷的鉴定;工程造价计价依据的编审;与工程造价业务有关的其他事项
	investment estimate, economic evaluation, cost control, calculating of claiming, etc.
注册咨询工程师 certified consulting engineer	从事经济社会发展规划、计划咨询;行业发展规划和产业政策咨询;经济建设专题咨询;投资机会研究;工程项目建议书的编制;工程项目可行性研究报告的编制;工程项目评估;工程项目融资咨询,绩效追踪评价,后评价及培训咨询服务;工程项目招投标技术咨询;国家发展计划委员会规定的其他工程咨询业务
	preliminary consultation on research investment opportunity and social appraisal, writing reports, project proposals, feasibility study report and amendments.

9.7　工程项目管理的信息化——BIM

9.7　Informationization of construction project management: BIM

信息化是指培养、发展以计算机为主的智能化工具为代表的新生产力,并使之造福于社会的历史过程。信息技术是实

Informationization refers to the extent, by which a geographical area, an economy or a society is becoming information-based, i.e., increase in size of its information labor force. Informationization to the Information

现信息化的重要技术支撑和手段。工程项目管理的信息化可以实现项目管理信息的大量储存、高效处理和快速传输；承担复杂的项目管理计算工作，如网络分析、资源和成本优化等；提供管理手段和方法的应用支持和途径，如系统控制方法，预测决策方法等。

　　建筑信息建模（building information modeling，BIM）是工程项目的物理特性与功能特性集成的数字化模型，如图9.10所示。它能够为从工程项目的最初概念设计开始的整个生命周期里的任何决策提供可靠的共享信息资源。实现BIM的前提是在工程项目生命周期的各个阶段，不同的项目参与方通过在BIM建模过程中插入、提取、更新及修改信息以支持和反映各参与方的职责。

Age is what industrialization was to the Industrial Age. When it comes to construction project management, information technology can help users to store quite a substantial amount of information, undertake complex calculations and support the management measures and methods.

Building information modeling（BIM）is a process involving the generation and management of digital representations of physical and functional characteristics of a facility（Fig. 9.10）. The resulted building information models become shared knowledge resources to support decision-making about a facility from earliest conceptual stages as well as design and construction, through its operational life and eventual demolition.

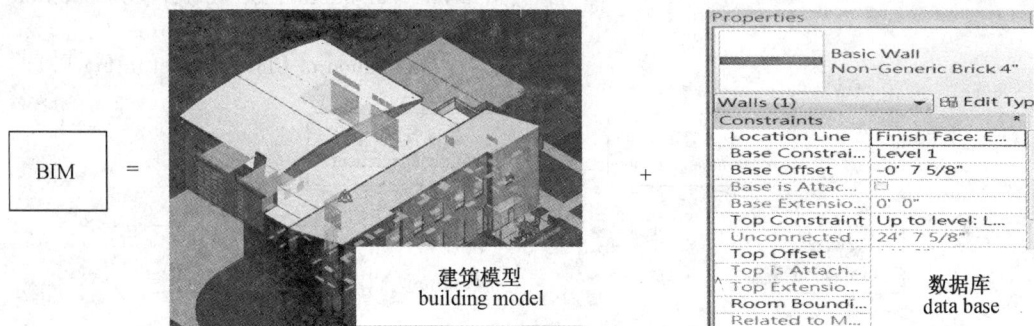

图 9.10　BIM 的概念
Fig. 9.10　Concept of BIM

　　传统的建筑设计依赖于2D建筑模型，其是用点、线、面元素模拟几何构件，只有长和宽的二维尺度。BIM则将这一模型扩展到了3D甚至更多的维度（nD），不同的维度对应于不同的功能。3D模型增加了高度，实现了空间化表达。当然，

Traditional building design is largely reliant on two-dimensional drawings（plans, elevations, sections, etc.）. BIM extends this beyond 3D, augmenting the three primary spatial dimensions（width, height and depth -X, Y and Z）with time as the fourth dimension and cost as the fifth. BIM therefore covers more than just geometry. It also

也可将时间作为第四维,费用作为第五维。

4D 的 BIM 模型是在 3D 模型的基础上增加了进度管理功能,可以研究项目的可施工性、进度安排、进度优化、精益化施工等,提高经济性与时效性。

5D 的 BIM 模型是在 4D 模型的基础上增加了费用管理功能,可以实现全寿命期的费用分析与优化。

6D(Green BIM)及 nD 的 BIM 模型是在 5D 模型的基础上进一步增加了全寿命周期管理功能,将更大化地满足业主和社会的需求,如舒适度模拟及分析、耗能模拟、可持续化分析等。

BIM 的应用框架如图 9.11 所示。

covers spatial relationships, light analysis, geographic information, and quantities and properties of building components.

4D BIM refers to the intelligent linking of individual 3D CAD components or assemblies with time or schedule-related information. 4D BIM can make schedule plan lean construction, and improve economy and timeliness.

5D BIM refers to the intelligent linking of individual 3D CAD components or assemblies with schedule (time) constraint and cost-related information. 5D BIM can be employed to do life cycle cost analysis and optimizing.

6D BIM (Green BIM) refers to the intelligent linking of individual 3D CAD components or assemblies with all aspects of project life-cycle management information. It is an innovative construction methodology which helps Contractor to Manage, Optimize, Review and Feedback 3 additional dimensions, such as comfort simulation and analysis, simulation of energy consumption and sustainable analysis.

Application of BIM is shown in Fig. 9.11.

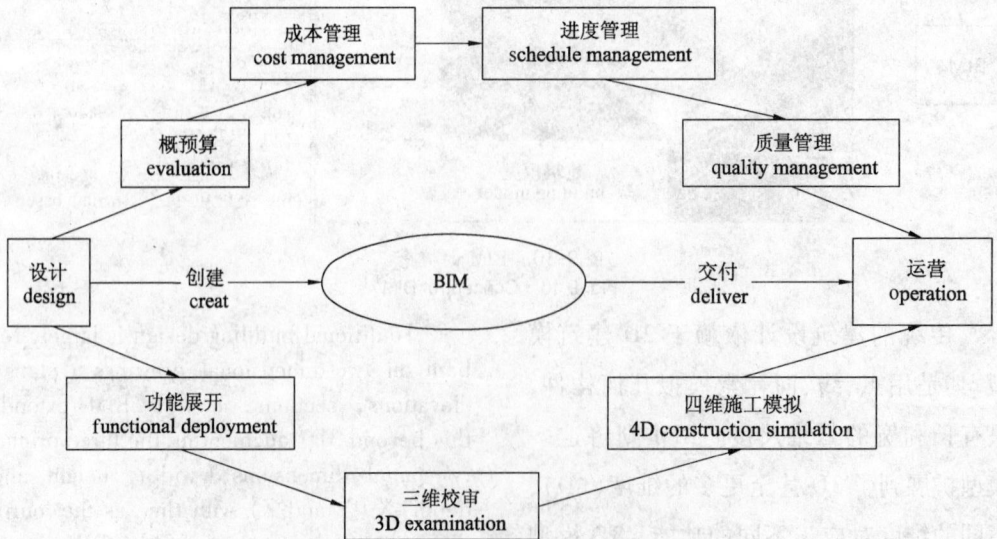

图 9.11　BIM 的应用框架
Fig. 9.11　Application of BIM

建设工程是分阶段展开的,各阶段涵盖不同的工作,具有不同的工作重点,而 BIM 能够实现对各建设阶段的全面覆盖。设计阶段,BIM 技术所输出的可视化效果可以为业主校核是否满足要求提供平台,且利用 BIM 技术可实现耗能与可持续发展设计与分析,能为提高建筑物、构筑物等的性能提供支持。实施阶段,BIM 技术可以加强各参与方之间的协作与信息交流的有效性,改善传统的工程项目管理模式和实施流程,通过加强伙伴协作、建筑工业化、优化供应链等方式,缩短工期、提高质量。运营阶段,将 BIM 竣工模型作为设备管理与维护的数据库能够显著提高运营和维护的工作水平。

9.8 工程项目的国际化——国际工程

国际工程是指一个工程项目具有国际化的工程要素,各个方面的参与者来自于多个国家,并且按照国际通用的工程项目管理模式和规则进行管理的工程。例如,鲁布格水电站是我国第一个真正意义上的国际工程,如图 9.12 所示。

国际工程通常包含国际工程咨询和国际工程承包两个领域。国际工程咨询包括对工程项目前期的投资机会研究、可行性研究、项目评估、勘测设计、招标评标、工程监理、项目管理、项目后评价等。国际工程承包涉及对工程项目进行投资、施工、设备采购、安装调试、项目分包、合

The BIM concept envisages virtual construction of a facility prior to its actual physical construction, in order to reduce uncertainty, improve safety, work out problems, and simulate and analyze potential impacts. Sub-contractors from every trade can input critical information into the model before beginning construction, with opportunities to pre-fabricate or pre-assemble some systems off-site. Waste can be minimized on-site and products delivered on a just-in-time basis rather than being stock-piled on-site. BIM can bridge the information loss associated with handing a project from design team, to construction team and to building owner/operator, by allowing each group to add to and reference back to all information they acquire during their period of contribution to the BIM model. This can yield benefits to the facility owner or the operator.

9.8 Internationalization of construction project: International project

International project has international project element, international participants and international convention of management. For example, Lubuger hydropower station (Fig. 9.13) is the first typical international project in China.

International project includes consulting project and contracting project. Consulting includes feasibility study, project appraisal, design planning, prepare the tender documents and supervision of the work. Contracting includes execution of the work, procurement of equipment, sub-construction of work and labor service.

作劳务等。

图 9.12　鲁布格水电站
Fig. 9.12　Lubuger hydropower station

国际工程主要有以下 3 类：

① 基础设施和土木工程,如水利、公路和铁路建设、地下建筑工程以及桥梁、码头工程；

② 以资源为基础的工程,如石油开发、炼油厂、矿山钢铁厂、化工厂、化肥厂；

③ 制造业工程,如发电厂、造纸厂、纺织厂和机器制造厂等。

国际工程是国际经济合作的最高层次。它能够输出资金、物资、技术和劳动力,实现全球范围的资源最优配置和合作双方的双赢。

进入 21 世纪后,经济发展和技术进步取得了巨大的成就,经济的全球化、信息化与网络化等技术的发展,管理理念的变革,人类追求可持续发展的目标,这些变化与发展趋势对国际建筑业的发展产生巨大的影响,带来了国际工程和建筑市场的新的变化。这些变化是相互关联的,并可以用

International project is divided into petroleum, general building, transportation, industry, power supply, manufacture industry, water supply, hazards treatment, sewage water and others according to *Engineering News Records*.

International project is a form of top level of international economic collaboration. It includes export of capital, goods, technology, and labor. It optimizes combination of the productive element around the world and makes double win.

The development of economy and technology facilitates globalization, informationization, networknization, and helps people achieve sustainable development, which strongly influences the development of international construction industry. These changes are shown as a chain in Fig. 9.13.

土木工程导论

一个发展变化链来描述,如图 9.13 所示。

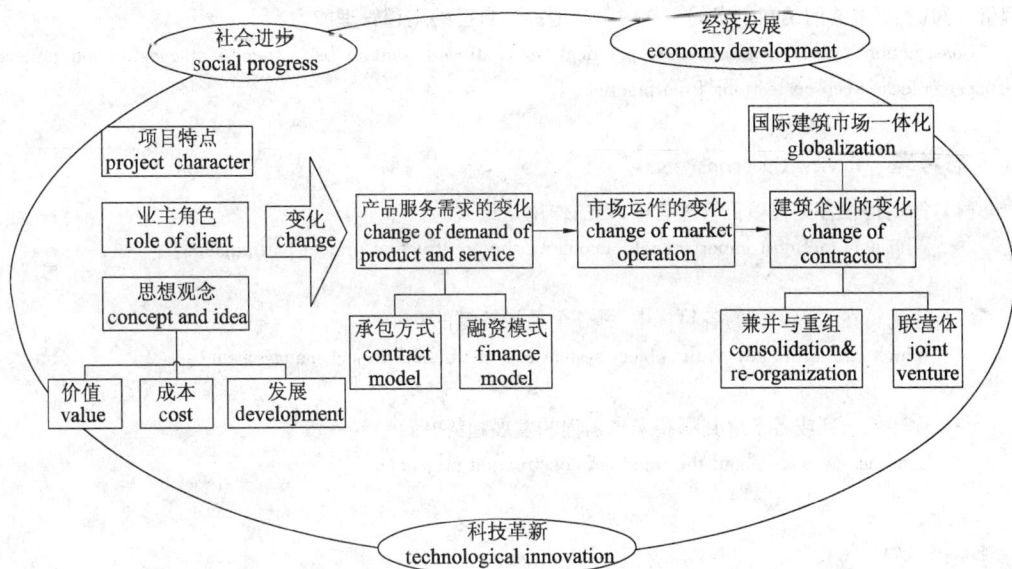

图 9.13　国际工程市场的变化与发展
Fig. 9.13　Change and development of international construction market

注:本章图片均来源于网络。
Note：In this chapter, all pictures are from webs.

知识拓展
Learning More

相关链接　Related Links

如果想了解工程项目管理课程及详细知识,可访问重庆大学的工程项目管理精品课程网站和同济大学的工程项目管理精品课程网站。

If you want to get more information about construction project management, please visit http://202.202.0.17/index. jsp/ and http://jpkc. tongji. edu. cn/jpkc/gcxmgl/web/.

如果想了解政策信息,可访问住房与城乡建设部网站 www. mohurd. gov. cn。

If you want to get some policy information, please visit www. mohurd. gov. cn.

如果想了解国际工程市场发展的最新动向,可访问工程新闻纪要网站 www. enr. com。

If you want to learn about the international construction market, please visit www. enr. com.

小贴士　Tips

(1) 工程管理本身就是一个本科专业,在许多学校都与土木工程专业共同设置,相辅相成。很多具有土木工程专业背景的人士,未来都走上了从事工程管理工作的道路。

Construction management is an undergraduate major, too. Many professionals with the civil engineering background work on construction project management.

(2) 工程项目管理是一项高度重视实践的工作,希望大家树立正确的工程管理意识,储备足够的工程管理知识,更重要的是在未来的工作中不断地提高自己的工程管理能力。

Construction project management is practical work. If you want to be a successful construction project manager, please keep on learning from practice.

思考题　Review Questions

(1) 你觉得在各工程项目管理者中,谁的作用是最大的?

Who play the most important role among all the construction project participators?

(2) 你觉得在工程项目的各目标中,哪一个是最关键的?

Which one is the key in the object system of construction project management?

(3) 你还能想到或者了解到哪些未来工程的发展趋势?

Can you say more about the trends of construction project?

参考文献
References

[1] 成虎,陈群:《工程项目管理》(第3版),中国建筑工业出版社,2009年。

[2] 成虎:《工程管理概论》(第2版),中国建筑工业出版社,2010年。

[3] 成虎:《建设工程合同管理与索赔》(第4版),东南大学出版社,2008年。

[4] 李启明,邓小鹏:《工程项目采购模式与管理》,东南大学出版社,2011年。

[5] 李启明:《土木工程合同管理实务》,东南大学出版社,2009年。

[6] 陆惠民:《工程项目管理》,东南大学出版社,2002年。

[7] 李启明:《国际工程管理》,东南大学出版社,2010年。

[8] Fredrick E. Gould, Nancy E. Joyce. Construction Project Management (Professional Edition). Tsinghua University Press,2004.

[9] Keith Collier. Construction Contracts. 3rd ed. Tsinghua University Press,2004.

[10] Frank R. Dagostino, Leslie Feigenbaum. Estimating in Building Construction. 5th ed. Tsinghua University Press,2004.

土木工程导论

第10章 工程合同管理

Chapter 10　Construction Contract Management

合同管理在现代建设工程项目管理中具有十分重要的地位,并发挥着重大的作用,它已成为与成本管理、质量管理、进度管理和信息管理等并列的一项管理职能。合同管理对项目的成本管理、质量管理和进度控制有总控制和总协调的作用,是工程管理的核心和灵魂。本章主要围绕工程招投标合同管理、工程索赔管理和合同风险管理等核心知识展开。

In modern construction project management,contract management plays a great role, and it has become a function coordinating with cost management, quality management, schedule management, information management,ect. Contract management plays a role of total control and total coordination in cost management, quality management, schedule control, and therefore, it is the core and soul of engineering management. We will carry out a discussion about the contract management in construction bidding and tendering, construction claim management, contract risk management,etc.

10.1　工程合同管理概述

10.1　Introduction to construction contract management

10.1.1　工程合同管理基本概念

10.1.1　Basic concept of construction contract management

(1) 建设工程合同概念

《中华人民共和国合同法》第 269 条:"<u>建设工程合同</u>是承包人进行工程建设,发包人支付价款的合同。"它可简单表示为如图 10.1 所示的形式。

(1) The concept of construction contract

The 269th article in the *Contract Law of the People's Republic of China* prescribes that <u>construction engineering contract</u> is the contract by which the contractor performs engineering construction and the employer pays the price. It may be expressed simply as Fig. 10.1.

图 10.1　建设工程合同关系
Fig. 10.1　Relationship in construction contract

在建设工程中,主要的建设合同体系如图 10.2 所示。

In the construction project, the main contract system of construction engineering is shown in Fig. 10.2.

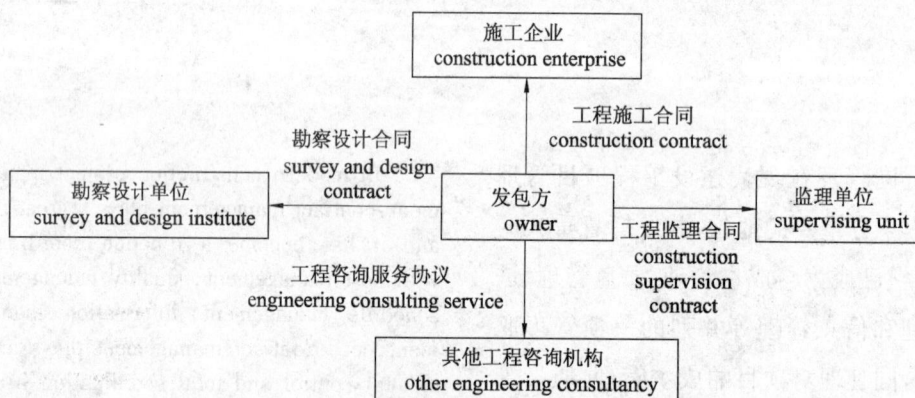

图 10.2　建设工程合同体系
Fig. 10.2　The system of construction contracts

建设工程项目是一个极为复杂的社会生产过程,而在工程中维系这种关系的纽带是合同,所以就产生了各式各样的合同。工程项目的建设过程实质上是一系列经济合同签订和履行的过程。

Construction project is an extremely complex social producing process, in which contract plays an important role. In fact, project construction process contains a series of economic contract signing and fulfilling.

(2) 工程合同管理概念

工程合同管理是对工程项目中相关合同的组织、策划、签订、履行、变更、索赔和争议解决的管理。根据合同管理的对象,工程合同管理可分为单项合同管理和整个项目的合同管理;而根据合同管理主体的不同,又可分为业主方合同管理和工程承包方合同管理。

(2) Concept of construction contract management

Construction contract management refers to organizing, planning, signing, performing, changing, claiming and dispute-resolving about the construction contracts. According to the management object, construction contract management can be divided into single contract management and comprehensive contract management. According to the management subject, it can be divided into owner's contract management and contractor's

10.1.2 工程合同管理的特点

（1）工程合同管理对工程经济效益的影响

由于工程项目建设投资大，合同价款高，使得合同管理的效益显著，合同管理对工程经济效益影响很大。若合同管理得好，合同主体可避免亏损，赢得利润，否则，将要承受较大的经济损失。

（2）工程合同管理工作极其复杂、繁琐

由于参建单位众多和项目之间接口复杂等特点，使得工程合同管理工作极其复杂、繁琐。在合同签订和履行过程中，涉及众多不同主体之间的各种错综复杂关系，处理好各方的关系极为重要，同时也很复杂和困难，稍有疏忽就可能导致巨大的经济损失。

（3）工程合同管理周期长、时间跨度大

这是由工程合同的生命期长决定的。工程合同的生命期不仅包括施工期，还包括招投标和合同谈判以及保修期，一般至少两年，有的长达五年或更长，难以预测此间的外界环境变化，因此工程合同管理周期长、跨度大，受外界环境的影响大、风险大。

10.1.2 Characteristics of the construction contract management

(1) The economic effects of construction contract management on project

Due to large project investment and high contract price, it makes the benefits of the contract management remarkable, so contract management has great influence on the economic benefits of the project. If the contract management is made well, the parties can avoid losses and win the profits. Otherwise, they will bear larger economic losses.

(2) The complicated and cumbersome work of construction contract management

Due to many participants in construction and complex engagements in the projects, the construction contract management is extremely complex and cumbersome. The process of signing and fulfilling contract includes various intricate relationships among many different subjects, and it is extremely important to handle the relationships properly among the parties, and it is very complex and difficult at the same time, so a slight negligence may lead to large economic losses.

(3) The long duration and large time span of construction contract management

This characteristic depends on the long lifecycle of a construction contract. The lifecycle of a construction contract not only includes construction period, but also includes the bidding, contract negotiations and the warranty period, and then there are generally two years at least; nevertheless, some projects are up to five years or longer. Therefore, it is difficult to predict the external environment changes in such a long time. The construction contract management cycle is long, and its span is large, and then the

工程合同管理 Construction Contract Management

(4) 工程合同管理是动态的

工程过程中内外的干扰事件多,合同变更频繁,这就要求合同管理必须是动态的,合同实施过程中合同变更管理显得尤为重要。因此,项目管理人员必须加强对合同变更的管理,做好记录,作为索赔、变更或终止合同的依据。

10.1.3　工程合同管理的主要工作

合同管理是工程项目管理中的一个重要职能,贯穿于工程项目的决策、计划、实施和结束的整个过程,如图 10.3 所示。

10.1.4　工程合同的发展趋势

纵观世界各国或地区,国际工程合同的趋势是建立和发展合同主体之间的合作伙伴关系,合同模式从"对抗型"和"敌对型"转向"合作型",思维模式从"如何解决发生的问题"转向"如何合作实现合同目标"。在早期,国际工程一般采用非契约型合作伙伴关系,各方只签署无法律约束力的宪章或意向书,而近年来普遍趋向于采用契约型合作伙伴关系。

而我国工程合同的发展趋势则是追求公平性,如通过电脑随机选择投标人甚

construction contract management is affected by the external environment. Its risk is large.

(4) The dynamic construction contract management

Due to plentiful interferences inside and outside the construction process and frequent contract changes, the contract management must be dynamic, and it seems particularly important to manage contract changes. Therefore, project managers must strengthen the contract change management and make records, which serve as the basis for lodging a claim, modifying or terminating a contract.

10.1.3　Main work of construction contract management

Contract management is an important function of the construction project management, which is throughout the entire project process of decision, plan, implementation and termination, and as is shown in Fig. 10.3.

10.1.4　Development trend of construction contract

Throughout various countries or regions in the world, the developmental trend of construction contract is the establishment and development of cooperative partnership among the contract subjects. Contract mode makes the transition from confrontation and hostility to the cooperation, and thinking mode transfers from how to solve the problem into how to realize the target. In the earlier days, international engineering commonly adopts cooperation by the non-contractual relationship that the parties only sign the charter or letter of intent without legally binding, but contractual relationship generally tends to be adopted in recent years.

Yet the developmental trend of the construction contract in China is the pursuit of fairness. For example, the bidder is selected

至抽签授予合同,这就损害了合作伙伴关系的基础。并且我国工程合同范本及其风险分配往往也只是以管理需要而不是国际惯例为准则,这容易造成彼此对立,因而需要强化独立工程师的作用甚至引入 JCT 的独立测量师,以便独立、公平、公正地履行合同。然而,越是追求公平,合同双方的对抗性就越强,而对抗性越强又越需要公平,以至陷入一种非理性循环状态。

randomly by the computer, and even contract is awarded by lottery. This will damage the foundation of cooperative partnership, and the construction contract model and its risk distribution are only ruled by managerial demand rather than the international convention, so it is likely to create a contrary relationship. The independent engineer's role needs to be strengthened and even to introduce the independent surveyor of Joint Contract Tribunal (JCT), so that the contract can be fulfilled independently, fairly and justly. However, the more the fairness is pursued, the stronger the antagonism of both parties is; nevertheless, the stronger the confrontation is, the more fairness we need, so that it is in an irrational circulation situation.

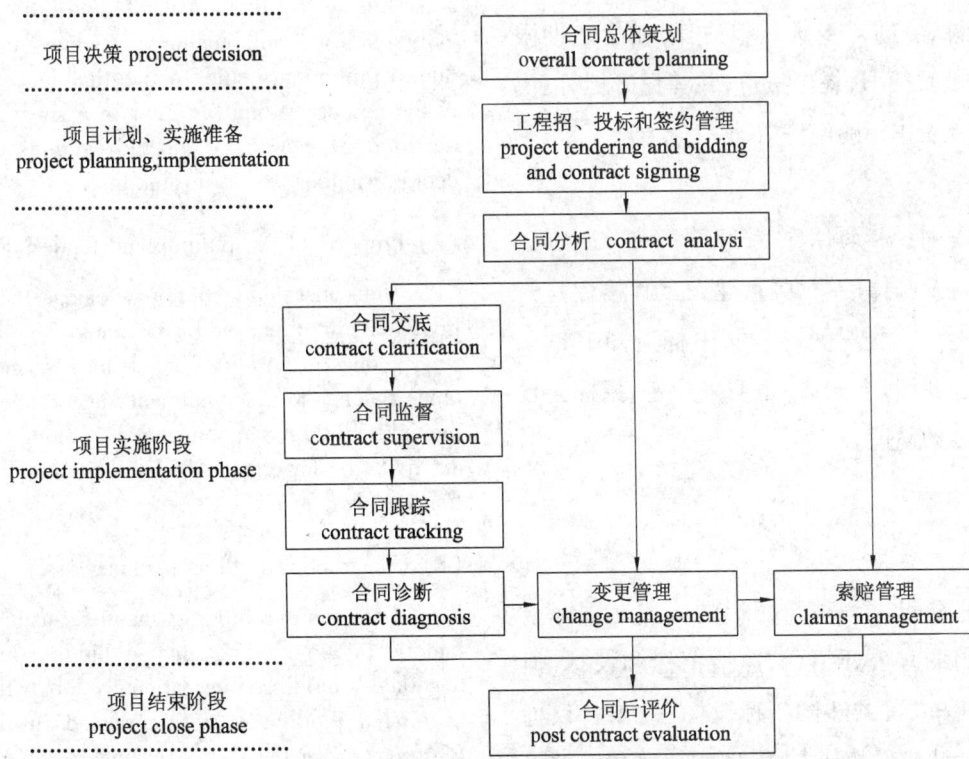

图 10.3　工程合同管理过程

Fig. 10.3　Construction contract management process

10.2.1　工程招投标概述

(1) 工程招投标的概念

　① 工程招标是指建设单位对拟建的工程项目通过法定的程序和方式吸引建设项目的承包单位竞争,并从中选择条件优越者来完成工程建设任务的法律行为。

　② 工程投标是工程招标的对称概念,指具有合法资格和能力的投标人根据招标条件,经过初步研究和估算,在指定期限内填写标书,提出报价,并等候开标,以决定能否中标的经济活动。

(2) 招投标的性质

　从法律的角度看,建设工程招标是要约邀请,而投标是要约,中标通知书是承诺。我国《合同法》也明确规定,招标公告是要约邀请。

(3) 工程招标方式

　建设工程项目招标的方式在国际上通行的有公开招标、邀请招标和议标,但《中华人民共和国招标投标法》未将议标作为法定的招标方式,即法律所规定的强制招标项目不允许采用议标方式,这主要是因为我国国情与建筑市场的现状条件,不宜采用议标方式,但法律并不排除议标

10.2.1　Summary of construction bidding and tendering

(1) Concept of construction bidding and tendering

　① Construction bidding means that the construction unit attracts many contractors to compete for the proposed project through the legal procedures and methods, and thus select the superior contractor to complete the construction task.

　② Construction tendering corresponds with construction bidding, which means that according to the bidding conditions, the legally qualified and competent bidders conduct a preliminary study and estimation, fill in the tender, submit an offer in a specified deadline, then wait for the bid opening and decide whether to win the bidding.

(2) Property of the bidding and tendering

　The construction bidding belongs to an invitation for offer, and construction tendering belongs to an offer, and letter of acceptance belongs to a commitment, according to the law. Bidding announcement belongs to an invitation for offer, which is stipulated in *Contract Law of China*.

(3) Construction bidding manners

　Construction bidding manners usually include public bidding, invited bidding and negotiated bidding internationally, but the negotiated bidding is not stipulated in the *Bidding Law of the People's Republic of China*. That is to say, the mandatory bidding projects, stipulated in the law, are not allowed to bid by negotiated bidding. Mainly owing to the national conditions and the present situation of the construction market in

方式。3 种工程招标方式的比较见表 10.1。

China, it is not suitable to adopt negotiated bidding, but the law is not opposed to using negotiated bidding. Comparison of the three bidding and tendering manners is shown in Table 10.1.

表 10.1 工程招标方式比较
Table 10.1 Comparison of the three bidding manners

招标方式 Bidding manners	含义 Implication	优点 Advantages	缺点 Disadvantages	适用范围 Applied ranges
公开招标 public bidding	招标人以招标公告的方式邀请不特定的法人或其他组织投标 bid inviter invites unspecified corporation-or or other organizations to bid by announcement.	投标竞争激烈,择优率更高,有利于保证建设工程的质量并降低工程的造价 fierce competition, higher preferential rate, ensuring quality, reducing cost	招标的工作量大,所需招标时间长、费用高 big workload, long time, high cost	标的额较大的项目 projects of high total price
邀请招标 invited bidding	招标人以投标邀请书的方式邀请特定的法人或其他组织投标 bid inviter invites specified some organizations to bid by invitation for bids.	简化了招标程序、能节省招标费用和时间;承包商违约风险降低 simplify procedure, save time and money, reduce contractor's default risk	竞争的激烈程度相对较差,有可能提高中标的合同价,也有可能排除了某些在技术上或报价上有竞争力的承包商参与投标 relatively weak competition, likely increase contract price, likely excludes some competitive contractors from bidding on technics or quotation.	标的额较小的项目和专业性强或需要在短时间内完成投标任务等的项目 projects of lower total price, projects strongly specialized or bidding needed completing in a short time, etc.
议标 negotiated bidding	由招标人直接与选定的承包单位就发包项目进行协商的一种招标方式 bid inviter negotiates directly with the selected contractors.	节省时间,容易达成协议,可迅速开展工作 save time, reach easily an agreement, work quickly	一对一谈判,竞争程度最差 one-on-one negotiations, the weakest competition	非国家法定招标项目、不宜进行公开或邀请招标的项目 statutory bidding project, project unsuited to public or invited bidding.

工程合同管理 Construction Contract Management

10.2.2 工程招投标管理工作流程

工程招标管理的工作流程如图 10.4 所示。

10.2.2 Construction bidding management process

Construction bidding management process is shown in Fig. 10.4.

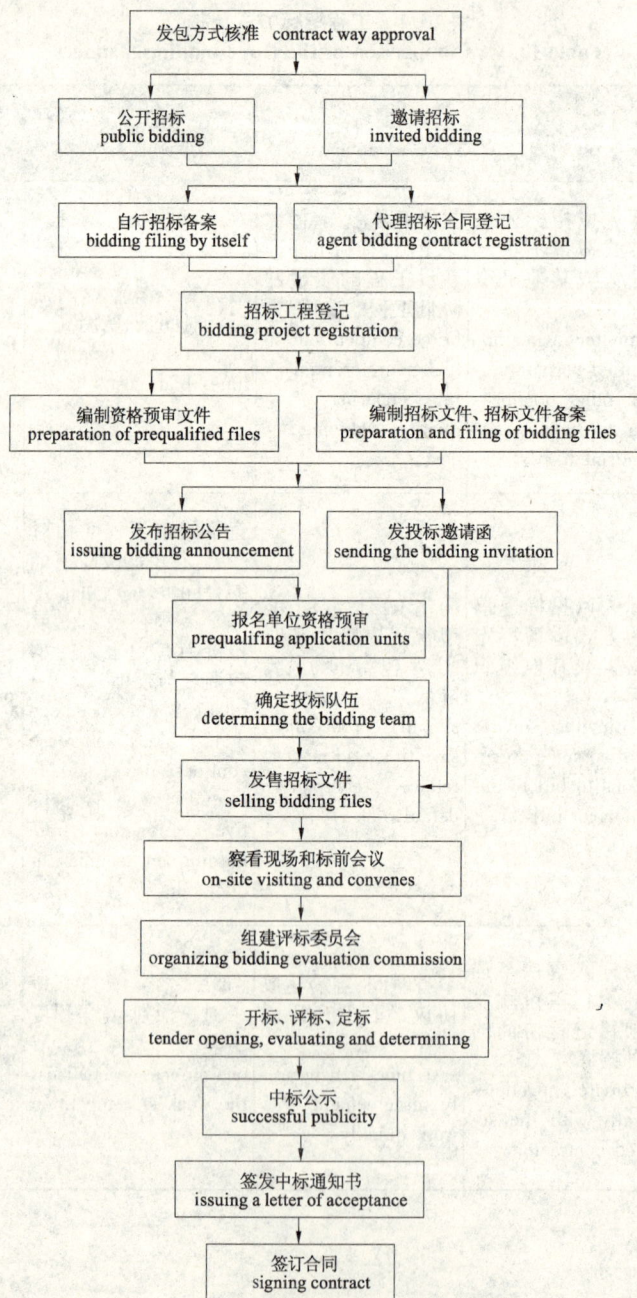

```
                    发包方式核准  contract way approval
                              |
              ┌───────────────┴───────────────┐
        公开招标                          邀请招标
        public bidding                  invited bidding
              |                               |
        自行招标备案                    代理招标合同登记
        bidding filing by itself        agent bidding contract registration
                              |
                    招标工程登记
                    bidding project registration
                              |
              ┌───────────────┴───────────────┐
        编制资格预审文件                 编制招标文件、招标文件备案
        preparation of prequalified files   preparation and filing of bidding files
              |                               |
        发布招标公告                     发投标邀请函
        issuing bidding announcement    sending the bidding invitation
              |
        报名单位资格预审
        prequalifing application units
              |
        确定投标队伍
        determinng the bidding team
              |
        发售招标文件
        selling bidding files
              |
        察看现场和标前会议
        on-site visiting and convenes
              |
        组建评标委员会
        organizing bidding evaluation commission
              |
        开标、评标、定标
        tender opening, evaluating and determining
              |
        中标公示
        successful publicity
              |
        签发中标通知书
        issuing a letter of acceptance
              |
        签订合同
        signing contract
```

图 10.4　工程招标管理流程

Fig. 10.4　Construction bidding management process

工程投标管理的工作流程如图 10.5 所示。

Construction tendering management process is shown in Fig. 10.5.

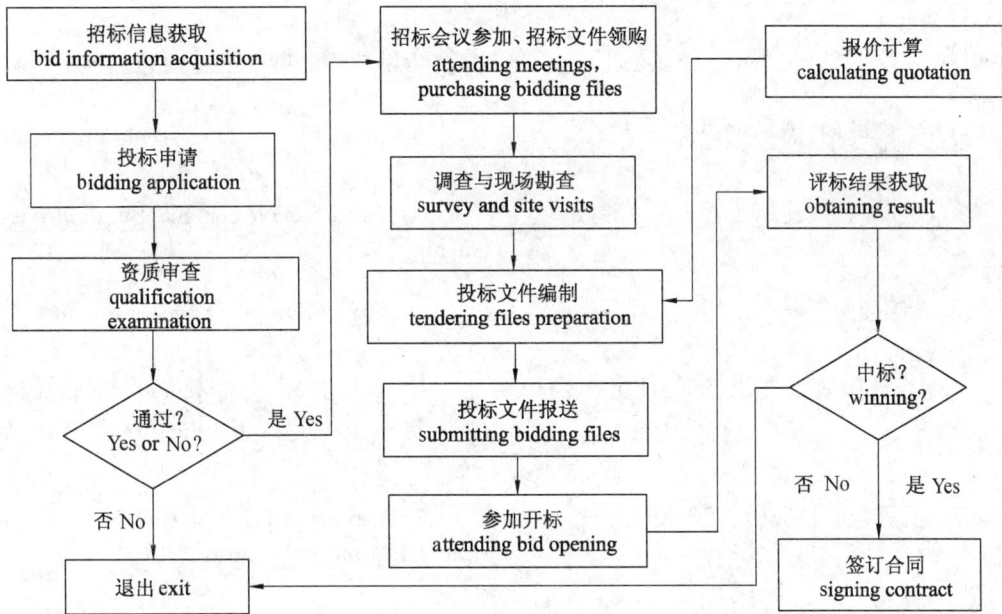

图 10.5 工程投标管理流程
Fig. 10.5 Construction tendering management process

10.2.3 工程招标公告示例

10.2.3 A case of construction bidding announcement

江苏省＊＊大厦建设项目

招 标 公 告

编号：CNJ130022-＊＊＊＊

江苏＊＊股份有限公司建设项目江苏＊＊大厦项目已经立项部门批准建设。工程所需资金来源自筹，现已落实。本工程对投标申请人的资格审查，采用资格预审方法选择合格的投标申请人参加投标。

一、江苏省＊＊股份有限公司（招标代理）受招标人委托负责本工程的招标事宜。

＊＊ building construction project in Jiangsu province
Bidding Announcement
Serial number：CNJ130022-＊＊＊＊

Jiangsu ＊＊ building construction project of Jiangsu ＊＊ Co. , LTD. has been approved. The construction funds need to be self-raised and have been implemented. This project adopts the prequalification method to review the applicant's qualifications of bidding, and selects qualified tendering applicants to participate in tendering.

Ⅰ. Jiangsu ＊＊ Co. , LTD（bidding agent）was entrusted by the bidder to be responsible for the construction bidding.

二、本招标工程概况:

1. 标段名称:江苏＊＊大厦项目工程施工;

2. 工程地点:南京市＊＊大街;

3. 工程类型:大型工程;

4. 建设内容:建筑工程;

5. 结构类型:框架结构;

6. 工期及项目要求:600 日历天,要求省优;

7. 工程合同估算价:15 000 万元;

8. 划分标段:本次招标共分一个标段;

9. 单位工程及招标范围说明:江苏＊＊大厦项目工程施工,招标文件及施工图纸所包含的所有内容;

10. 工程规模:占地总面积为 15 009 平方米,建筑面积约 65 000 平方米(其中:地上 2 幢 60 米高 15 层建筑)

三、投标申请人资格条件:

1. 申请人资质等级及范围:＊＊＊;

2. 项目经理资质类别和等级:＊＊＊;

3. 投标申请人的单位名称必须与企

Ⅱ. The summary of bidding project

1. Project name: Jiangsu ＊＊ building construction;

2. Project site: Nanjing ＊＊ street;

3. Project type: Large-scale project;

4. Construction content: Building engineering;

5. Structure type: Frame structure;

6. Period and requirements for project: 600 days, provincial fine-quality;

7. Estimated price of construction contract: 150 million yuan;

8. Divided into bid lots: Only one bid lot;

9. Introduction to construction unit and bidding scope: Jiangsu ＊＊ building construction, all of the contents contained in bidding documents and construction drawings;

10. Project scale: A total land area of 15 009 square meters, a gross building area of approximately 65 000 square meters (two buildings of 60 meters high and 15 layers on the ground).

Ⅲ. Qualifications for tendering applicant:

1. Level and scope of applicant's qualification: ＊＊＊;

2. Type and level of project manager's qualification: ＊＊＊;

3. Company name of the tendering applicant must be in line with company name

业资质证书上的单位名称一致;

4. 本次招标不接受联合体投标。

四、报名及获取资格预审文件方法: * * * *。

五、资格审查办法: * * * *。

六、投标报价评审方法:使用综合评估法。

七、其他。

八、有下列行为之一的投标人,本工程不接受其参加投标: * * * *。

九、本公告发布法定媒体: * * * *。

招标代理机构地址:南京市 * * 路 * * 号 * * 室

联系人: * * * *

电话: * * * *

传真: * * * *

10.3 工程索赔管理

10.3.1 工程索赔概述

(1) 工程索赔的概念

工程索赔是工程建设项目中常见的一种合同管理的内容。所谓工程索赔,是指在工程合同履行过程中,对于并非自己的过错,而是应由对方承担责任的情况造

in enterprise quality certificates;

4. The bidding does not accept a consortium bid.

Ⅳ. Manners of application and obtaining prequalification documents: * * * *.

Ⅴ. Qualification reviewing methods: * * * *.

Ⅵ. Reviewing methods of tender offers: Comprehensive evaluation method.

Ⅶ. Others.

Ⅷ. The tenderer, who has one of following behaviors, cannot participate in bidding of this project: * * * *.

Ⅸ. The media legally issuing this announcement: * * * *.

Bidding agent address: * * Room * * Road in Nanjing

Contact person: * * * *
Tel: * * * *
Fax: * * * *

10.3 Construction claim management

10.3.1 Summary of construction claim

(1) Concept of construction claim

Construction claim belongs to one usual kind of construction contract management. Construction claim refers to it that in the process of the contract fulfilling, the actual losses, not caused by his own fault but borne by the other party, are requested to get the

成的实际损失向对方提出经济补偿和（或）时间补偿的要求。

economic compensation and（or）time compensation for the other.

（2）常见的工程索赔分类

从不同的角度，按不同的标准，工程索赔通常有以下几种分类方法（见表10.2）。

（2）Usual classification of construction claims

From different angles or according to different standards, claims are usually classified according to Table 10.2.

表10.2　常见的工程索赔分类

划分标准	索赔类型	说　明
干扰事件的性质	工期拖延索赔	因发包人未按合同要求提供施工条件，或因发包人指令工程暂停或不可抗力事件等原因造成工期拖延
	不可预见的外部障碍或条件索赔	如在施工期间，在现场遇到一个有经验的承包商通常不能预见的外界障碍或条件
	工程变更索赔	由于业主或工程师指令修改设计、增加或减少工程量、增加或删除部分工程、修改实施计划、变更施工次序，造成工期延长和费用损失
	工程终止索赔	由于不可抗力因素或业主违约等原因，使工程被迫在竣工前停止实施，并不再继续进行，使承包商蒙受经济损失
	其他索赔	如货币贬值、汇率变化、物价、工资上涨、政策法令变化、业主推迟支付工程款等原因
合同类型	总承包合同索赔	即承包商和业主之间的索赔
	分包合同索赔	即总承包商和分包商之间的索赔
	联营合同索赔	即联营成员之间的索赔
	劳务合同索赔	即承包商与劳务供应商之间的索赔
	其他合同索赔	除上述以外的索赔
索赔要求	工期索赔	即要求业主延长工期
	费用索赔	即要求业主补偿费用损失
索赔的起因	业主违约	包括业主和监理工程师没有履行合同责任；没有正确地行使合同赋予的权力，工程管理失误，不按合同支付工程款等
	合同错误	如合同条文不全、错误、矛盾、有二义性，设计图纸、技术规范错误等
	合同变更	如双方签订新的变更协议、备忘录、修正案，业主下达工程变更指令等
	工程环境变化	包括法律、市场物价、货币兑换率、自然条件的变化等
	不可抗力因素	如恶劣的气候条件、地震、洪水、战争状态、禁运等

划分标准	索赔类型	说　明
索赔所依据的理由	合同内索赔	即发生了合同规定给承包商以补偿的干扰事件
	合同外索赔	指工程过程中发生的干扰事件的性质已经超过合同范围
	道义索赔	指承包人在合同内或合同外都找不到可以索赔的依据，因而没有提出索赔的条件和理由，但承包人认为自己有要求补偿的道义基础，而对其遭受的损失提出具有优惠性质的补偿要求
索赔的处理方式	单项索赔	即针对某一干扰事件提出的索赔
	总索赔	又叫一揽子索赔或综合索赔

Table 10. 2　Usual classification of construction claims

Classification standards	Claim types	Explanation
nature of the interference	delaying claim	delaying caused by the owner, irresistible factors, ect.
	claim of unforeseen external disorders or conditions	such as external disorders or conditions that an experienced contractor cannot foresee in the construction site
	project change claim	delaying and expense caused by project change due to the owner or the engineer
	project termination claim	economic loss caused by project termination due to the owner, irresistible factors, ect.
	other claims	such as depreciation, exchange rates change, prices and wage rising
contract types	total contract claim	between the contractor and the owner
	subcontract contract claim	between the general contractor and the subcontractor
	joint contract claim	among joint members
	labor contract claim	between the contractor and the labour supplier
	other contract claim	in addition to the above
claim requirements	claim for extension of time	request the owner to extend the duration
	cost claims	ask the owner for compensation costs
claim causes	owner defaults	caused by the default responsibility of the owner, supervision engineer, etc.
	contract error	such as incomplete, false, contradictory, ambiguous provisions, and errors in designing or technical specification
	contract change	such as a new change agreement, memorandum, amendment and the owner's engineering change instructions
	environment change	including law, the change of market prices, currency exchange rates, natural conditions, etc.
	irresistible factors	such as harsh climatic condition, earthquakes, floods, war and embargo

Classification standards	Claim types	Explanation
claim basis	contractual claim	the interference, in which the contractor can get compensation stipulated in the contract, happens
	non-contractual claim	the interference which happened in construction is beyond contract
	ex-gratia payment	when the contractor cannot find basis for claims in the contractual or non-contractual provisions, there are no conditions and reasons to claim compensation, but the contractor thinks he has the moral basis to propose some requirements for compensating his losses
the methods of handling claims	single claim	for a certain interference
	total claim	also known as blanket claim or comprehensive claim

10.3.2 工程索赔工作流程

工程索赔工作流程如图 10.6 所示。

10.3.2 Construction claim process

Construction claim process is shown in Fig. 10.6.

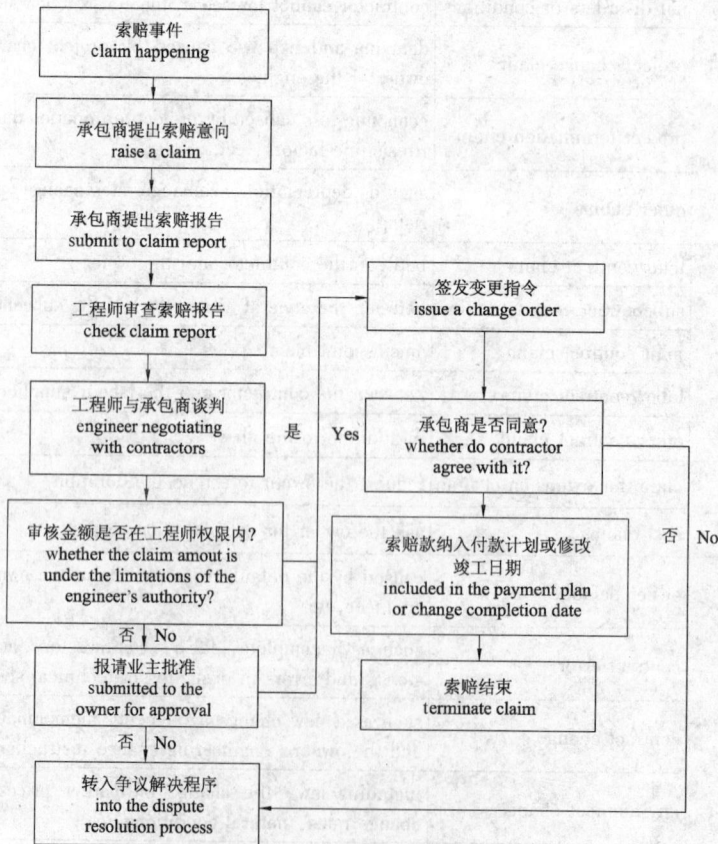

图 10.6 工程索赔工作流程

Fig. 10.6 The construction claim process

10.3.3 工程索赔争议处理

（1）工程索赔处理的方式

① 和解——解决纠纷的最佳选择。

和解是指纠纷发生后，当事人本着自愿、互谅、协商一致的原则解决问题的一种方式。纠纷产生后，当事人应当首先考虑通过和解的方式解决纠纷。

② 调解——解决纠纷的温和手段。

调解是指在履行合同的过程中，双方当事人对合同权利义务产生纠纷，通过非国家司法机关的社会团体的主持，促使双方在相互妥协的基础上达成谅解，从而解决纠纷的一种方式。

③ 仲裁——解决纠纷的必要补充。

仲裁是指合同双方当事人事前或者事后达成协议，自愿将发生争议的事项提交仲裁委员会进行仲裁，并依据仲裁裁决履行义务的一种争端解决方式。

④ 诉讼——解决纠纷的最后保障。

诉讼是指合同当事人在纠纷发生后，请求人民法院对纠纷予以裁决的活动，实际上是人民法院行使审判权的活动。

（2）工程索赔争议处理流程

工程索赔争议的处理基本流程如图10.7所示。

10.3.3 Claim dispute resolution

(1) Claim resolving manners

① Reconciliation is the best choice for dispute resolution.

Reconciliation refers to that the parties resolve disputes, based on their own willing, mutual understanding and consensus after the disputes occur. The parties should firstly choose reconciliation to resolve the disputes.

② Mediation is the more moderate manner for dispute resolution.

Mediation refers to that in the process of the contract fulfilling, the parties resolve the disputes, caused by contract rights and obligations, by non-state judicial organizations prompting both sides to arrive at an understanding on the basis of mutual compromise.

③ Arbitration is the necessary supplement to resolve disputes.

Arbitration refers to that the parties voluntarily submit to arbitration committee for arbitration of disputes, and perform obligations on the basis of arbitration results.

④ Litigation is the last guarantee to resolve disputes.

Litigation refers to that the parties request the court for vindicating disputes after the disputes occur. That is to say, the court exercises jurisdiction in fact.

(2) Claim dispute resolving process

The basic process of claim dispute resolving is shown in Fig. 10.7.

提出索赔要求 make a claim	报送索赔资料 submit claim files	谈判解决索赔争端 negotiate to resolve claim	协调解决索赔争端 coordinate to resolve claim	提交仲裁或诉讼 submit to arbitration or litigation

图 10.7　工程索赔处理的流程
Fig. 10.7　The claim resolving process

简言之,上述 5 个工作程序可归纳为两个阶段,即友好协商解决和诉诸仲裁或诉讼。

10.3.4　案例分析

2002 年 8 月 27 日,某光电有限公司(发包人)与某建设有限责任公司(承包人)签订工程建设合同一份,约定发包人所有的厂房工程发包给承包人承建,合同工期 250 天,合同价款暂定 3 000 万元。结算价按重庆市建设工程基价表(99)定额下浮 2%。工程进度款按承包人当月完成量的 85% 支付。工程结算款在发包人审查完结算后半月内付清。承包人交给被告工程保证金 240 万元。合同签订后,承包人按发包人要求于 2002 年 11 月进场施工,在施工过程中,由于发包人资金缺口太大,根本不能按合同约定给付工程进度款,因承包人已无力继续垫资,到 2003 年 5 月 15 日全面停工。经承、发包人多次协商,将已做部分工程办理结算,并于 2004 年 4 月 20 日签署了结算书。工程款结算后,承包人多次催取工程款,发包人至今未付,只退还工程保证金 126 万元。本索赔事件由承包人向人民法院起诉,通

In short, the above five working procedures can be divided into two stages, namely, friendly consultation and resort to arbitration or litigation.

10.3.4　A case study

On August 27th, 2002, a certain photoelectric Co., LTD. (the owner) and a certain construction Co., LTD. (the contractor) signed a consruction contract, which stipulated all of the owner's factory engineering need to be contracted by the contractor. The contract period is 250 days, and the price is 30 million yuan. Based on Chongqing construction project quota in benchmark table (99), settlement price fell by 2%. Progress payment was paid in accordance with 85% of the contractor's completed quantities for the month. Engineering settlement money was paid within half a month after the owner had checked the settlement, and the contractor paid the defendant for security deposits of 2.4 million yuan. After signing the contract, the contractor began to construct in November 2002 in accordance with the requirements of the owner. However, due to serious lack of funding in the process of construction, the owner could not radically pay the progress payment according to the contract agreement, and the contractor was unable to continue loaning, so all of constructions had to be stopped completely on May 15th, 2003. After repeated negotiations between the contractor and the

过诉讼方式解决。

问题分析：

法院认为，工程的停工是由于发包人的资金缺口过大而导致的，而在双方签署结算书之后，发包人理应支付剩余的已完成工程的工程款并且退还工程保证金。因此，法院最终判决：发包人支付拖欠的工程款以及退还剩余的保证金，并承担相应的诉讼费。

10.4　工程合同风险管理

10.4.1　工程合同风险管理概述

（1）工程合同风险管理的概念

①工程合同风险是指工程合同签订和履行过程中可能发生的不确定性事件对当事人造成的经济损失。

工程合同风险管理是指通过风险识别、风险分析和风险评价去认识工程合同的风险，并以此为基础合理地使用各种风险应对措施、管理方法、技术和手段对合同的风险实行有效地控制，妥善处理风险事件造成的不利后果，以最少的成本保证

owner, the accounts of part of completed projects had been settled, and the statement was signed on April 20th, 2004. After the payment being settled, the contractor urged the owner to take payment for many times. However, the owner did not pay it, and only refunded security deposits of 1. 26 million yuan. The contractor brought a suit to court for the claim, and thus the claim was settled through litigation.

Problem analysis：

The court found that the stoppage was caused by the owner due to lack of funding, so after the statement is signed, the owner should pay for the rest of the completed project and return the project security deposits. At last, the court made the following judgement: the owner should pay the contractor the overdue payment, refund the rest of the security deposits, and bear corresponding litigation costs.

10.4　Risk management of construction contract

10.4.1　Summary of construction contract risk management

（1）Basic concept of construction contract risk management

① Construction contract risk refers to the economic losses of the parties due to uncertain events which occur in the process of contract signing and contract fufilling.

Construction contract risk management refers to understanding construction contract risk by risk identification, risk analysis and risk assessment, and on this basis, controlling effectively contract risk by the various and reasonable use of risk measures, management methods, techniques and means, to properly handle the adverse effects of risk

合同总体目标实现的管理工作。

(2) 工程合同的风险来源

工程合同风险主要来源于以下的几个方面：

① 由合同类型所决定的风险；

② 合同中明确或隐含规定应由一方来承担的风险(免责条款)；

③ 合同缺陷导致的风险,如条文不全面、不完整,表达意思模糊甚至错误、矛盾、歧义。

(3) 工程合同的风险分析

工程合同风险在很大程度上取决于合同签订时所确定的类型,而按照不同的标准可以将合同划分成不同的类型。这里只根据风险来源的特点,按计价方式和管理模式两种标准来划分并比较各自的风险大小。

① 不同计价方式的工程合同风险比较。

工程合同按照项目计价方式来划分,主要有总价合同、单价合同和成本加酬金合同。其中,总价合同是指业主支付给承包人的施工款项在承包合同中是一个固定的金额；单价合同是指整个合同期内执行同一个单价,而工程量则按实际完成的数量进行计算；成本加酬金合同指业主向承包方支付工程项目的实际成本,并按事先约定的某一种方式支付酬金。

events, thus ensuring the overall goals to be realized by the minimal cost.

(2) The source of construction contract risks

Construction contract risks mainly come from the following several aspects:

① The risks are determined by the contract type;

② The risks are explicitly or implicitly included by the provisions (exception clause);

③ The risks are caused by contract defects, such as incomplete or imperfect provisions, vague expressions or even wrong and ambiguous expressions.

(3) Risk analysis of construction contract

Construction contract risk, to a great extent, depends on the determined contract type. In accordance with the different standards, contracts can be divided into many different types. However, according to the characteristics of risk source, the contracts are divided by pricing manner and management pattern, and are compared with them.

① Comparisons with construction contract risk of different pricing manners.

According to the project pricing manners, construction contracts mainly consist of lump-sum contract, unit price contract and cost plus compensation contract. A lump-sum contract refers to the fixed price giving by the owner. A unit price contract means that the owner should pay the contractor the same unit price in the entire contract period. A cost plus compensation contract means that the owner should pay the contractor the actual project costs and the predetermined rewards in a certain way.

不同计价方式的工程合同风险比较见表 10.3。

Comparisons with construction contract risk of different pricing manners are shown in Table 10.3.

表 10.3 不同计价方式的工程合同风险比较
Table 10.3 Comparisons with construction contract risk of different pricing manners

合同类型 Contract type	风险分配 Risk allocation	
	业 主 owner	承包方 contractor
总价合同 lump-sum contract		
单价合同 unit price contract		
成本加酬金合同 cost plus compensation contract		

② 不同管理模式的工程合同风险比较。

工程合同按照施工管理模式来划分，主要有传统合同管理模式（DBB）、设计 - 建造合同管理模式（DB）、设计 - 采购 - 施工合同管理模式（EPC）和建造 - 运营 - 移交合同管理模式（BOT）。不同管理模式的工程合同风险比较见表 10.4。

② Comparisons with construction contract risk of different management modes.

According to the construction management modes, construction contracts mainly consist of Design-Bid-Build (DBB) contract, Design-Build (DB) contract, Engineering-Procure-Construct (EPC) contract and Build-Operate-Transfer (BOT) contract. The comparisons with construction contract risk of different management modes are shown in Table 10.4.

表 10.4 不同管理模式的工程合同风险比较
Table 10.4 Comparisons with construction contract risk of different management modes

合同类型 Contract type	风险分配 Risk allocation	
	发包方 owner	承包方 contractor
DBB 合同管理模式 Design-Bid-Build		
DB 合同管理模式 Design-Build		
EPC 合同管理模式 Engineering-Procure-Construct		
BOT 合同管理模式 Build-Operate-Transfer		

10.4.2 工程合同风险管理的主要工作

工程合同风险贯穿于合同的签订、履行、变更或转移、终止等各个环节。因此，工程合同风险管理是一个贯穿于工程合同管理的全过程管理，具体工作内容见表10.5。

10.4.2 Main management work of construction contract risk

Construction contract risk exists in the every phase of signing, fulfilling, changing or transferring, terminating, etc. Therefore, construction contract risk management is a life-cycle management throughout the whole process of construction contract management, and its specific content is shown in Table 10.5.

表 10.5 工程合同风险管理的主要工作

阶 段	风险管理重点与内容	风险管理应对策略
项目投标前	(1) 项目风险调查与分析:项目环境、业主资信、项目资源条件的调查分析等; (2) 投标决策:在上述基础上,做出投标或放弃投标的决策。	(1) 回避策略; (2) 合同风险管理策略。
项目投标与合同谈判	(1) 合同风险审查:审查合同风险。 (2) 报价策略:确定工程投标报价策略。 (3) 合同风险分配:争取业主给予有利的合同条件。 (4) 合同风险转移:确定工程保险或分包方案。 (5) 风险管理方案:制定完整的、系统的风险管理方案。	(1) 报价策略; (2) 合同谈判争取有利的合同条件; (3) 工程保险转移风险; (4) 与其他单位共同承担风险。
合同实施至工程竣工	(1) 实施并调整风险管理方案; (2) 工程索赔管理。	(1) 加强项目管理; (2) 加强工程索赔管理。
缺陷责任期	(1) 加强对工程尾款和保留金的催讨; (2) 进行合同风险管理总结; (3) 建立风险管理档案。	建立合同风险管理案例库,为后续项目合同风险管理策略的制定提供依据。

Table 10.5 Main management works of construction contract risk

Phase	Focuses and contents of risk management	Risk management strategy
before the bidding	(1) project risk investigation and analysis: project environment, owner's credit, project resource conditions, etc; (2) bidding decision-making: on the basis of the above, decide to bid or abandon bidding.	(1) avoidance strategy; (2) contract risk management strategy.

Phase	Focuses and contents of risk management	Risk management strategy
project bidding and contract negotiation	(1) checking contract risk; (2) pricing strategy; (3) contract risk allocation: best obtaining the beneficial contract conditions from the owner; (4) contract risk transferring: determining the engineering insurance or the subcontract plan; (5) risk management plan: making a complete and systematic plan of risk management.	(1) pricing strategy; (2) best obtaining the beneficial contract conditions by contract negotiation; (3) transferring risk by engineering insurance; (4) taking risks together with the others.
from contract implementation to project completion	(1) implementing and adjusting the plan of risk management; (2) claim management.	(1) strengthening the project management; (2) strengthening the engineering claim management.
defects liability period	(1) strengthening to dun for the fund of tail and retention money; (2) summarizing contract risk management; (3) building a profile of risk management.	The contract risk management is established, so as to provide references for developing a strategy of subsequent construction contract risk management.

10.4.3 工程合同风险管理流程

工程合同风险管理流程如图 10.8 所示。

10.4.3 Management process of construction contract risk

Management process of construction contract risk is shown in Fig. 10.8.

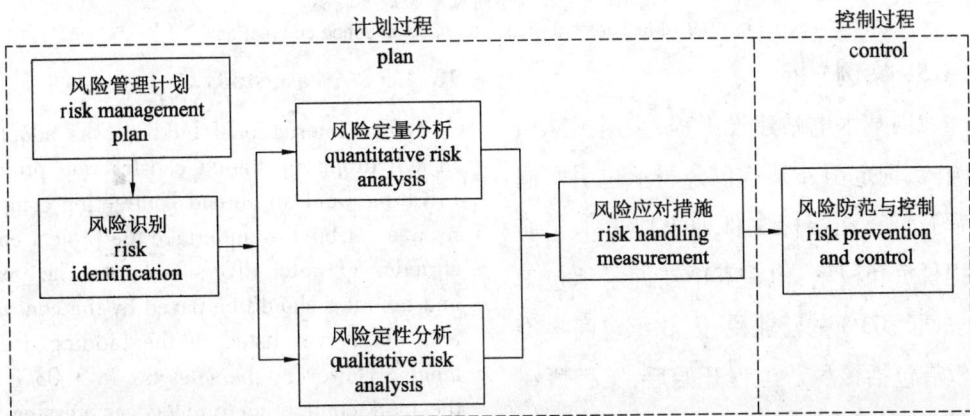

图 10.8 工程合同风险管理流程
Fig. 10.8 Management process of construction contract risk

10.4.4 工程合同风险管理的措施

工程合同风险管理的措施如图10.9所示。

10.4.4 Management measures of construction contract risk

Management measures of construction contract risk are shown in Fig. 10.9.

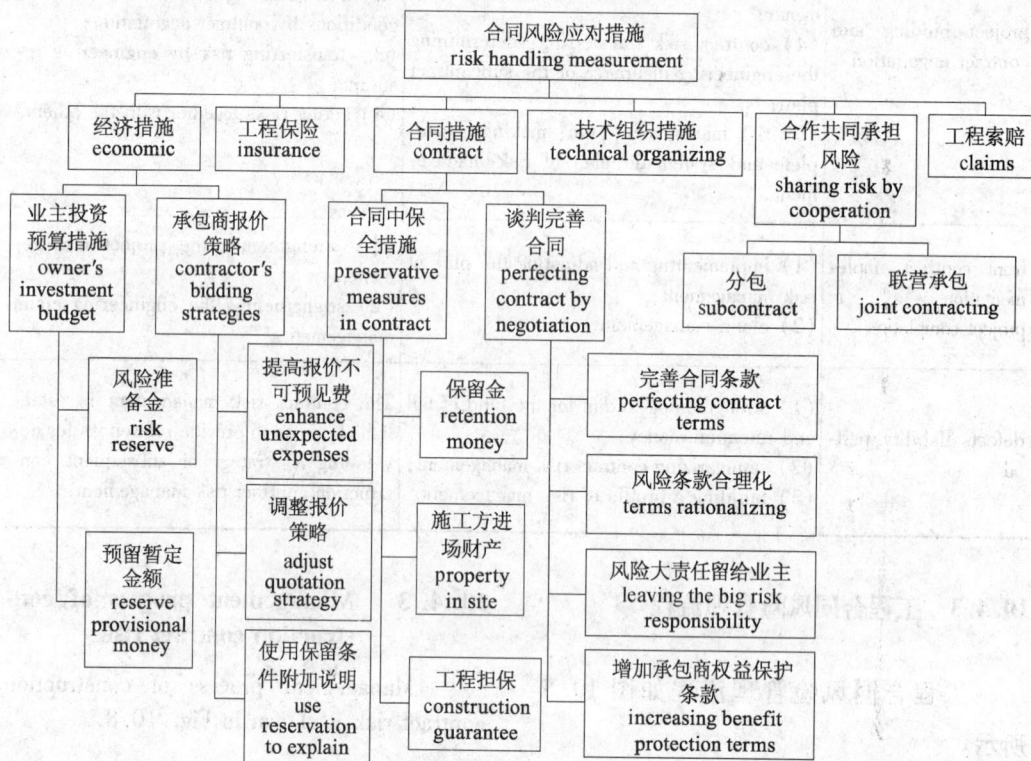

图 10.9 工程合同风险管理的措施

Fig. 10.9 Management measures of construction contract risk

10.4.5 案例分析

我国某水电站建设工程,采用国际招标方式,选定国外某承包公司承包引水洞工程施工。在招标文件列出应由承包商承担税负和税率,但在其中遗漏了承包工程总额3.03%的营业税,因此承包商报价时没有包括该税。工程开始后,工程所在地税务部门要求承包商交纳已完工程的营业税92万元,承包商按时缴纳,同时向业主提出索赔要求。

10.4.5 A case study

The international bidding was adopted in a hydropower station construction project in China, and an abroad contracting company was selected to undertake the project construction of water diversion hole. The taxes and tax rates should be payed by the contractor, which was listed in the bidding documents. However, the sales tax in 3.03% of the total contracting project was missing in it, so the sales tax was not included in the contractor's price. After the construction work started, local tax authority required the contractor to pay sales tax of 920 000 yuan

问题分析：

如果业主在招标文件中仅列出几个小额税种，而忽视了大额税种，是招标文件的不完备或者是有意的误导行为，业主应该承担责任。但是，如果招标文件中没有给出任何税收目录，而承包商报价中遗漏税赋，则索赔要求是不能成立的。这属于承包商环境调查和报价失误，应由承包商负全责。因为合同明确规定："承包商应遵守工程所在国一切法律"，"承包商应交纳税法所规定的一切税收"。

for the completed project, so the contractor paid it on time, and at the same time, the contractor lodged a claim against the owner.

Problem analysis：

If the owner only lists a few small tax types in the bidding documents and ignores the large tax types, which belong to the imperfectness of the bidding documents or deliberately misleading behavior, the owner should bear the corresponding responsibility. However, if any tax directory is not given in the bidding documents, and the taxes are missing in the contractor's quotation, this claim can not be supported. This belongs to the contractor's lapses in the environmental investigation and quotation, so the contractor should be responsible for it. As clearly stated in the contract, the contractor should comply with all of local laws, and the contractor should pay all of the taxes stipulated in the tax law.

注：本章图片均来自网络。
Note：In this chapter, all pictures are from webs.

知识拓展
Learning More

相关链接　Related Links

（1）工程合同管理师考试中心 http://www.cettic.co/

（2）工程合同管理师职业培训 http://www.bogohr.com/

（3）全国合同管理网 http://www.htgls.com/

（4）百高教育工程合同管理师培训 http://www.htgls.com/

（5）恒润基合同管理系统 http://www.woorich.cn/channels/175.html

小贴士　Tips

FIDIC 是国际咨询工程师联合会（Fédération Internationale Des Ingénieurs Conseils）的法文缩写，有人称 FIDIC 是国际承包工程的"圣经"。可以说，FIDIC 是集工业发达国家土木建筑业上百年的经验，把工程技术、法律、经济和管理等有机结合起来的一个合同条件。国际承包工程行业涉及的 FIDIC，主要有土木工程方面的红色封皮 FIDIC，海外通常称红皮 FIDIC；机电工程方面的黄色封皮 FIDIC，常称黄皮 FIDIC；设计咨询方面的白色封皮 FIDIC，也称白皮 FIDIC。

FIDIC is short for the International Federate Association of Consulting Engineers in French, and it is said to be a bible of international contracted construction. FIDIC has, as it were, hundreds of years of civil engineering experience in industrial developed countries, and combines organically with engineering technology, law, economy, management, ect. FIDIC mainly includes the red-covered FIDIC about civil engineering, as is referred to the red-covered FIDIC overseas; the yellow-covered FIDIC about electromechanical engineering, as is referred to the yellow-covered FIDIC; and the white-covered FIDIC about design consultancy, as is referred to the white-covered FIDIC.

FIDIC 施工合同文件的主要组成内容,包括:

(1) 工程施工合同协议书;

(2) 中标通知书;

(3) 投标书及其附件;

(4) 工程施工合同专用条款;

(5) 工程施工合同通用条款;

(6) 标准、规范及有关技术文件;

(7) 图纸;

(8) 工程量清单;

(9) 工程报价单或预算书;

(10) 其他。

The construction contract documents in FIDIC mainly include the following:

(1) Construction contract agreement;

(2) Bid-winning notice;

(3) Book of tender and its accessories;

(4) Special terms of construction contract;

(5) Universal terms construction contract;

(6) Criterion, specification and relevant technical documents;

(7) Drawings;

(8) Bill of quantities;

(9) Engineering quotation or budget;

(10) Others.

除此之外,现在世界上使用范围较广的还有美国的 AIA 系列合同条件和 ICE 合同条件。

In addition, the series of the AIA contracts in USA and ICE contracts in UK are now widely used in the world.

思考题 Review Questions

(1) 我国建设工程合同管理有哪些特征?

What are the characteristics of the construction contract management in China?

(2) 如何理解招投标的含义?

How do you understand the meaning of the construction bidding and tendering?

(3) 试分析索赔的程序包括哪几个方面的内容?

What content does the construction claim procedure include?

（4）工程建设中业主和承包商遇到的风险有哪几个方面？

What risks do owners and contractors encounter in the construction?

参考文献
References

［1］成虎，虞华：《工程合同管理》（第2版），中国建筑工业出版社，2011年。

［2］王俊遐：《建筑工程招标与合同管理》，机械工业出版社，2011年。

［3］杨平，丁晓欣，等：《工程合同管理》，人民交通出版社，2007年。

［4］高显义：《工程合同管理教程》，同济大学出版社，2009年。

［5］何佰洲：《工程合同法律制度》，中国建筑工业出版社，2003年。

［6］黎广军：《21世纪国际工程合同的发展趋势——反思中国工程合同的发展趋势》，中国论文下载中心，2007年。

［7］张尚：《国际工程合同的发展趋势研究》，《项目管理技术》，2010年第8卷第11期。

［8］姜兴国，张尚：《工程合同风险管理理论与实务》，中国建筑工业出版社，2009年。

［9］《中华人民共和国合同法》，1999年3月15日第九届全国人民代表大会第二次会议通过。

［10］《中华人民共和国招标投标法》，1999年8月30日第九届全国人民代表大会常务委员会第十一次会议通过。

［11］刘黎红：《工程招投标与合同管理》，机械工业出版社，2008年。

［12］陈正：《工程招投标与合同管理》，东南大学出版社，2009年。

［13］郝永池：《工程招投标与合同管理》，机械工业出版社，2011年。

［14］吴冬平：《工程招投标与合同管理》，机械工业出版社，2012年。

［15］刘黎虹：《工程招投标与合同管理》，机械工业出版社，2012年。

［16］王晓林：《合同管理在项目管理中的地位和作用》，《铜业工程》，2005年第2期。

［17］王潇洲：《工程招投标与合同管理》，华南理工大学出版社，2009年。

［18］何红锋：《工程建设中的合同法与招标投标法》，中国计划出版社，2002年。

［19］朱宏亮，成虎：《工程合同管理》，中国建筑工业出版社，2006年。

［20］Cao Xiaolin, Song Yang. International Infrastructure Contract Management of Chinese Construction Company. Science&Technology Progress and Policy, 2011(13).

［21］Mourad El Meziane, Khadija El Meziane(eds.). Knowledge Management：Developing System for Construction Contract Management and Facing the Challenge of Managing Tacit. Business Management Dynamics, 2011(6).

［22］D Bryan Morgan. International Construction Contract Management. Ashgate Pub Co, 2005.

第11章 工程造价管理

Chapter 11 Engineering Cost Management

工程造价管理是对建设项目总投资实施的规划和控制活动，是保证建设项目在既定投资额度内实现预期目标的重要管理工作。本章主要介绍工程造价管理的基本概念和思想、建设程序过程中的造价管理工作以及工程造价管理的发展趋势。

Engineering cost management is the overall planning and control of engineering cost. It is important to achieve the desired objectives in the given cost requirements. This chapter introduces the basic ideas of engineering cost management, specific cost management work during the construction process and the tendency of engineering cost management.

11.1 工程造价管理概述

11.1 Overview introduction to engineering cost management

11.1.1 工程造价

建设项目是指在一定的约束条件下（时间、质量、成本等），按照一定的设计和规划进行施工的一个或几个单项工程总体，例如一所医院、一座工厂或是一所学校均为一个建设项目。建设项目按照实体构成分为单项工程、单位工程、分部工程和分项工程，如图11.1所示。

11.1.1 Engineering cost

Construction project is one or several units projects constructed according to the design and planning in certain constraints (time, quality, cost, etc.), such as a hospital, a factory or a school. It is divided into units projects, engineering units, branch works and subdivision works (Fig. 11.1)

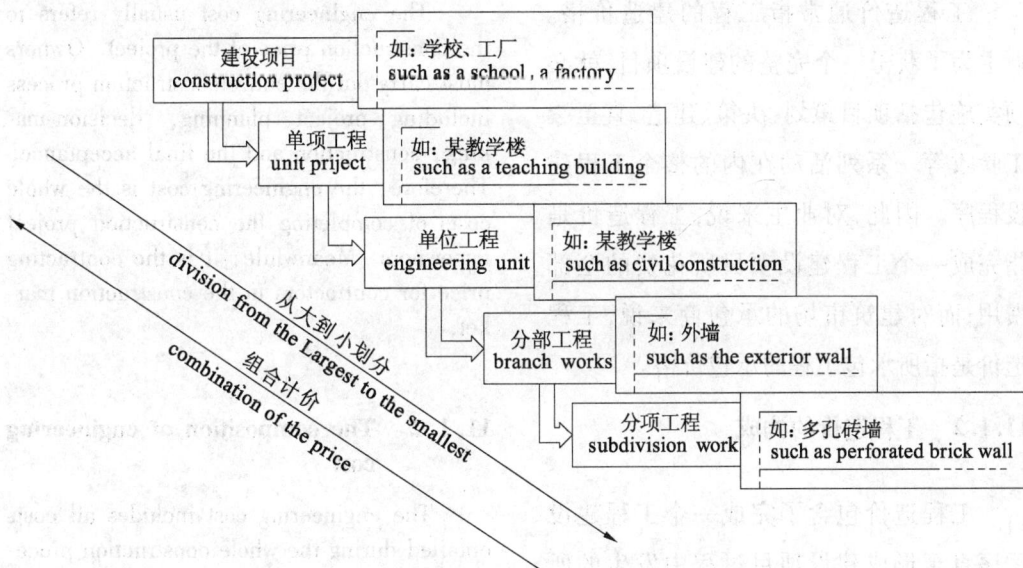

図 11.1 建设项目的构成
Fig. 11.1 The components of engineering project

由于建设项目一般具有周期长、投资规模大等特殊性,所以在实施过程中要遵循一定的工程建设程序,如图11.2所示。

The construction process must follow certain procedures (Fig. 11.2), owing to its long construction cycle and large-scale investment.

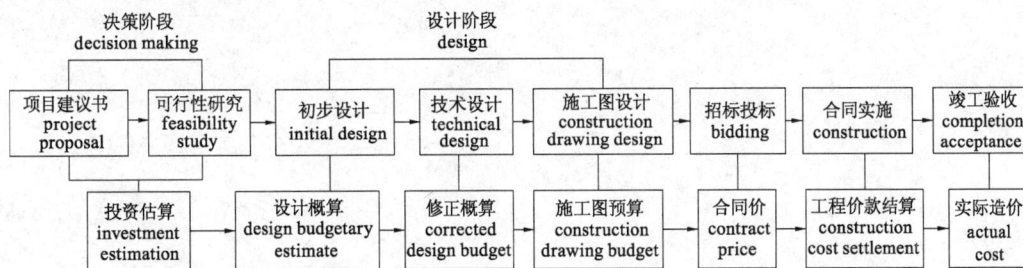

图 11.2 我国的工程建设程序
Fig. 11.2 The whole construction procedure in China

工程建设程序是指完成一个建设项目必经的环节以及各个环节必须遵循的先后顺序。工程建设程序是由国家主管部门制定颁布的,反映了建设工作客观的规律性,严格遵循和坚持按建设程序办事是提高工程建设经济效果的必要保证。

The construction procedure is necessary links of the construction, and its orders must be followed so as to complete the construction project. The construction procedure which reflects the objective construction work regularities is promulgated by national authorities. Strictly following and adhering to the construction procedure is necessary to improve the economic effect of the construction work.

工程造价通常指工程的建造价格。业主为了获得一个完整的建设项目，就必须实施包括项目策划、决策、建造，直至竣工验收等一系列活动在内的整个工程建设程序。因此，对业主来说，工程造价是指完成一个工程建设项目所花费的全部费用；而对建筑市场的承包商来说，工程造价是指所承包工程的承包价格。

11.1.2　工程造价的构成

工程造价包含了完成一个工程建设程序直至形成建设项目过程中发生的所有费用，具体包括设备及工器具购置费、建筑安装工程费、工程建设其他费、预备费、建设期贷款利息以及固定资产投资方向调节税。工程造价的构成如图 11.3 所示。

The engineering cost usually refers to the construction price of the project. Owners must carry out the entire construction process including project planning, decision-making, construction and the final acceptance. Therefore, the engineering cost is the whole costs of completing the construction project for owners. Meanwhile, it is the contracting price for contractors in the construction market.

11.1.2　The composition of engineering cost

The engineering cost includes all costs entailed during the whole construction procedure. Specifically, there are purchasing costs for equipment and tools, construction and installation costs, other construction costs, reserve funds, interest for loans during construction and adjustment tax for fixed assets investment. The composition of engineering cost is shown in Fig. 11.3.

- 建设项目工程造价 engineering cost
  - 设备及工器具购置费 purchasing costs for equipment and tools
    - 设备购置费 equipment cost
    - 工具器具及生产家具购置费 tools cost
  - 建筑安装工程费 construction and installation costs
    - 直接费 direct cost
    - 间接费 indirect cost
    - 利润 profits
    - 税金 taxes
  - 工程建设其他费用 other construction costs
    - 土地使用费 land-use fees
    - 与项目建设有关的其他费用 other costs associated with construction
    - 与未来企业生产经营有关的其他费用 costs related to the future enterprise operation
  - 预备费 reserve funds
    - 基本预备费 basic reserve funds
    - 涨价预备费 pricing reserve funds
  - 建设期贷款利息 interest for loans during construction
  - 固定资产投资方向调节税 adjustment tax for fixed assets investment

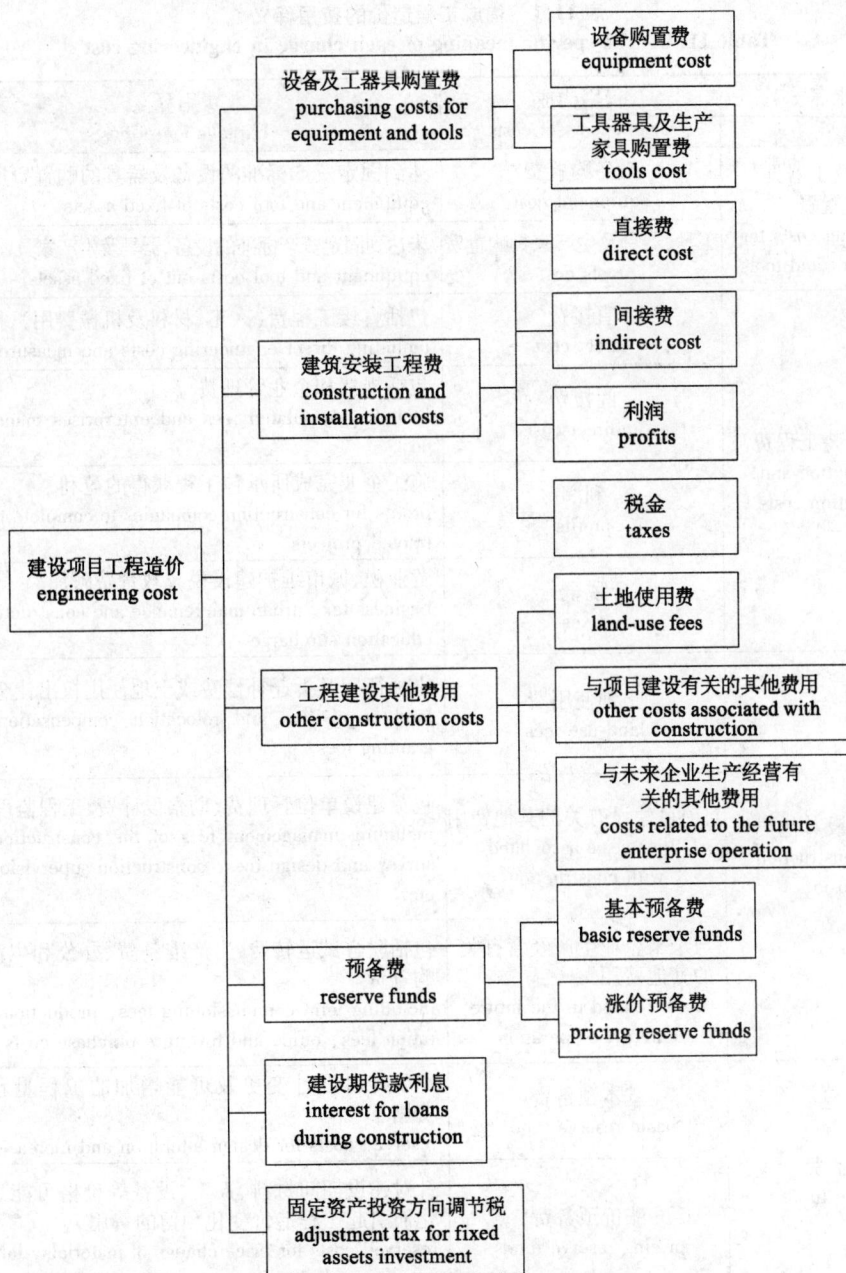

图 11.3　我国现行工程造价的构成

Fig. 11.3　The components of engineering cost in China

工程造价形成固定资产投资,与流动资产投资共同构成了我国建设项目的总投资,每项费用的具体含义见表 11.1。

The engineering cost which form the fixed assets and current assets constitute the total investment. The specific meaning of each charge is shown in Table 11.1.

表 11.1　构成工程造价的费用释义
Table 11.1　The specific meaning of each charge in engineering cost

费用名称 Name of costs	费用构成 Components of costs	费用释义 Explanation of costs
设备及工器具购置费 purchasing costs for equipment and tools	设备购置费 equipment cost	达到固定资产标准的设备及器具的购置费用 equipment and tool costs of fixed assets
	工具、器具及生产家具购置费 tools cost	未达到固定资产标准的设备、器具及生产家具购置费用 equipment and tool costs out of fixed assets
建筑安装工程费 construction and installation costs	直接费 direct cost	包括直接工程费(人工、材料及机械费用)、措施费 including direct engineering costs and measure fees
	间接费 indirect cost	包括规费和企业管理费 including stipulated fees and enterprises management cost
	利润 profits	施工企业完成所承包工程获得的盈利 profits for construction companies to complete the contracted projects
	税金 taxes	营业税、城市维护建设税及教育费附加 business tax, urban maintenance and construction tax, education surcharge
工程建设其他费 other construction costs	土地使用费 land-use fees	土地征用及拆迁补偿费或土地使用权出让金 land acquisition and relocation compensation, land granting fees
	与项目建设有关的其他费用 other costs associated with construction	包括建设单位管理费、勘察设计费、工程监理费等 including management fees of the construction unit, survey and design fees, construction supervision fees, etc.
	与未来企业生产经营有关的其他费用 costs related to the future enterprise operation	包括联合试运转费、生产准备费、办公和生活家具购置费 including joint commissioning fees, production preparation fees, office and furniture purchase costs
预备费 reserve funds	基本预备费 basic reserve funds	主要针对设计变更及可能增加的工程量预留的费用 reserved costs for design alteration and increased work
	涨价预备费 pricing reserve funds	针对建设期间材料、人工、设备等价格可能发生变化而引起工程造价变化预留的费用 reserved costs for price change of materials, labor and equipment during the construction period.
建设期贷款利息 interest for loans during construction	包括向银行以及其他金融机构的贷款等在建设期间应计的借款利息 including interest from banks and other financial institutions accrued during the construction period	
固定资产投资方向调节税 adjustment tax for fixed assets investment	2000 年 1 月 1 日起暂停征收 suspended since January 1, 2000	

11.1.3 工程造价管理

工程造价管理是利用技术、经济、法律等手段对工程建设活动中的造价进行规划和控制，以达到在预期投资额内完成项目的目标。

工程造价管理的基本内容包括合理规划和有效控制工程造价（如图 11.4 所示）。合理规划工程造价是指在工程建设各个阶段，采用科学的计价方法和切合实际的计价依据，对建设程序的各个阶段进行计价，具体包括在决策阶段进行的投资估算、设计阶段实施的设计概算和施工图预算、招投标阶段形成的合同价、工程实施过程中的结算价以及竣工验收后形成的实际造价。合理规划工程造价是有效控制工程造价的前提和先决条件，没有合理的工程造价规划，就无法进行工程造价的有效控制。

11.1.3 Engineering cost management

Engineering cost management is to plan and control costs of the whole construction activities using the technology, economy, legal and other measures, in order to achieve the desired objectives in the given cost requirements.

The basic engineering cost management includes rational planning and effective control of engineering costs (Fig. 11.4). Rational planning includes estimating costs of all the stages using scientific method of valuation and realistic valuation basis. Specifically, it includes investment estimation during decision making, design budgetary estimation and construction drawing budget during design stage, contract price during bidding stage, settlement price during construction and the actual cost. Rational planning is the prerequisites of effective control for the engineering cost. It is impossible to carry out effective control without rational planning.

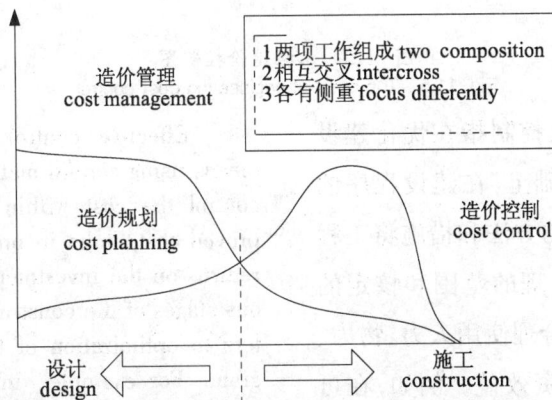

图 11.4　工程造价管理的含义
Fig. 11.4　Meanings of engineering cost management

造价规划和造价控制贯穿于建设项目实施的整个过程，并在不同的阶段各有

Cost planning and cost control have different emphases at different stages in the

侧重。一般在项目实施的前期主要实行造价规划工作,在项目具体实施过程中根据造价规划对造价进行控制。工程造价控制的作用主要体现在建设程序前一个阶段形成的估算投资额要控制下一个阶段的工作,不能出现超出前一个阶段投资额的现象,如图 11.5 所示。

whole process of construction. Usually in early stage, cost planning is the major work, and cost control should be carried out based on cost planning. Cost control is mainly reflected in the construction phase that the previous estimation takes control of the next stage and excess of the investment is not allowed (Fig. 11.5).

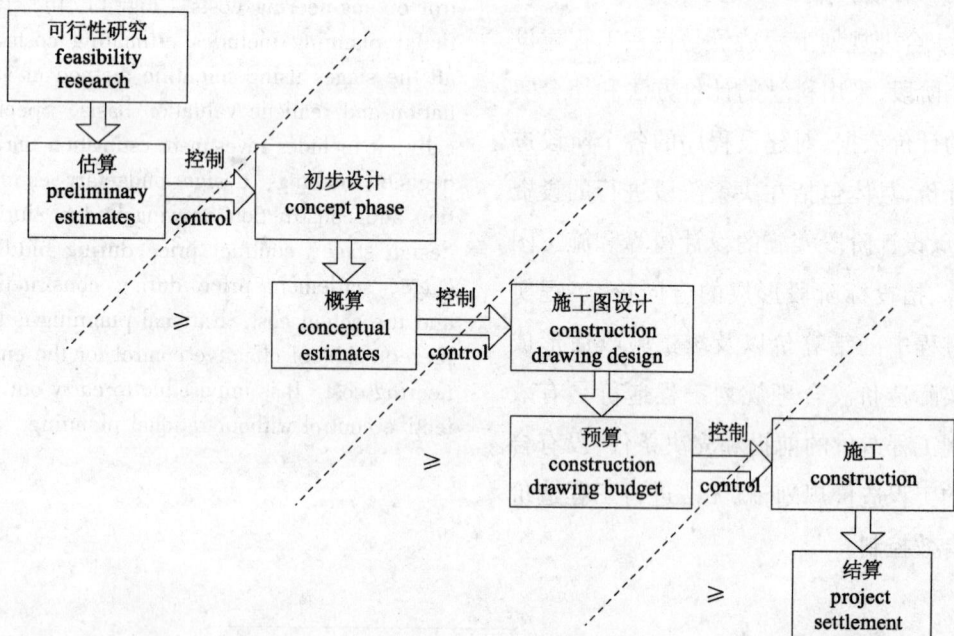

图 11.5　工程造价控制图
Fig. 11.5　The chart of engineering cost control

　　工程造价的有效控制指在优化建设方案、设计方案的基础上,在建设程序的各个阶段,采用一定的方法和措施将工程造价的发生控制在合理的范围和核定的造价限额以内,以求合理使用人力、物力、财力,取得较好的投资效益。例如,在可行性研究阶段形成的投资估算要控制初步设计,初步设计阶段的概算额要控制在估算额之内,而设计概算要控制施工图设计阶段的预算额,最终的结算费用不能超

　　Effective control of engineering cost means using certain methods and measures to control the cost within reasonable and approved cost limits in order to achieve better returns on the investment. It occurs in various stages of the construction process and refers to optimization of the construction program. For example, investment estimate in the feasibility study stage should control the preliminary design; the budget estimates cannot exceed the investment estimation and the final settlement costs cannot exceed the estimated cost.

出预算费用。

建设项目决策及设计阶段的
造价管理

Engineering cost management of decision-making and design stage

11.2.1 建设项目决策与投资估算

建设项目决策是选择和决定投资行动方案的过程,是对拟建项目的必要性和可行性进行技术经济论证,对不同建设方案进行技术经济比较以做出判断和决定的过程。项目决策正确与否,直接关系到项目建设的成败,关系到工程造价的高低及投资效果的好坏。因此,正确决策是合理确定和控制工程造价的前提。

投资估算是指在建设项目立项之前的决策阶段,依据现有的资料和一定的方法,对将来进行该项目建设可能要花费的各项费用的事先匡算。建设项目的投资决策过程一般要经历一个逐步详细的技术经济论证过程,通常把项目的投资决策过程划分为项目规划阶段、项目建议书阶段、预可行性研究阶段和可行性研究阶段,投资估算工作也相应分为4个阶段。

不同决策阶段拟建项目的要求明确程度不同,对应的估算条件和资料的掌握程度不同,因而投资估算的准确程度不同,每个阶段投资估算所起的作用也不同。随着阶段的不断发展,掌握的资料越来越丰富,投资估算将逐步准确,作用也

11.2.1 Decision making and investment estimation

Making decisions for construction project is a process of selecting and deciding the investment actions, including technical and economic argumentation about the necessity and feasibility of proposed project, comparison of different construction schemes and the following judgment and decisions. The quality of a project decision is directly related to the success of the project construction, the cost as well as the investment effect. Therefore, a correct decision is the premise of the reasonable determination and control of the engineering cost.

Investment estimation is to calculate costs of the project construction in the future based on available data and certain methods. Generally, the decision making procedure will go through a detailed technical and economic argumentation step by step, which can be divided into planning stage, proposal stage, pre-feasibility and feasibility study stage, while the investment estimation work features four stages correspondingly.

Different decision stages of a proposed project require different definite degrees, and the acquisition of corresponding estimate condition and information differs, meaning a different accuracy of investment estimation and different roles of investment estimation at each stage. With the project going on and the collected data getting richer, the accura-

工程造价管理 Engineering Cost Management

越来越重要。投资估算的阶段划分情况见表11.2。

cy of the investment estimation is improved, and its role becomes more important. Investment estimation stages are shown in Table 11.2.

表 11.2　投资估算的阶段划分
Table 11.2　Stages of the investment estimation

阶段名称 Name of stages	主要作用 Main role	投资估算误差 Deviation
项目规划阶段 programming stage	根据国民经济、地区和行业发展规划粗略估算建设项目所需要的投资额 rough estimating of the total investment according to national, regional and industry development planning	$\geqslant \pm 30\%$
项目建议书阶段 proposal stage	依据产品方案、项目建设规模、产品主要生产工艺、企业车间组成等估算所需投资额 estimating the total investment based on product solutions, project size, main production process and enterprise workshops composition	$\leqslant \pm 30\%$
预可行性研究阶段 pre-feasibility study stage	在项目方案初步明确的基础上做出投资估算,为项目进行技术经济论证提供依据 estimating the total investment based on the initial project plan and providing a basis for technical and economic feasibility	$\leqslant \pm 20\%$
详细可行性研究阶段 detailed feasibility study stage	全面、详细、深入的技术经济分析论证,是决定项目可行性、编制设计文件、控制初步设计概算的依据 comprehensive, detailed, in-depth technical and economic analysis providing basis for the feasibility of the project, preparation of design documents, and control of the preliminary design estimates	$\leqslant \pm 10\%$

　　为适应市场价格运行机制的要求,建设项目的工程造价区分为静态投资和动态投资,以使投资的计划、估算、控制更加符合实际。静态投资包括设备及工器具购置费、建筑安装工程费、工程建设其他费以及基本预备费(如图 11.6 所示)。现行静态投资部分估算的方法,主要以类似工程对比为主要思路,利用各种数学模型和统计经验公式进行估算。

　　In order to meet the requirements of the operation mechanism of market price, construction costs of the project are divided into static investment and dynamic investment. This kind of classification can make the investment plan, estimation and control become more practical. Static investment includes equipment and tools purchase costs, installation costs, other construction costs and basic reserve funds (Fig. 11.6). Current estimation methods for static investment use various mathematical models and statistic empirical formula based on what are used in similar projects.

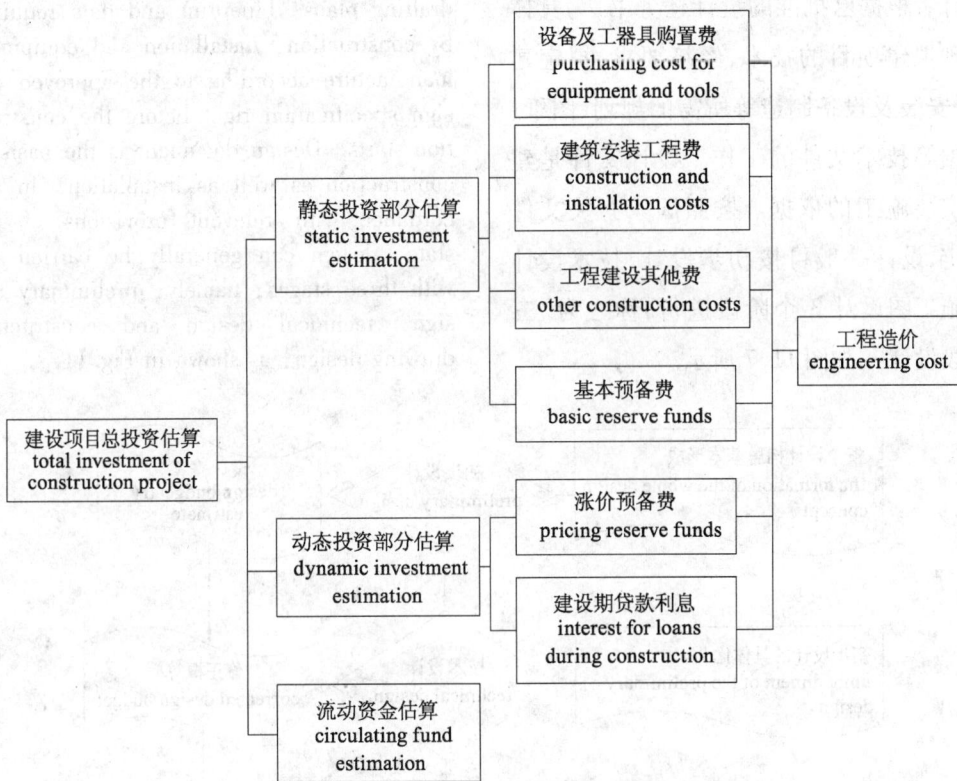

图 11.6　投资估算结构图

Fig. 11.6　Investment estimation structures

动态投资部分主要包括涨价预备费和建设期贷款利息,这两部分的估算以基准年静态投资估算额为基础,采用国家或行业的具体规定计算。静态投资部分和动态投资部分之和为建设项目的工程造价,建设项目总投资除了包括工程造价之外,还包括流动资金部分。

流动资金指生产经营项目建成后,为保证项目正常生产或服务运营所必需的周转资金,对流动资金的估算主要采用扩大指标估算法和分项详细估算法。

Dynamic investment mainly includes the pricing reserve fund and interest for loans during construction, which can be estimated on the basis of the static investment estimates in the benchmark year and calculated according to provisions of state or industry. Sum of those two parts mentioned above are called the construction cost, together with the circulating fund, making up the total investment of construction project.

Circulating fund refers to the working cash to ensure the regular product or service operation after the completion of project. The estimation of circulating fund mainly adopts expand index estimation and itemized detailed estimation.

11.2.2　设计概算

工程设计是指在工程开始施工之前,

11.2.2　Design budgetary estimate

Project design refers to the work of

设计者根据已批准的设计任务书,为具体实现拟建项目的技术、经济要求,拟定建筑、安装及设备制造等所需的规划、图纸、数据等技术文件的工作。设计文件是建筑安装施工的依据。按照国家有关文件规定,设计一般可按初步设计、技术设计和施工图设计 3 个阶段进行,称之为"三阶段设计",如图 11.7 所示。

drafting plans, blueprint and data required by construction, installation and equipment manufacture according to the approved design specification right before the construction starts. Design document is the basis of construction as well as installation. In accordance with relevant provisions of the state, design can generally be carried out with three stages, namely, preliminary design, technical design and construction drawing design, as shown in Fig. 11.7.

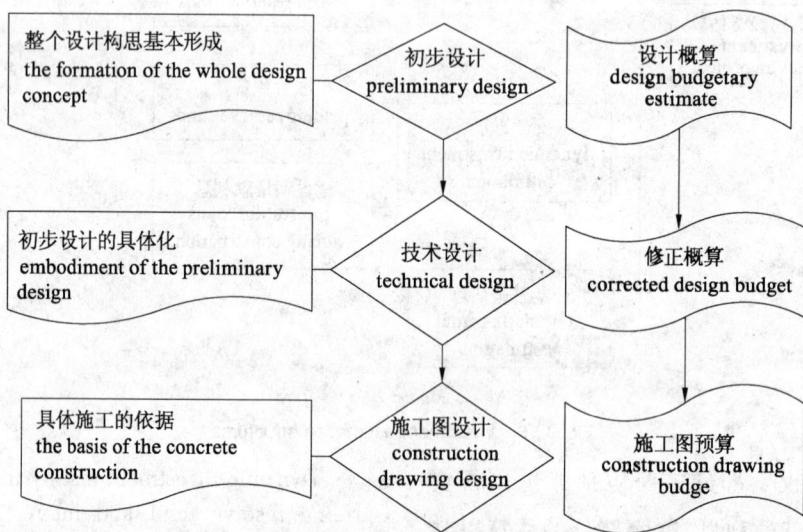

图 11.7 三阶段设计模式
Fig. 11.7 Three-stage design pattern

设计的每个阶段对应不同的造价管理内容,在初步设计阶段主要编制设计概算;在技术设计阶段,依据深入的设计资料对设计概算进行修正;在施工图设计阶段主要编制施工图预算,以便进行工程造价的计价和控制从而使造价构成更合理,提高资金使用效率。

设计概算是由设计单位根据设计图纸、说明和造价管理部门颁发的计价依据

Each stage of the design corresponds to different content of cost management. The preliminary design stage focuses on working out budgetary estimate. In the technical design stage, modifications of budget estimate is made based on more detailed design data. While at the last stage, construction drawing budget is established in order to control the engineering cost valuation and make the construction cost more reasonable, this improving the utilization efficiency of funds.

Design budgetary estimate files are compiled by the design unit, according to design drawings, instructions and price foundation

等资料编制的建设项目从筹建到竣工交付使用所需全部费用的文件。设计概算可分单位工程概算、单项工程综合概算和建设项目总概算三级。各级概算之间的相互关系如图 11.8 所示。

issued by cost management department, which cover the total expense from planning to completion of a project. Design budgetary estimate can be further divided into budgetary estimate of unit project, composite budgetary estimate of single construction and overall budgetary estimate of construction project. The relationship among three levels is shown in Fig. 11.8.

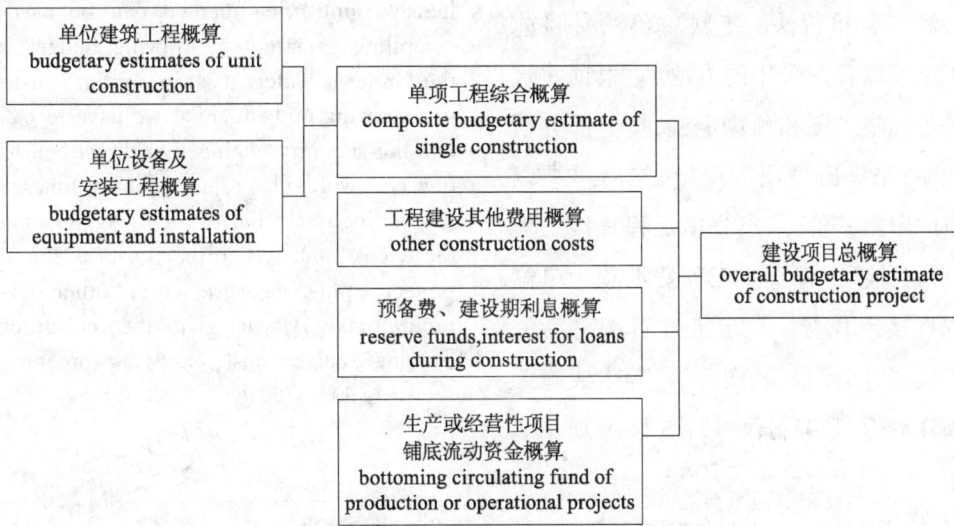

图 11.8　工程造价三级概算
Fig. 11.8　Three-grades budgetary estimate of the construction cost

单位建筑工程和设备及安装工程概算的编制可采用概算定额法、概算指标法以及类似工程预算法。单位建筑工程和单位设备及安装工程概算汇总后即得到单项工程综合概算,它包括了工程造价构成中的设备及工器具购置费、建筑安装工程费,然后汇总工程建设其他费用概算,计取相应的预备费、建设期贷款利息以及铺底流动资金得到建设项目总概算。

Budgetary estimate of the unit construction project and equipment as well as installation can be compiled with quota method, index method or comparison with the similar project. Aggregation of those two above gives composite budgetary estimate of single construction. That, plus other costs, reserve funds, interest loans during construction and circulating fund, make up the overall budgetary estimate of a construction project.

11.2.3　施工图预算

施工图预算是由设计单位在施工图设计完成后,工程开工前,根据已批准的

11.2.3　Construction drawing budget

Construction drawing budget is compiled by the design unit between the finish of

施工图纸、现行预算定额、费用定额以及地区设备、材料、人工、施工机械台班等资源价格,在施工方案或施工组织设计已大致确定的前提下,编制和确定的建筑安装工程造价文件。

施工图预算的编制可以采用工料单价法和综合单价法。工料单价法是目前施工图预算普遍采用的方法,它根据建筑安装工程施工图和预算定额,按分部分项的顺序,先算出分项工程量,然后再乘以对应的定额基价,求出分项工程直接工程费,再将分项工程直接工程费汇总为单位工程直接工程费。直接工程费汇总后另加措施费、间接费、利润、税金等生成施工图预算造价,其编制程序如图 11.9 所示。

construction drawing design and the commencement of the works. It is made based on approved construction drawings, current budget quota, expense standards and prices of local equipment, materials, labors, construction machinery while premises of construction scheme and organization design are being roughly determined.

Quantity unit price method and comprehensive unit price method can be used in compiling construction drawing budget, and the former is widely used in current construction drawing budget. First we have to calculate the subentry engineering direct construction cost with sub-project quantity times corresponding quota basic price. Aggregation of those cost, namely direct cost of the unit project, plus measure cost, other costs, profit, taxes, will give the construction drawing budget cost, and its program is shown in Fig. 11.9.

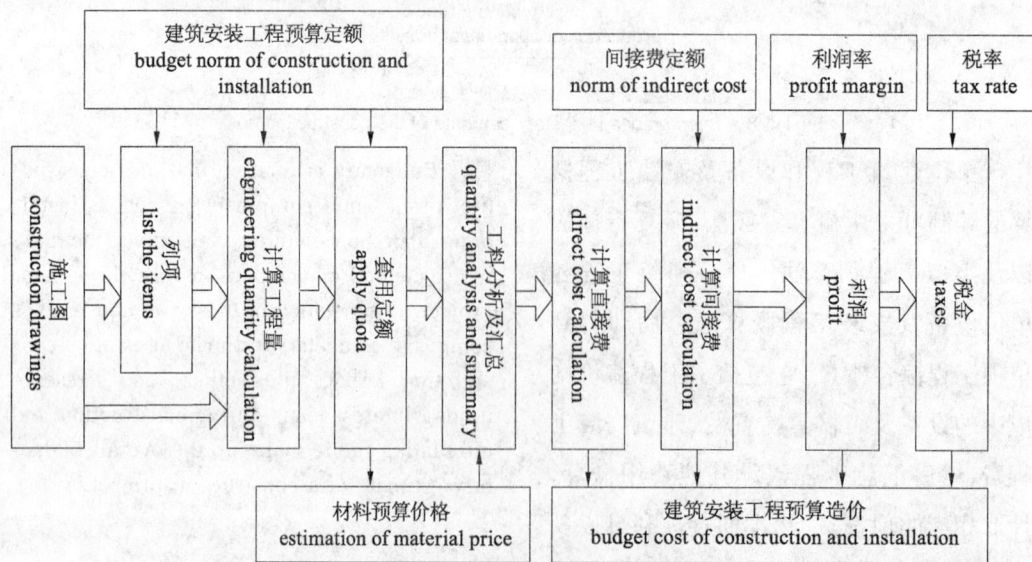

图 11.9　施工图预算编制程序
Fig. 11.9　The procedure of construction drawing budget

以直接费为计费基数的施工预算编制方法见表 11.3。

Methods of construction drawing budget based on direct cost are shown in Table 11.3.

序号 Number	费用项目 Expense item	计算方法 Computational method
①	直接工程费 direct engineering cost	按预算表 according to the budget table
②	措施费 measure fees	按规定标准计算 calculation according to the standard
③	直接费小计 sum of the direct cost	①+②
④	间接费 indirect cost	③×相应费率 ③ × the corresponding rate
⑤	利润 profit	[③+④]×相应利润率 [3 + 4] × the corresponding rate
⑥	合计 total	③+④+⑤
⑦	含税造价 cost including taxes	⑥×(1+相应税率) ⑥ × (1 + the corresponding rate)

综合单价综合了人工费、材料费、机械费,有关文件规定的调价、利润、税金,现行取费中有关费用、材料的差价以及部分风险费用。综合单价法与工料单价法相比较,主要区别在于:间接费和利润是用一个综合管理费率分摊到分项工程单价中,从而组成分项工程全费用单价,某分项工程单价乘以工程量即为该分项工程的完全价格。

Comprehensive unit price method is different from quantity unit price method. It includes labor cost, material cost, mechanical cost, and the provisions of the price, profit, taxes in the relevant documents as well as the risk cost. Indirect costs and profits are assessed to the project unit by comprehensive management rate. Therefore, the unit price multiplied by the quantity is the cost of the entire project.

11.3 建设工程结算与竣工决算

11.3 Construction price settlement and final settlement of account

11.3.1 建设工程发承包造价管理

建设单位完成工程建设准备工作并具备开工条件后,应按照法律规定进行工程施工发包,寻找最佳的施工单位实施工

11.3.1 Cost management in contracting stage

The employer should award constructions in accordance with laws and regulations and seek the best constructor to execute the work with proper commencement conditions. Project contracting is a trading activity in

程。工程发承包是建筑市场的交易活动，而发承包最主要的实现形式为工程招投标，通过招投标确定最合适的承包方，并签订合同形成建设工程合同价。

工程发承包阶段造价管理的主要任务是合理确定合同价，也就是合理确定一个发承包双方均满意的工程实施价款，作为工程实施过程中结算管理的依据。在工程招投标过程中，发包人的主要造价管理工作为编制招标控制价。招标控制价是招标人依据设计图纸编制的招标工程最高限定造价，其主要内容如图 11.10 所示。

construction market and the main practical form to find the best constructor is tendering and bidding with prices determined by contracts.

The main task of cost management at contracting stage is to find a reasonable contract price for both sides and to take it as the basis for balance of accounts management during project implementation. During the bidding procedure, cost management for employers is to determine the tender control price, which is the highest cost of bidding project compiled with design drawings. Its main content is shown in Fig. 11.10.

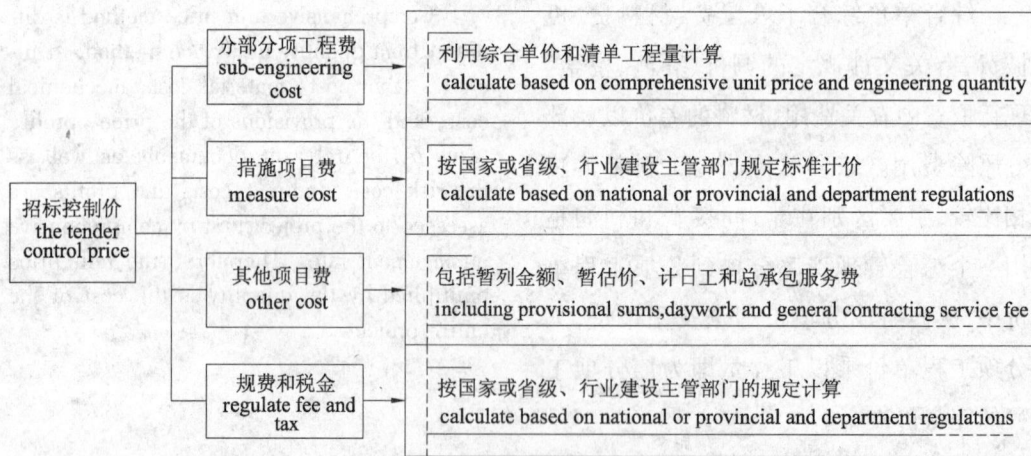

图 11.10　招标控制价的主要内容
Fig. 11.10　The main contents of the tender control price

在工程发承包阶段，投标人的造价管理工作表现为进行最优的投标报价以获得招标工程项目。任何建设项目的投标报价都是一项复杂的系统工程，需要周密思考，统筹安排，并遵循一定的程序，如图 11.11 所示。

Correspondingly, cost management work for bidders is to offer the optimal price to win the bidding. The bidding part is a complicated systematic work in any construction project which needs careful thinking and overall arrangement with certain procedures, as shown in Fig. 11.11.

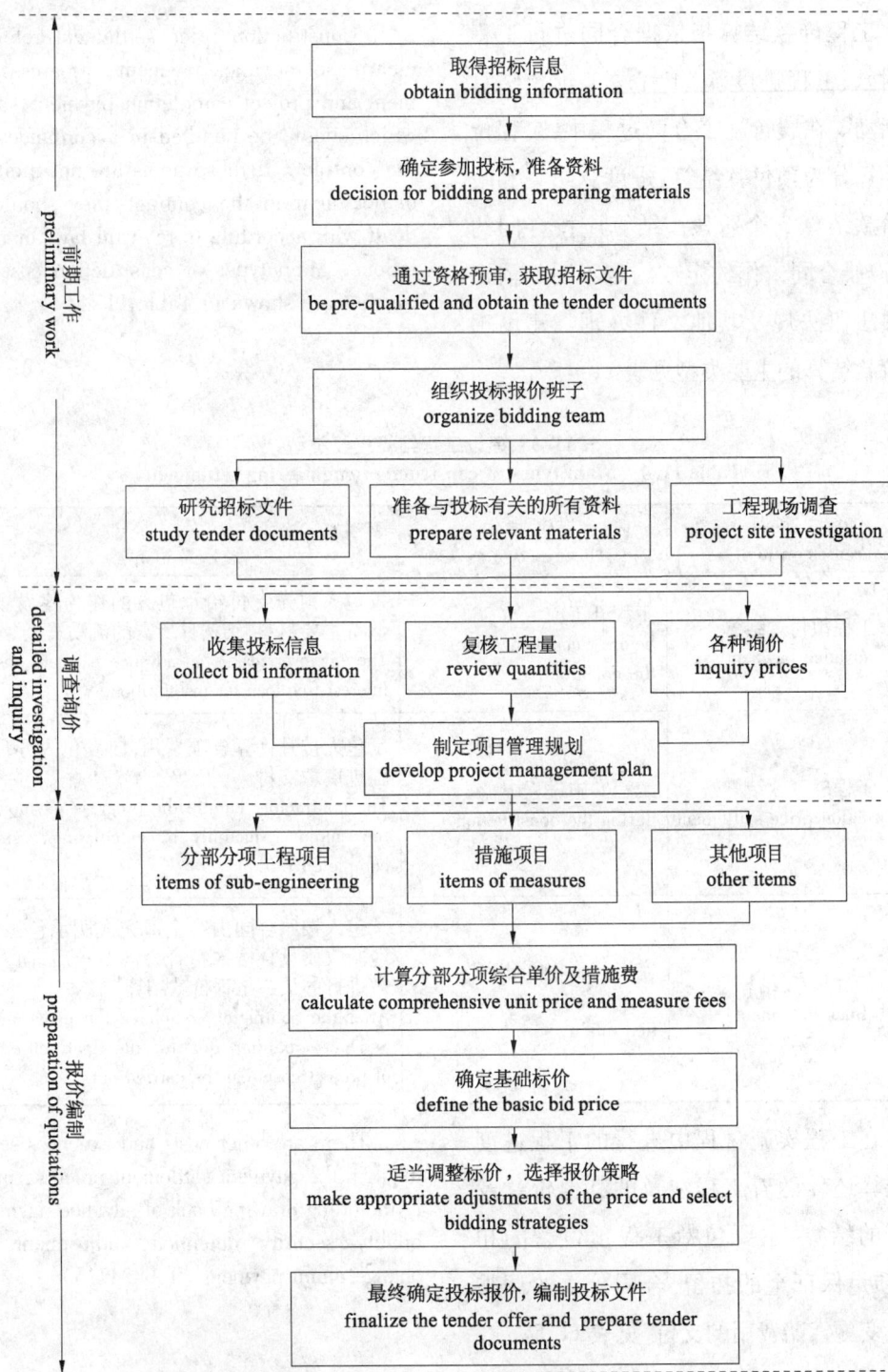

preliminary work
前期工作

取得招标信息
obtain bidding information

确定参加投标,准备资料
decision for bidding and preparing materials

通过资格预审,获取招标文件
be pre-qualified and obtain the tender documents

组织投标报价班子
organize bidding team

研究招标文件
study tender documents

准备与投标有关的所有资料
prepare relevant materials

工程现场调查
project site investigation

detailed investigation and inquiry
调查询价

收集投标信息
collect bid information

复核工程量
review quantities

各种询价
inquiry prices

制定项目管理规划
develop project management plan

preparation of quotations
报价编制

分部分项工程项目
items of sub-engineering

措施项目
items of measures

其他项目
other items

计算分部分项综合单价及措施费
calculate comprehensive unit price and measure fees

确定基础标价
define the basic bid price

适当调整标价,选择报价策略
make appropriate adjustments of the price and select bidding strategies

最终确定投标报价,编制投标文件
finalize the tender offer and prepare tender documents

图 11. 11　施工投标工程量清单报价程序

Fig. 11.11　Construction tender BOQ program

11.3.2 建设工程价款结算

工程价款结算指依据合同进行工程预付款、工程进度款、工程竣工价款结算的活动。在履行施工合同过程中，工程价款结算分为预付款结算、进度款结算和竣工价款结算 3 个阶段。结算应按合同约定办理，合同未作约定或约定不明的依据有关法律法规及其他文件办理。建设工程价款结算的主要类型见表 11.4。

11.3.2 Construction price settlement

Construction price settlement refers to clearing of advance payment, progress payment and project completion payment, all of which should be handled in accordance with the contract. If those items are not included or not clear in the contract, they should be dealt with according to relevant laws or regulations. Main types of construction cost settlement are shown in Table 11.4.

表 11.4 建设工程结算主要类型
Table 11.4 Main types of construction engineering settlement

结算类型 Settlement type	发生时间 Occurrence time	定义 Definition
工程预付款结算 advance payment	工程开工前 before commencement of the construction	发包人预先支付给承包方的作为建设项目储备主要材料、结构件所需的流动资金。 The employer makes an advance payment, as an interest-free loan for mobilization.
工程进度款结算 construction price settlement	工程实施过程中 during the construction	承包人按月计算各项费用,向发包人办理工程进度款支付。 The contractors handle the progress payment to the employer monthly by calculating costs of completed works.
工程竣工结算 final settlement	完工且验收质量合格后 completion and get qualified quality	承包人按照合同内容全部完成所承包工程,经验收质量合格并符合合同要求后,向发包人进行的最终工程价款结算。 After the contractor completes the construction and gets qualified quality, the final settlement of project costs can be carried out.

在工程实施过程中进行的工程进度款结算还伴随着工程涉及的其他价款及费用的结算，主要包括工程预付款的扣回、质量保证金的扣留、合同价款的调整以及变更索赔费用的支付，见表 11.5。

There are other costs and expenses settlement in the payment settlement process, mainly including drawing back of advance payment, quality security detention, adjustment and change claim payment (Table 11.5).

表 11.5 进度款支付过程中涉及的其他费用结算

Table 11.5 Other costs and expenses settlements in the payment settlement process

结算类型 Settlement type	发生时间 Occurrence time	定义 Definition
工程预付款的扣回 Drawing back of advance payment	进度款支付过程中 in the payment settlement process	随着工程实施以抵充工程价款的方式陆续扣回,抵扣方式在合同中约定 gradual deduction with the implementation of the project according to the contract
质量保证金的扣留 quality security detention	进度款支付过程中 in the payment settlement process	从应付工程款中预留,用以保证承包人在缺陷责任期内对建设工程出现的缺陷进行维修的资金 reserved from the payment to guarantee repairing defects of the construction project during the defects liability period
合同价款的调整 adjustment of the contract	进度款支付过程中 in the payment settlement process	调整的方式方法在合同中约定,调整价款在进度款支付中增加或扣减 adjustment according to the contract, and the adjusted funds should be increased and decreased.
变更索赔费用的支付 change and claim payment	进度款支付过程中 in the payment settlement process	工程实施过程中产生的工程变更或索赔费用均在进度款支付中完成 change and claim payment during the construction should be paid during the payment settlement process

工程竣工验收符合合同要求之后,发包人和承包人要进行最终的工程价款结算,竣工结算由承包人编制,发包人审核后予以财务支付。在清单计价方式下,竣工结算的编制内容应包括工程量清单计价表中包含的各项费用内容,即分部分项工程费、措施项目费、其他项目费、规费和税金。

After completion of the project acceptance, the employer and the contractor should carry out the final settlement of the engineering cost. The contractors should work out the final settlement which will be examined by the employer. In the list valuation mode, completion of the settlement shall include sub-engineering cost, measure cost, other cost, stipulated fees and taxes.

11.3.3 竣工决算的编制

当建设项目按设计文件的规定内容和要求全部建成后,建设单位要组织工程竣工验收。竣工验收是投资成果转入生产或使用的标志,也是全面考核工程建设成果、检验设计和施工质量的重要步骤。工程竣工验收之后,还要进行竣工决算的

11.3.3 Final settlement of account

When the construction project is completed according to the contents and requirements of design documents, quality of the project should be checked by the construction unit. That is called completion acceptance. Completion acceptance is the sign of investment achievements transformation and an important step in test and assessment of construction achievement, design rationality and

编制和工程保修费用的处理。

construction quality. After completion acceptance, final settlement of the account and budget of maintenance cost should be made.

竣工决算由竣工财务决算说明书、竣工财务决算报表、工程竣工图和工程竣工造价对比分析四部分组成。前两部分又称建设项目竣工财务决算，是竣工决算的核心内容，根据大、中型建设项目和小型建设项目分别制定。

Final settlement of the account includes final accounts instruction of completion finance, annual final accounts of completion finance, drawings of completion, comparison analysis of completion cost. Final accounts instruction of completion finance and annual final accounts of completion finance are called final accounts of completion finance which is the core content of final settlement of account. They are formulated separately according to the scale of project.

建设项目竣工财务决算审批表是竣工决算上报有关部门审批时使用的。大中型项目概况表反应了建设项目的总投资、主材消耗以及主要技术经济指标等情况；大、中型建设项目竣工财务决算表反应建设项目的全部资金来源和资金运用情况，是考核和分析投资效果的依据（见表11.6）；使用资产总表和使用资产明细表反映项目建成后，新增固定资产、流动资产、无形资产和递延资产的全部价值。

Approval forms of the final accounts of completion finance are materials submitted to the relevant department for approval of the final accounts. The total investment of construction project, main material consumption and other main technical and economic indexes are reflected in the large and medium-sized project profile. All sources and use of the construction project funds are reflected in annual final accounts of completion finance of large and medium-sized project which is the basis evaluation and analysis of investment effect (Table 11.6). All value of newly increased fixed assets, circulating assets, immaterial assets of project completed is reflected in summary statement of assets employed and detail statement of assets.

表 11.6   建设项目竣工财务决算表

Table 11.6   Final accounting of the construction project

资金来源 Financial sources	金额/元 Amount/yuan	资金占用 Occupation of funds	金额/元 Amount/yuan
一、基建拨款 allocating funds	110 440	一、基本建设支出 capital construction expenditure	170 160
1. 预算拨款 budget allocation	52 000	1. 交付使用资产 delivery of assets	107 600
2. 基建基金拨款 infrastructure fund appropriation		2. 在建工程 project in construction	62 510

资金来源 Financial sources	金额/元 Amount/yuan	资金占用 Occupation of funds	金额/元 Amount/yuan
其中:国债专项资金拨款 treasury under special funding included		3. 待核销基建支出 infrastructure spending to be written off	50
3. 专项建设基金拨款 special construction funds		4. 非经营性项目转出投资 investment from non-operating items	
4. 进口设备转账拨款 imported equipment transfer funds		二、应收入生产单位投资借款 investment loans of production units	1 400
5. 器材转账拨款 equipment transfer funds		三、拨付所属投资借款 investment loans disbursed	
6. 煤代油专用基金拨款 special fund allocations of substitute coal for oil		四、器材 equipment	50
7. 自筹资金拨款 self-raised funds allocation		其中:待处理器材损失 pending equipment losses	16
8. 其他拨款 other fund sources		五、货币资金 monetary capital	470
二、项目资本 project fund		六、预付及应收款 prepayments and other receivables	18
1. 国家资本 national capital		七、有价证券 securities	
2. 法人资本 corporate capital		八、固定资产 fixed assets	50 528
3. 个人资本 personal capital		固定资产原价 original value of fixed assets	60 550
4. 外商资本 foreign capital		减:累计折旧 accumulated depreciation	10 022
三、项目资金公积 project funding reserves		固定资产净值 net value of fixed assets	50528
四、基建借款 capital construction loan		固定资产清理 disposal of fixed assets	
其中:国债转贷 state bond-turned loan		待处理固定资产损失 pending loss on fixed assets	
五、上级拨入投资借款 the investment loan from the superior			
六、企业债券资金 corporate bond funds			
七、待冲基建支出 infrastructure spending to be blunt			

资金来源 Financial sources	金额/元 Amount/yuan	资金占用 Occupation of funds	金额/元 Amount/yuan
八、应付款 account payable			
九、未交款 unpaid expenses			
1. 未交税金 　unpaid taxes			
2. 其他未交款 　other unpaid expenses			
十、上级拨入资金 the fund from the superior			
十一、留成收入 retained earnings			
合计 total	222 626	合计 total	222 626

11.4　全面造价管理

11.4.1　建设工程全面造价管理

　　全面造价管理(total cost management, TCM)是由曾任国际全面造价管理促进协会(The Association for the Advancement of Cost Engineering, AACE)会长的 Richard Westney 于 1991 年在西雅图年会上代表协会提出的。"全面造价管理就是有效地使用专业知识和专门技术去计划和控制资源、造价、盈利和风险"。

　　所谓建设工程全面造价管理,就是指政府主管部门、行业协会、业主、承包单位、设计单位、监理及咨询单位等,在建设工程策划决策、设计、招投标、施工、竣工验收的各个阶段,基于建设工程项目全寿

11.4　Total cost management

11.4.1　Total cost management of construction projects

　　Total cost management (TCM) is a process for applying the skills and knowledge of cost engineering. It was put forward by AACE firstly in the 1990s and the full presentation of the process was published in the *Total Cost Management Framework* in 2006. Total Cost Management is a company-wide systematic and structured approach, which provides a holistic framework to control, reduces and eliminates costs throughout the value chain. This process of managing the financial outcome of activities encompasses all operations, internal and external.

　　Total cost management of construction projects is some kind of integrated management of quality, time, and security based on the whole life of construction projects. It refers to government departments, industry associations, owners, contractors, design units, supervision and consulting units, and

命期,对建设工程本身的建造成本以及质量成本、工期成本、安全成本、环保成本等进行的集成管理。由此可以看出,建设工程全面造价管理的主体涉及建设工程管理有关各方,管理的指导思想是基于建设工程项目全寿命期,管理的纵向范围覆盖建设工程项目决策与建设实施的全过程,管理的横向范围涉及影响建设工程造价的各个要素。综上所述,建设工程全面造价管理是一个综合性概念,主要包括全寿命期造价管理、全过程造价管理、全要素造价管理和全方位造价管理 4 个方面。全面造价管理的体系如图 11.12 所示。

it occurs at various stages including decision-making, design, bidding, construction, completion and acceptance. It can be seen that total cost management of construction projects involves all the related parties. Its guiding ideology is the total life cycle of construction projects. The longitudinal extent covers the whole process from decision-making to building construction. The lateral extent refers to various elements of construction cost. In summary, total cost management is a comprehensive concept, including full life-cycle cost management, cost management of the whole process, cost management of all elements and cost management of omnibearing. Total cost management system is shown in Fig. 11.12.

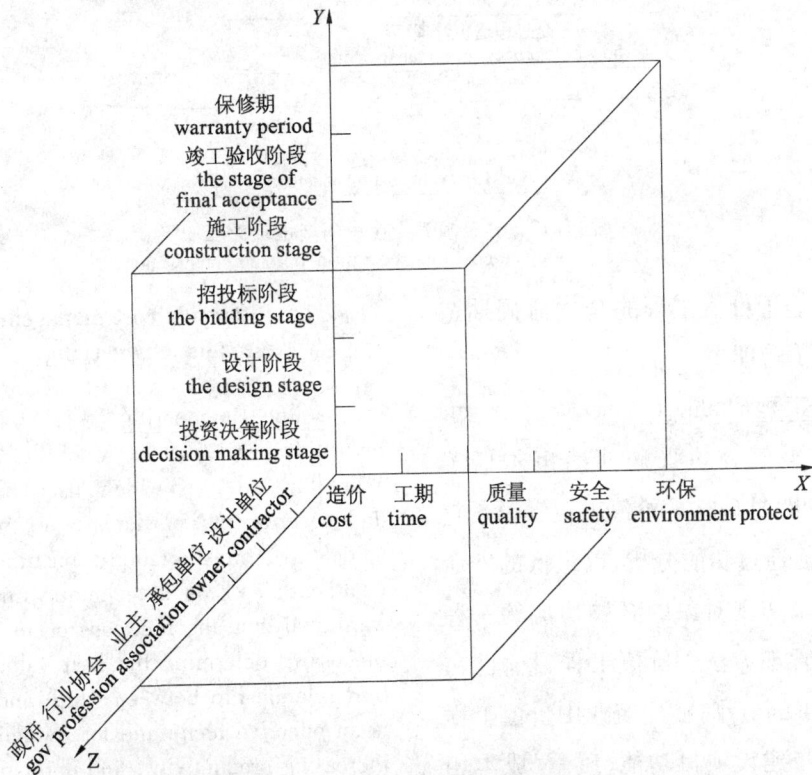

图 11.12 建设项目全面造价管理体系
Fig. 11.12 The total cost management system of building construction

建设项目全过程造价管理包括前期决策、设计、招投标、施工和竣工验收各个

Cost management of the whole process includes various stages of pre-decision

阶段,而全寿命周期造价管理将使用维护阶段纳入了管理范围。全寿命周期造价是指建设工程初始建造成本和建成后的日常使用成本之和,包括建设前期、建设期、使用期和拆除期各个阶段的成本费用。全过程造价管理和全寿命周期造价管理之间的关系如图 11.13 所示。

process, design, bidding, construction and acceptance. The full life-cycle cost management also gets the maintenance phase included. The full life-cycle cost refers to the initial construction costs and costs of daily use after completion of construction projects, including costs of various stages of pre-construction, construction, use and demolition. It is shown in Fig. 11.13.

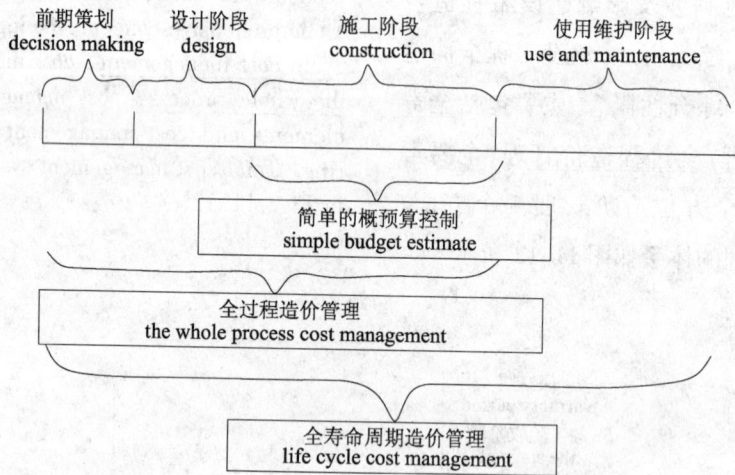

图 11.13 建设项目全生命周期造价管理
Fig. 11.13 Life-cycle cost management of building construction

11.4.2 基于价值工程的全生命周期造价管理

价值工程(Value Engineering, VE)由美国通用电气公司采购部的迈尔斯于 1961 年提出,并确立了价值工程学说。它是研究如何通过功能分析,以最低的寿命周期成本来实现对象的必要功能的一种技术经济管理方法。价值工程目前已在众多领域得到应用,在建筑业中价值工程被认为是与建设项目功能、质量、成本相关的项目优化决策方法。功能(或效用)、成本(或投入)、价值是 VE 的核心术语,它们间的关系可以表达为:

11.4.2 Life-cycle cost management based on value engineering

Value Engineering (VE) was developed at General Electric Corp during World War II, and it is widely used in industry and government, particularly in areas such as defense, transportation, construction, and healthcare. VE is a technique directed toward analyzing the functions of an item or a process to determine the best value, or the best relationship between worth and cost. It is an effective technique for reducing costs, increasing productivity, and improving quality. The relationship between them can be expressed as:

$$V = F/C,$$

wherein, V is the value, F is the function,

$$V = F/C,$$

式中,V 为价值;F 为功能;C 为成本。

　　在建设项目实施过程中,价值工程中的功能 F 是指建设项目能够满足某种特定要求的属性,可理解为用途、效能或是作用等;价值工程中的成本 C 一般为建设项目的全寿命周期造价,包含从项目决策、设计、施工、竣工验收直至运营维护的整个时期的成本,分为建造成本和运营维护成本两部分(如图 11.14 所示)。

C is the cost.

　　In the construction process, function F is the ability to meet the requirements of a particular property; the cost C is generally the entire life cycle cost including costs of decision-making, design, construction, operation and maintenance. All these costs can be divided into construction costs and operation costs (Fig. 11.14).

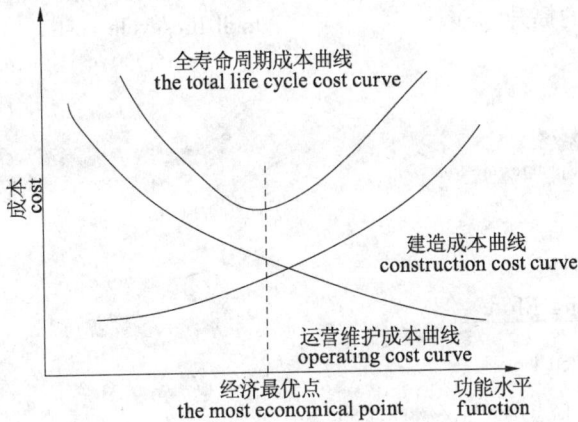

图 11.14　建设项目全寿命周期成本图
Fig. 11.14　The total life cycle cost curve of building construction

　　建设项目初期建造成本的提高必然带来运营维护成本的降低,建设项目要进行有效的造价管理就必须降低全生命周期成本,寻求建造成本和运营维护成本之和的最低点。

　　在建设项目造价管理中,将利用价值工程对项目的功能、质量、成本进行优化,实现价值最大化。建设项目价值提升包括 4 种途径。

　　① $F \rightarrow /C \downarrow = V$:在保持建设项目现

　　The added construction costs in the preliminary stage will inevitably reduce the operation and maintenance costs. The effective cost management means reducing total life cycle costs and seeking the economical point.

　　The engineering cost management tries to maximize value within the optimized costs and qualities by using value engineering. There are four ways to enhance the value of construction projects.

　　① $F \rightarrow /C \downarrow = V$: enhancing the value

有功能不变的情况下通过合理有效的造价管理手段降低成本,实现价值的提升。

② $F\uparrow/C\downarrow=V$:提升建设项目功能的同时通过一定的技术手段降低其全寿命期成本,实现价值的提升。

③ $F\uparrow/C\rightarrow=V$:在保持建设项目造价不变的情况下,提高其功能实现价值。

④ $F\uparrow/C\uparrow=V$:在大幅度提升建设项目功能的同时允许项目成本有所增加,但仍保证项目价值的提升。

by reducing costs through rational and effective cost management while maintaining the existing function of construction projects.

② $F\uparrow/C\downarrow=V$: enhancing the value by improving the function of construction projects while reducing their total life-cycle costs through certain technical means.

③ $F\uparrow/C\rightarrow=V$: improving function of construction projects while keeping their total life-cycle costs.

④ $F\uparrow/C\uparrow=V$: enhancing the value by greatly improving the function of construction projects while slightly increasing their total life-cycle costs.

注:本章图片均来自网络。
Note: In this chapter, all pictures are from webs.

知识拓展
Learning More

相关链接　Related Links

中国建设工程造价信息网 http://www.cecn.gov.cn/

http://www.aacei.org/

http://www.value-eng.org/

思考题　Review Questions

(1) 你觉得建设程序与工程造价有什么关系?

What is the relationship between the construction procedure and the engineering cost?

(2) 全面造价管理在造价管理中怎么应用?

How is the total cost management applied in engineering cost management?

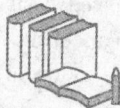

参考文献
References

[1] 尹贻林,严玲:《工程造价概论》,人民交通出版社,2009 年。

[2] 全国造价工程师执业资格考试培训教材编审组:《工程造价计价与控制》,中国计划出版社,

2007 年。

［3］［美］Dagostino,F. R. :《建筑工程估价(英文版)》,清华大学出版社,2004 年。

［4］沈杰:《工程造价管理》,东南大学出版社,2006 年。

［5］Apfelbaum, Adek. Construction Cost Management：Cost Engineering, Cost Controls and Controlled Bidding. Authorhouse, 2005.

［6］Phillip F. Ostwald, Timothy S. McLaren. Cost Analysis and Estimating for Engineering and Management. Pearson Education,2004.

工程造价管理 Engineering Cost Management